Smart Innovation, Systems and Technologies

Volume 445

The Smart Innovation, Systems and Technologies book series encompasses the topics of knowledge, intelligence, innovation and sustainability. The aim of the series is to make available a platform for the publication of books on all aspects of single and multi-disciplinary research on these themes in order to make the latest results available in a readily-accessible form. Volumes on interdisciplinary research combining two or more of these areas is particularly sought.

The series covers systems and paradigms that employ knowledge and intelligence in a broad sense. Its scope is systems having embedded knowledge and intelligence, which may be applied to the solution of world problems in industry, the environment and the community. It also focusses on the knowledge-transfer methodologies and innovation strategies employed to make this happen effectively. The combination of intelligent systems tools and a broad range of applications introduces a need for a synergy of disciplines from science, technology, business and the humanities. The series will include conference proceedings, edited collections, monographs, handbooks, reference books, and other relevant types of book in areas of science and technology where smart systems and technologies can offer innovative solutions.

High quality content is an essential feature for all book proposals accepted for the series. It is expected that editors of all accepted volumes will ensure that contributions are subjected to an appropriate level of reviewing process and adhere to KES quality principles.

Indexed by SCOPUS, EI Compendex, INSPEC, WTI Frankfurt eG, zbMATH, Japanese Science and Technology Agency (JST), SCImago, DBLP.

All books published in the series are submitted for consideration in Web of Science.

Shu-Chuan Chu · Lingping Kong ·
Chien-Ming Chen · Jeng-Shyang Pan
Editors

Advances in Intelligent Data Analysis and Applications

Proceedings of the 8th Euro–China Conference on Intelligent Data Analysis and Applications, December 07–09, 2024, Xiamen, China, Volume 2

Springer

Editors
Shu-Chuan Chu
School of Artificial Intelligence
Nanjing University of Information Science and Technology
Nanjing, China

Lingping Kong
Faculty of Electrical Engineering and Computer Science
VŠB—Technical University of Ostrava
Ostrava, Czech Republic

Chien-Ming Chen
School of Artificial Intelligence
Nanjing University of Information Science and Technology
Nanjing, China

Jeng-Shyang Pan
School of Artificial Intelligence
Nanjing University of Information Science and Technology
Nanjing, China

ISSN 2190-3018 ISSN 2190-3026 (electronic)
Smart Innovation, Systems and Technologies
ISBN 978-981-96-7276-9 ISBN 978-981-96-7277-6 (eBook)
https://doi.org/10.1007/978-981-96-7277-6

This Springer imprint is published by the registered company Springer Nature Singapore Pte Ltd.
The registered company address is: 152 Beach Road, #21-01/04 Gateway East, Singapore 189721, Singapore

Preface

This volume comprises the proceedings of the 8th Euro–China Conference on Intelligent Data Analysis and Applications (ECC 2024), which is hosted by Minnan Science and Technology College, China and co-sponsored by Nanjing University of Information Science and Technology, China. The conference is held in Xiamen, on December 7–9, 2024. The aim of this conference is to provide a premier interdisciplinary platform for researchers, practitioners, and industry experts to present and discuss the most recent innovations, trends, and challenges in intelligent data analysis and applications.

We received a total of 113 submissions for ECC 2024 and, after a rigorous peer-review process, 68 papers were accepted for inclusion in this proceeding. We extend our sincere appreciation to all authors for their invaluable contributions. We also express our deep gratitude to the program committee members and reviewers for their dedication and meticulous efforts in ensuring the quality and integrity of the review process.

Furthermore, we would like to acknowledge and express our heartfelt thanks to all sponsors and supporting institutions for their generous support and contributions, which have been instrumental in the successful organization of this conference.

We hope that the proceedings of 8th Euro–China Conference on Intelligent Data Analysis and Applications will serve as a valuable resource for researchers and practitioners worldwide, fostering new collaborations and inspiring further advancements in intelligent data analysis and applications.

Nanjing, China — Shu-Chuan Chu
Ostrava, Czech Republic — Lingping Kong
Nanjing, China — Chien-Ming Chen
Nanjing, China — Jeng-Shyang Pan
March 2025

Contents

Part III Artificial Intelligence

Part IV Intelligent Applications

Part I
Intelligent Data Analysis

Chapter 1
Time Effect of Anxiety Stimuli on EEG Data Discovered by Wavelet and Data Mining Analysis

Zhuohui Xu and Nanlin Jin

Abstract Anxiety is a prevalent mental state that commonly appears in daily life, and how electroencephalographic (EEG) signals respond to anxiety triggers remains poorly understood. This study aims to investigate the changes in EEG signals that respond to anxiety stimuli, at different levels of anxiety. The EEG data during exposure to the stimuli and the time immediately after exposure is analysed using both numerical computation and wavelet transform techniques. Interestingly the stimulus decreased anxiety in severely anxious participants but slightly increased anxiety in normal participants. This helps understand time effects of anxiety stimuli and aids future clinical assessments and interventions.

1.1 Introduction

About 30% of adults suffer from anxiety, making it one of the most prevalent psychological diseases [2, 7]. Electrical signals produced by brain neurons are known as electroencephalographic (EEG) signals [4]. Research has demonstrated that distinct EEG signals at various brain regions reflect changes in emotions [13]. Rich frequency components are present in EEG, and varied psychological, physiological and cognitive demands alter how energy is distributed at different frequencies in different areas of the brain. It should be noted that anxiety presents itself in varying ways among individuals subjected to identical external stimuli [3]. Significant questions concerning the processes behind this diversification are raised by this variability. A better understanding of the intricate brain dynamics associated with anxiety has been made possible by recent developments in EEG analysis, particularly wavelet analysis [8].

Z. Xu · N. Jin (✉)
Department of Computing, Xi'an Jiaotong-Liverpool University, Suzhou, China
e-mail: Nanlin.Jin@xjtlu.edu.cn

Z. Xu
e-mail: Zhuohui.Xu22@student.xjtlu.edu.cn

S.-C. Chu et al. (eds.), *Advances in Intelligent Data Analysis and Applications*, Smart Innovation, Systems and Technologies 445,
https://doi.org/10.1007/978-981-96-7277-6_1

Machine learning methods are invaluable tools for analysing and interpreting EEG data. They are excellent at spotting connections, patterns and trends. Machine learning methods, for example, categorize EEG data according to their correlation with anxiety states [9] and use feature selection [10]. However, not much research has been done to compare and explain how different types of people respond differently to the same stimuli in terms of changes in EEG data.

The purpose of this paper is to investigate how different levels of anxiety affect people's EEG data in reaction to various stimuli as the examples shown in Fig. 1.1. To find patterns linked to anxiety states, the EEG data were subjected to time-frequency analysis using wavelet transform techniques.

We organize this paper as follows. Section 1.2 outlines the related work on existing methods for anxiety analysis using EEG. Section 1.3 describes our method in detail. Section 1.4 analyses and discusses the results obtained. The last section summarizes our research findings.

1.2 Related Work

1.2.1 Fundamentals of Anxiety EEG Signal Analysis

With advances in technology, researchers have begun to use EEG signals more extensively, such as to identify emotional and anxiety states. Many of the research works try to recognize emotion. For example, EEG-based emotion recognition is validated by the DEAP dataset [4, 14], which was collected with 32 EEG channels. During the experiments for the Database of Anxiety States Based on Psychological Stimulation (DASPS), anxiety was introduced by face-to-face mental stimuli. EEG data was collected and participants' anxiety levels were labelled before and after these stimuli. This paper will analyse EEG data collected from DASPS [3].

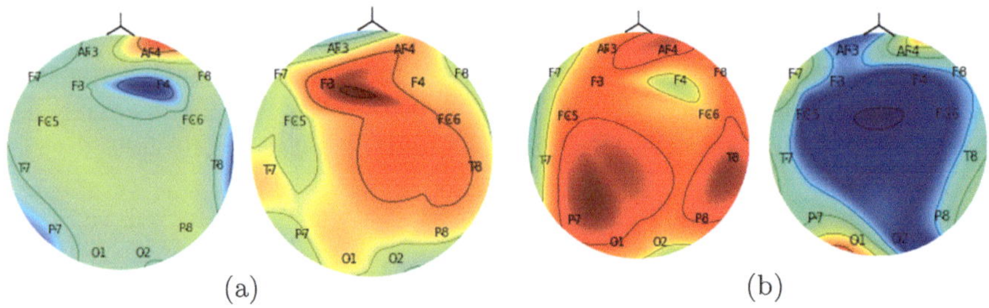

Fig. 1.1 EEG data on the tested channels **a** during the exposure to stimuli and during the recall immediately after the exposure, for a participant without anxiety (normal) and **b** during the exposure to stimuli and during the recall immediately after the exposure, for a participant with severe anxiety

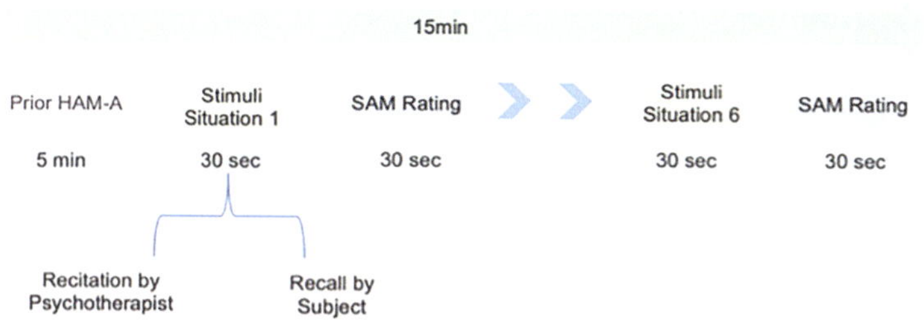

Fig. 1.2 Data acquisition procedure [3]

1.2.2 Wavelet Transform

Wavelet transform method is very effective in detecting brain disorders [1]. Discrete Dobsonian and harmonic wavelets were studied for analysing epileptic EEG data, suggesting physiological processes in the brain at the onset of epilepsy. The research in [5] aims to understand the potential role of heart rate variability (HRV) in predicting different levels of anxiety and affect in non-clinical samples through wavelet transform tools. It was shown that resting HRV was effective in predicting anxiety traits and effective traits and states. The performance of the FBDAWT method was evaluated [12] in an EEG database [6], which consists of ECG and RSP data. Data collected were done by playing non-anxious and anxiety inducing video clips to the participants.

1.2.3 Dataset Description

The DASPS dataset [3] contains EEG signals recorded during face-to-face psychological stimulation. The unit of EEG data is voltage μV. Their EEG data was collected on 14 channels, shown in Fig. 1.1.

Before and after the experiment, anxiety was measured using HAM-A. As shown in Fig. 1.2 the psychotherapist began by reciting one scenario at the first 15 s, and then the participant recalled the second 15 s. This process repeats for six scenarios. Their EEG data were recorded at a frequency of 128 Hz and an impedance of 7 kOhm. The DASPS dataset retained the first 30 s of each experiment, for a total of 180 s across the six experiments, with 23040 records for each experimenter. Dataset was partitioned into 12 sets per experimenter, no additional processing was needed.

Two sets of comparative examples of the EEG data are visualized in Fig. 1.1. The colour coding scheme of the heat map was then applied to these values. Higher values are translated to brighter colours (red), representing areas of engaged activity, while lower values are mapped to darker colours (blue), which represent areas of low activity.

1.3 Methodology

1.3.1 Exploratory Data Analysis

This study utilized pre-processed EEG signals from 14 channels in the DASPS dataset, selected for the first of six experiments.

Anxiety Classification The study labelled the EEG data for anxiety status based on the Self-Assessment Manikin (SAM). Russell defined anxiety as low mood and high arousal [11], and categorized anxiety into the following four conditions based on mood scores versus arousal scores. We have then classified the participants as in Table 1.1.

Data Normalization EEG normalization mitigates disparities, reduces noise, enhances stability, aligns with analysis and facilitates comparisons. This study employs Min-Max normalization to scale data within the range of 0–1, ensuring uniformity of all feature values. The formula for each feature is

$$x' = \frac{x - x_{min}}{x_{max} - x_{min}}. \tag{1.1}$$

In this case, x represents the original data, while x_{min} and x_{max}, respectively, indicate the minimum and maximum values of the channel. Using Min-Max normalization, it is guaranteed that the resulting normalized values lie between 0 and 1.

1.3.2 Wavelet Transform Analysis

EEG signals' complexity makes extracting frequency and transient features challenging with traditional methods. The wavelet transform, with multi-resolution characteristics, accurately characterizes local signal information in both domains, ideal for extracting fleeting EEG information.

The wavelet transform (WT) is a linear combination of a number of wavelet functions that, as shown in Fig. 1.3. It approximates a signal by means of a family

Table 1.1 Anxiety level and number of participants

SAM	Valence in [5–9] and Arousal in [0–5]	Valence in [4–5] and Arousal in [5–6]	Valence in [2–4] and Arousal in [6–7]	Valence in [0–2] and Arousal in [7–9]
Anxiety	Normal	Light	Moderate	Severe
Number	4	6	6	7

of functions known as wavelet bases. These wavelet bases are created by translating and scaling the parent wavelet functions to create 'subwaves'.

In wavelet analysis, selecting an appropriate wavelet function aligned with the signal is crucial for computing the continuous wavelet transform. Different wavelet analyses yield varied results for the same problem, thus the choice of wavelet depends on the situation. Common wavelet functions include Daubechies (dbN), Morlet and Gaussian wavelets. Morlet and Gaussian wavelets are the primary wavelets chosen for analysis in this publication, db3.

For that instant in time, the continuous wavelet transform coefficients *C* are computed. The value of C, which represents how similar the chosen wavelet is to the signal's waveform over the course of the analysis period. These coefficients, directly produced by the wavelet transform, reveal the signal's properties at a specific frequency and time. The choice of wavelet function affects the coefficients, impacting the description of the original signal.

The continuous wavelet transform produces XW, a complex (len(S), len(x)) matrix showing wavelet amplitude maps reflecting coefficients of input signal X at various scales. XW helps identify brainwave patterns in anxiety states via time-frequency analysis. Scale values (s) correspond to distinct frequency components. Each element in S (a row in XW) represents wavelet coefficients at a scale. Briefly, C are individual coefficients, and XW is a matrix encompassing signal transformation details across scales and time intervals.

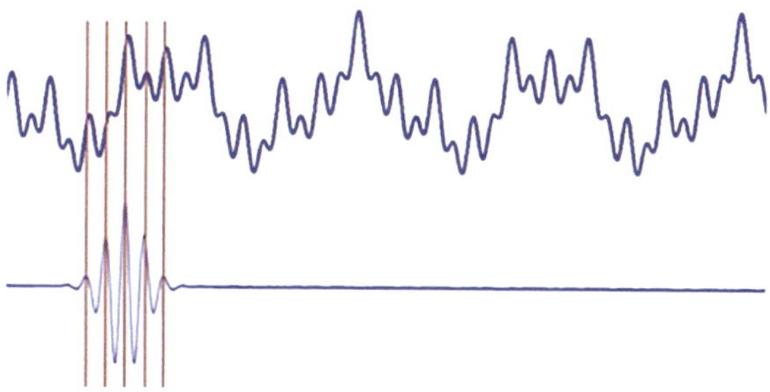

Fig. 1.3 Mother wavelet function translation and scaling

In conclusion, the wavelet transform is used in this study to analyse the input signal X. Specifically, through XW and S, Gaussian wavelets are utilized to expose signal fluctuations at various frequencies and offer comprehensive details about these variations. The input signal X is analysed using the wavelet coefficients of XW and S.

1.3.3 Statistical Test

To evaluate outcome precision and statistical relevance of EEG signal variations across anxiety states, we conducted t-tests on wavelet-transformed data (XW) from 14 channels per participant, comparing stimulatory and recollective phases. Wavelet transformation was applied per channel, followed by statistical evaluations. T-tests per frequency band yielded p-values; those <0.05 indicated statistical significance.

1.4 Experimental Results

1.4.1 Visualization for Comparison

Two participants with varying degrees of anxiety were chosen and their EEGs were used for visualization when recalling and receiving stimuli, respectively, to visualize the differences in brain waves between individuals.

Figure 1.1 are the electroencephalographic patterns recorded during the HAM assessment, deemed representative of an individual with average anxiety. Upon experiment initiation, the participant's EEG showed a transition from blue to red and yellow hues, indicating a possible anxiety increase. Compared to highly anxious individuals, the EEG of an average anxious person maintained a balanced colour distribution with blue patches. These variations suggest that anxiety in average individuals may stabilize or decrease post-stimulus. Notably, blue areas were concentrated in the cranium's superior regions, especially at AF4, T8, F7 and O1, potentially indicating reduced EEG activity. Figures 1.4 and 1.5 show the thermogram of a participant with moderate anxious and one with light anxiety.

Figure 1.5 shows an EEG of an average-anxiety subject in the HAM evaluation. The EEG initially shifted from blue to red/yellow, suggesting anxiety increase. This pattern, with balanced colours and blue segments, contrasts high-anxiety patterns. Blue spots, mainly in the upper head (AF4, T8, F7, O1), indicate reduced EEG activity. This supports the hypothesis that lower baseline anxiety may lead to higher anxiety post-stimulation.

Fig. 1.4 EEG data on the tested channels **a** during the exposure to stimuli and **b** during the recall immediately after the exposure, for a participant with moderate anxiety

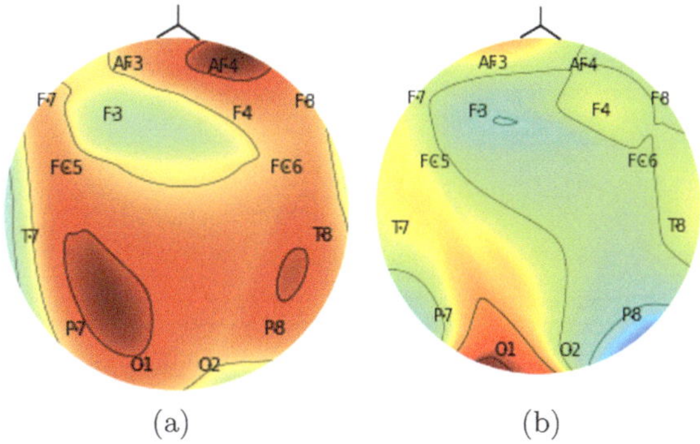

Fig. 1.5 EEG data on the tested channels **a** during the exposure to stimuli and during the recall immediately after the exposure, for a participant with light anxiety

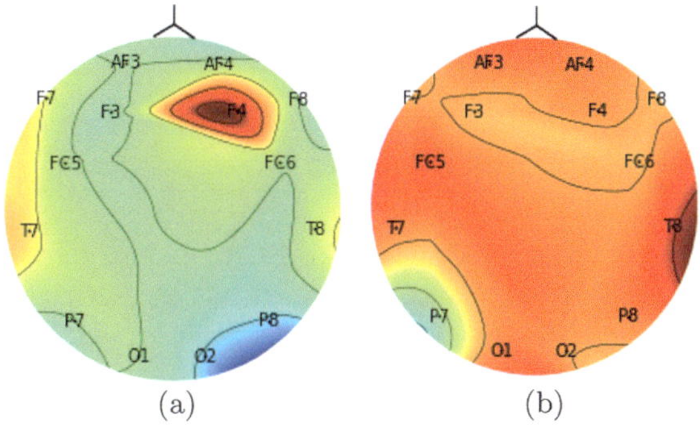

1.4.2 Aggregated Changes

The difference results for each experimenter were used to calculate the mean degree of change μV for each anxiety level (severe, moderate, light and normal). The mean indicates the entire difference between the two sets of data.

Table 1.2 shows the average changes for each of participants.

Take, for instance, Participant S03, categorized under Severe Anxiety, for whom the change value was registered at 231. This denotes that, relative to the subsequent session, there was a pronounced elevation in EEG signal intensity for S03 during the initial experiment. Conversely, for Participant S10, classified as Normal, a change value of –750 was observed, signifying a substantial diminution in EEG signal strength in the latter experiment when juxtaposed with the former.

After stimuli, 5/7 severely anxious individuals felt less anxious, suggesting psychological stimulation helps reduce anxiety. Conversely, 3/4 normally anxious individuals felt more anxious, suggesting cues may upset low-anxiety people. Moderate and light anxiety sufferers had mixed responses to stimuli, likely due to individual differences.

Table 1.2 Aggregated changes of each participant

Participant ID	S10	S14	S15	S23			
Normal	–750	–73	–429	306			
Participant ID	S17	S16	S13	S07	S02	S01	
Moderate	–296	–507	330	41	500	342	
Participant ID	S22	S21	S18	S11	S09	S08	
Light	–745	202	223	–210	–144	–52	
Participant ID	S03	S04	S05	S06	S12	S19	S20
Severe	231	28	–170	170	88	–491	10

1.4.3 *Wavelet Analysis*

The XW matrix is composed of 1920 columns, correlating to the length of the experimental input signal X, which extends to 1920 data points, in conjunction with a sampling rate (fs) of 128 Hz. This indicates that there will be 100 scale values for a Gaussian wavelet, and as a result, the XW matrix's size is (100, 1920). Where a matrix comprising 15 coefficient matrices is split into each second in the manner shown in Fig. 1.6.

Based on the Gauss wavelet approach, the following was selected as the Gauss wavelet coefficient matrix for the AF3 channel of the Severe Anxiety Provider.

Based on the Gauss wavelet approach, the following was selected as the Gauss wavelet coefficient matrix for the AF3 channel of the Severe Anxiety Provider. Each matrix row and column represents unique scales (or frequencies) and time points, respectively. It details wavelet coefficients for each frequency at a given time, revealing frequency composition, intensity and phase, offering a localized time-frequency view of the signal.

The XW matrix reveals varied patterns across different scales and times. Notably, at lower scales (corresponding to shorter time scales), the low-frequency components

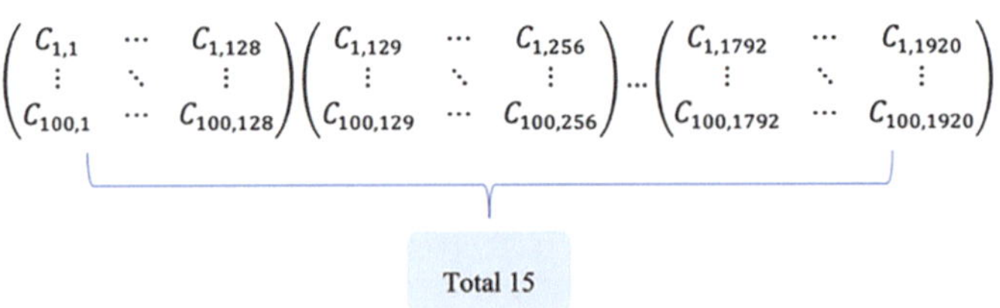

Fig. 1.6 An example of the wavelet coefficient matrix

display a higher energy distribution and notable transient changes, characterizing their dynamics.

a

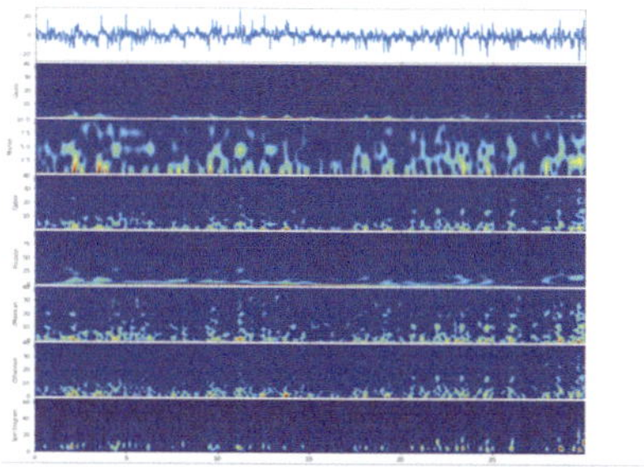

Wavelet transforms over the stimulation and recall for a participant without anxiety (normal).

b

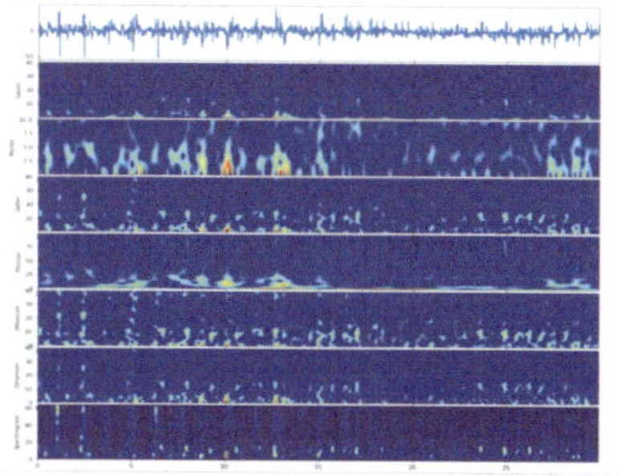

Wavelet transforms over the stimulation and recall for a participant with of severe anxiety.

During stimulation, EEG amplitudes increased significantly in severely anxious individuals. Low-frequency components were more pronounced, with higher wavelet transform amplitudes at lower scales, indicating more transient changes. Additionally, wavelet transform amplitudes were more pronounced in the low-scale range (the lower vertical region), indicating that EEG signals in severely anxious individuals were subject to more transient changes at these lower scales (shorter time scales). This observation underscores the dynamic response of severely anxious individuals to stimuli, with significant alterations in signal intensity and variability at finer time scales.

Before the experiment, general anxiety patients were calm (low EEG amplitude). During recall, EEG amplitude increased, showing anxiety rise. This change underscores the impact of the recall phase on elevating anxiety levels, reflecting a discernible physiological response associated with their anxiety.

1.5 Conclusion

We have investigated the effects of the stimuli on participants at different anxious levels. These stimuli were purposely designed by the experts to trigger anxiety. A rich body of research studies the EEG patterns of different types of individuals in terms of their anxiety levels. It is also known that EEG data bears personal patterns so that generalized or shared characteristics do not always apply to individuals. We investigate a new research question: how different EEG data is before and after exposure to anxiety stimuli.

Their EEG data, including it before and after exposure, was used for our analysis. Our research methods are aggregated EEG signals, wavelet transforms and spectrograms. It is interesting to find out that (1) for the severely anxious individuals, after anxiety stimulations, their EEG shows a decrease in the energy of the high-frequency signal and, at the same time, an increase in the low-frequency signal. This indicates that relaxation responses were produced after stimulation. In addition, our localized feature analysis, including transform and successive Shannon entropy, has observed a reduction in the instantaneous peak of anxiety after stimulation. On the other hand, (2) for those who were classified as normal, such a stimulation does trigger some levels of anxiety, observed in their EEG.

Our study offers initial, intriguing evidence; however, to confirm the generalizability and reliability of these findings, it necessitates further validation through studies involving larger sample sizes and more diverse stimuli. This step is crucial for reinforcing the significance and applicability of our observations across broader contexts.

Acknowledgements This work was partly funded by Xi'an Jiaotong-Liverpool University's RDF-23-01-015.

References

1. AlSharabi, K., Salamah, Y.B., Abdurraqeeb, A.M., Aljalal, M., Alturki, F.A: EEG signal processing for Alzheimer's disorders using discrete wavelet transform and machine learning approaches. IEEE Access **10**, 89781–89797 (2022)
2. Arif, M., Basri, A., Melibari, G., Sindi, T., Alghamdi, N., Altalhi, N., Arif, M.: Classification of anxiety disorders using machine learning methods: a literature review. Insights Biomed. Res. **4**, 95–110 (2020)
3. Baghdadi, A., Aribi, Y., Fourati, R., Halouani, N., Siarry, P., Alimi, A.M.: DASPS: a database for anxious states based on a psychological stimulation (2019). CoRR arXiv:1901.02942
4. Baghdadi, A., Aribi, Y., Fourati, R., Halouani, N., Patrick, S., Alimi, A.: Psychological stimulation for anxious states detection based on EEG-related features. J. Ambient. Intell. Hum. Comput. **12**, 08 (2021)
5. Bajaj, N.: Wavelets for EEG analysis. In: Mohammady, S. (ed.) Wavelet Theory, Chapter 5. IntechOpen, Rijeka (2020)
6. Elgendi, M., Galli, V., Ahmadizadeh, C., Menon, C.: Dataset of psychological scales and physiological signals collected for anxiety assessment using a portable device. Data **7**(9), (2022)
7. Fancher, T.L., DeMartini, J., Patel, G.: Generalized anxiety disorder. Ann. Intern. Med. **170**(7), ITC49–ITC64 (2019)
8. Grobbelaar, M., Phadikar, S., Ghaderpour, E., Struck, A.F., Sinha, N., Ghosh, R., Ahmed, M.Z.I.: A survey on denoising techniques of electroencephalogram signals using wavelet transform. Signals **3**(3), 577–586 (2022)
9. Jin, N., Wu, Y., Park, J., Qin, Z., Liang, H.-N.: Brain-metaverse interaction for anxiety regulation. In: 2023 9th International Conference on Virtual Reality (ICVR), pp. 385–392 (2023)
10. Park, J., Jin, N., Chen, J., Qi, J.: EEG-based anxiety detection with feature selection. In: 2023 International Conference on Cyber-Enabled Distributed Computing and Knowledge Discovery (CyberC), pp. 32–39 (2023)
11. Russell, J.: A circumplex model of affect. J. Pers. Soc. Psychol. **39**, 1161–1178 (1980)
12. Tripathy, R.K., Dash, D.K., Ghosh, S.K., Pachori, R.B.: Detection of different stages of anxiety from single-channel wearable ECG sensor signal using Fourier–Bessel domain adaptive wavelet transform. IEEE Sens. Lett. **7**(5), 1–4 (2023)

13. WHO. Depression and other common mental disorders. https://www.who.int/publications-detail-redirect/depression-global-health-estimates. Accessed 15 Oct 2024
14. Yang, L., Wang, Z., Wang, G., Liang, L., Liu, M., Wang, J.: Brain-inspired modular echo state network for EEG-based emotion recognition. Front. Neurosci. **18**, 1305284 (2024)

Chapter 2
Integrating SMOTE and Heterogeneous Ensemble Methods for Online Fraud Detection

Shilpa Srivastava, Niharika Patni, Meenu Singh, Millie Pant, and Václav Snášel

Abstract In the continuous evolving digital era, the escalation of online fraud demands a robust and efficient mechanism for its detection and prevention. In the recent years there has been a significant increase in the online bank transactions. The research delves into the integration of different machine learning algorithms and to enhance the model's adaptability, Synthetic Minority Oversampling Technique (SMOTE) has been utilized. The approach addresses the challenges of data imbalance and also strengthens the overall detection performance. Through an extensive literature review the study highlights the limitations in the existing issues in online financial fraud. The proposed model employs a heterogeneous ensemble model consisting of K-Nearest Neighbors (KNN), Random Forest, and XGBoost. KNN functions as an anomaly detector, identifying irregularities in transactional data. Simultaneously, Random Forest assesses feature significance and detects intricate patterns, contributing to a comprehensive understanding of fraudulent activity.

S. Srivastava
Christ University, Bengaluru, India
e-mail: shilpa.srivastava2015@gmail.com

N. Patni
IMT Ghaziabad, Ghaziabad, India

M. Singh (✉) · V. Snášel
Department of Computer Science, VŠB-Technical University of Ostrava, Ostrava, Czech Republic
e-mail: meenu.singh@vsb.cz

V. Snášel
e-mail: vaclav.snasel@vsb.cz

M. Pant
Department of Applied Mathematics and Scientific Computing, Indian Institute of Technology, Roorkee, Uttarakhand, India

Mehta Family School of Data Science and Artificial Intelligence, Indian Institute of Technology, Roorkee, Uttarakhand, India

M. Pant
e-mail: pant.milli@as.iitr.ac.in

S.-C. Chu et al. (eds.), *Advances in Intelligent Data Analysis and Applications*,
Smart Innovation, Systems and Technologies 445,
https://doi.org/10.1007/978-981-96-7277-6_2

XGBoost, known for its computational efficiency, ensures real-time responsiveness by adapting to emerging fraud tactics. The system also introduces a soft voting mechanism that seamlessly integrates individual algorithm predictions, resulting in a robust and highly accurate ensemble fraud detection system. Validation on an authentic bank fraud dataset underscores the framework's prowess, showcasing superior fraud detection capabilities and a significant reduction in false positives. The purpose of adopting this approach is to enhance the financial security and safeguard the consumer's assets.

2.1 Introduction

The threat of financial fraud is on the rise in the current scenario. It has been a serious threat for an individual and as well for the organizations. Online finance has undoubtedly transformed the saga of global economic integration, placing a tremendous ease on cross-border transactions. The possibility of fraudulent activities has taken a great height with the rapid growth of businesses in various sectors which including Banks, media and telecommunication, customer, retail and manufacturing. The digital transformation during the COVID-19 pandemic has significantly altered the landscape of fraud risks. As per the Financial and Cyber Fraud Survey Report 2024 [1] every second the organization observes one or more fraud incidents with the top three categories belong to cyber incidents, diversion of assets and regulatory. The reason for this includes work from home culture and challenges related to technology. However, this process also brings to the fore a new wave of advanced digital fraud- and that is a great threat to financial security. Traditional rule-based systems, although interpretable, cannot keep the pecker up with the changing tactics of fraudsters. With the advancement of the artificial intelligence (AI) approach, machine learning and data mining have been utilized to detect fraudulent activities in the financial sector [2]. The machine learning (ML) algorithms have a high potential to adapt to the emerging threats and search for intricate patterns in large datasets. They offer more advanced techniques to analyze the vast amount of data and study the susceptible behavior of the data produced during the online financial transactions. The key machine learning approaches include supervised learning, ensemble methods, neural networks and boosting algorithms.

This study proposes a novel multi-stage ensemble learning framework for bolstering bank fraud detection. Drawing inspiration from recent advancements in ensemble learning literature, the framework leverages the distinct strengths of K-Nearest Neighbors (KNN), Random Forest, and XGBoost algorithms. Each algorithm within the ensemble excels in specific areas of fraud detection. The purpose of using KNN is to excel at anomaly detection, effectively identifying transactions that deviate significantly from established patterns within the data. KNN acts as a seasoned detective, readily flagging suspicious individuals amidst a bustling crowd [3]. Random Forest extracts hidden relationships and identifies important features within the data, building a comprehensive understanding of fraudulent activity. Random Forest can be envisioned as a team of expert investigators, meticulously analyzing seemingly disparate clues to construct a compelling case against the perpetrators [4]. The XGBoost algorithm boasts rapid model training capabilities and computational efficiency, enabling swift adaptation to emerging fraud tactics. It can be considered as a skilled tactician, constantly updating defensive strategies based on real-time intelligence to outmaneuver the enemy [5].

By strategically combining these algorithms through a multi-stage approach, our framework aims to outperform single-algorithm frameworks. The initial stage focuses on data pre-processing and balancing, addressing class imbalance using techniques like Synthetic Minority Oversampling Technique (SMOTE). This mitigates biases towards majority class transactions and ensures effective detection of fraudulent activities.

Each algorithm independently analyzes pre-processed data, generating individual predictions for each transaction. This collaborative approach mitigates potential biases and fosters a robust and accurate detection system. By harnessing the power of ML and fostering collaborative intelligence among diverse algorithms, this research aims to significantly contribute to combating bank fraud and enhancing the overall financial ecosystem's security.

The paper is divided into eight sections. The recent research work is discussed in the Sect. 2.2. The problem in online transaction and the proposed solution is discussed in the Sect. 2.3. The methodology is explained in the Sect. 2.4. Section 2.5 presents results and discussion. The ethical considerations are provided in the Sect. 2.6 followed by the conclusion in Sect. 2.7.

2.2 Literature Review

This section provides the recent research work done in the domain of online fraud detection using machine learning and SMOTE. A range of studies focusing the technologies and methods for detecting bank payment frauds are explored. The detailed literature review and its implications are presented in the Table 2.1.

These studies collectively showcase the immense potential of machine learning for tackling fraud detection in financial transactions. The integration of explainable AI techniques and the exploration of advanced deep learning architectures hold significant promise for further enhancing the accuracy and efficacy of fraud detection systems.

2.3 Statement of the Problem

Financial institutions and payment systems suffer significant losses due to fraudulent transactions. Traditional fraud detection methods often struggle to keep pace with the evolving tactics of fraudsters, leading to missed detections and financial losses. This

Table 2.1 Overview of approaches used in literature across financial sectors

References	Description of the paper	Technique/approach	Application area
[6]	Explores machine learning and explainable AI (XAI) methods to address imbalanced data during fraud detection	SMOTE for handling imbalanced data	Bank fraud detection
[7]	Utilizes ensemble machine learning and PCA to enhance fraud detection accuracy	PCA (Principal Component Analysis) for dimension reduction	Financial fraud detection
[8]	Investigates encoding techniques, early stopping, and hyperparameter tuning to optimize machine learning models for fraud detection	Machine learning models	Bank fraud detection
[9]	Combines multiple machine learning models and anomaly detection methods to improve fraud resilience	Support vector machine, Logistic regression, Neural network, Decision tree and isolation forests	Fraud detection
[3]	Applies unsupervised learning algorithms to detect anomalies in fraud detection	Unsupervised learning algorithm	Credit card fraud detection
[5]	Develops machine learning techniques for unstructured supplementary service data	Machine learning techniques based on short-term memory	Telecom fraud detection
[10]	Examines automated methods for detecting fraudulent bank payments	Machine learning	Bank payment fraud detection
[11]	Applies machine learning techniques to detect anomalies	Machine learning	Fintech applications
[12]	Review of machine learning and data mining techniques for detecting fraud in financial statements	Machine learning and data mining methods, and Kitchenham methodology for identifying gaps, and limitations	Financial statements fraud detection
[13]	Analyzes and compares the performance of various machine learning algorithms	Comparation of three machine learning algorithms	Credit card fraud detection
[14]	Evaluates the effectiveness of SMOTE and AdaBoost techniques to handle imbalanced data in fraud detection	Machine learning with SMOTE and AdaBoost	Credit card fraud detection
[15]	Investigation of data mining techniques for fraud detection	14 data mining techniques for high fraud coverage with advantage and disadvantage analysis	Credit card fraud detection

(continued)

Table 2.1 (continued)

References	Description of the paper	Technique/approach	Application area
[16]	Proposes a deep learning framework for detecting fraudulent in mobile during COVID-19	Deep learning with Recurrent Neural Network (RNN)	Mobile based transactions fraud detection
[17]	Develops a quantitative fraud detection model on large e-commerce datasets	Deep learning model	E-Commerce fraud detection
[18]	Detection of small anomalies in large datasets	Data mining and supervised algorithms	Anomaly detection in large datasets

necessitates the development of robust and intelligent systems that can effectively identify and prevent fraudulent transactions in real-time.

Specifically, our research addresses the following challenges:

- **High number of false positives**. Existing fraud detection systems often trigger unnecessary alerts and block legitimate transactions due to their inability to accurately distinguish between genuine and fraudulent activities. This results in user inconvenience and potential loss of business.
- **Evolving fraud tactics**. Fraudsters constantly adapt their methods to bypass traditional detection methods. Machine learning-based solutions are needed to identify and adapt to these ever-changing patterns in real-time.
- **Imbalanced data**. Fraudulent transactions are significantly less frequent than legitimate ones, creating an imbalanced dataset that can hinder the effectiveness of traditional machine learning models [15].

2.4 Methodology

This research endeavors to address this critical problem by developing a machine learning model capable of accurately identifying fraudulent transactions within a financial dataset. By leveraging advanced analytical techniques and algorithms, the aim is to enhance the detection capabilities of financial institutions and regulatory bodies, thereby safeguarding financial systems and restoring trust in digital transactions. Through rigorous analysis and experimentation, this study seeks to contribute to the ongoing efforts to combat financial fraud and mitigate its adverse effects on the global economy. Here is a systematic flow of the model (Fig. 1).

The research design encompasses five phases—Data Collection, Research Design, Exploratory Data analysis, Model Building and Model evaluation.

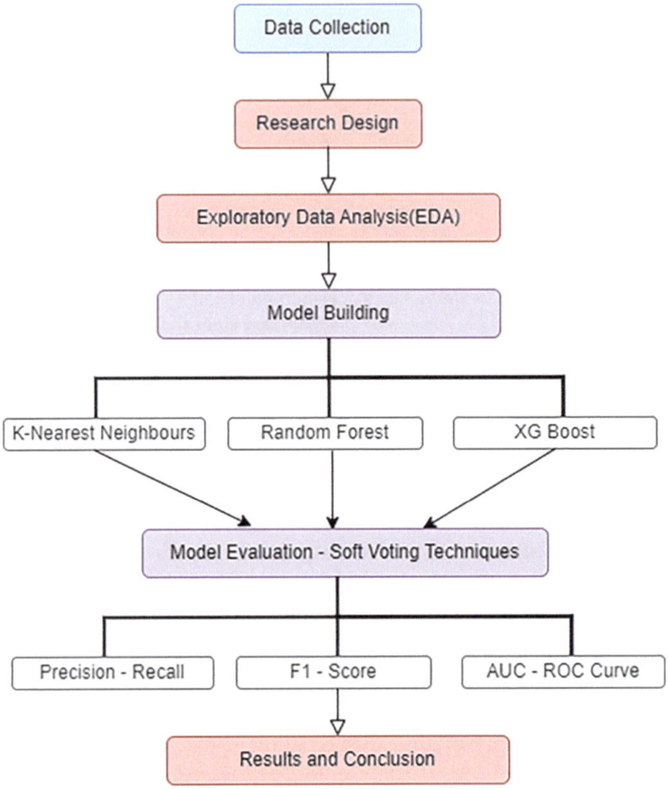

Fig. 2.1 Representing flowchart of methods used

2.4.1 Data Collection

The dataset is synthetically generated dataset called BankSim Data which consists of payments from various customers made in different time periods and with different amounts. BankSim is an agent-based simulator of bank payments based on a sample of aggregated transactional data provided by a bank in Spain. The dataset includes historical USSD transaction data with the following features:

- Customer ID: Unique identifier for each customer.
- Transaction Date/Time: Timestamp of the transaction.
- Merchant ID: Unique identifier for the merchant involved in the transaction.
- Amount: Transaction amount.
- Category: Category of the transaction (e.g., food, travel, leisure)
- Step/Day: Day of the transaction within the simulation period.
- Location: Location information associated with the transaction (optional)
- Feature-derived variables: Additional features created from existing data like time-based features, aggregated values, or interaction features

Fig. 2.2 Fraudulent and non-fraudulent payments

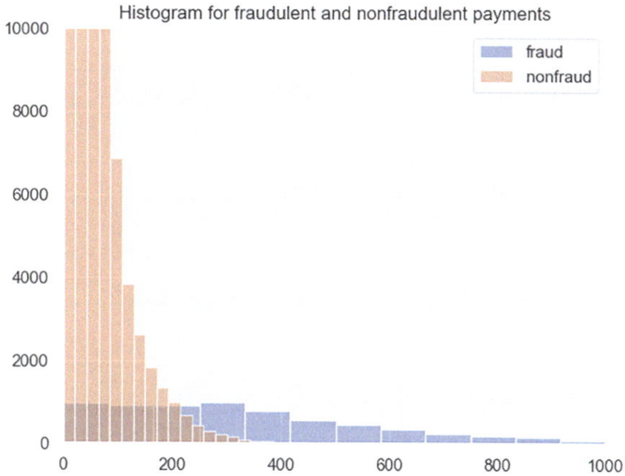

- Target variable: Label indicating whether the transaction is fraudulent (1) or non-fraudulent (0).

2.4.2 Research Design

The research design encompasses a structured approach to conducting experiments and analyzing results in a rigorous and systematic manner. Random sampling techniques are employed to partition the dataset into training and testing sets, ensuring the representative nature of the data subsets used for model training and evaluation. Additionally, advanced techniques are applied to address class imbalance issues, thereby enhancing the models' ability to generalize to real-world scenarios [14].

Model validation is conducted using both traditional train-test splits and cross-validation techniques to ensure the reliability and generalizability of the findings. Exploratory Data Analysis (EDA) which is the first Stage of our model, plays a crucial role in gaining insights into the underlying patterns and distributions within the dataset [18]. Figure 2.2 shows the initial segregation of data based on fraud and non-fraud payment entries. Figure 2.3 represents the count of fraud payments.

Figure 2.4 presents the percentage of fraud in each category whereas Fig. 2.5 represents the boxplot illustrating fraud in each category. Figure 2.6 presents the fraud based on the age of card.

2.4.3 Model Building

Model building being the second stage of the model is the heart of the research which lies in the construction of machine learning models capable of accurately detecting

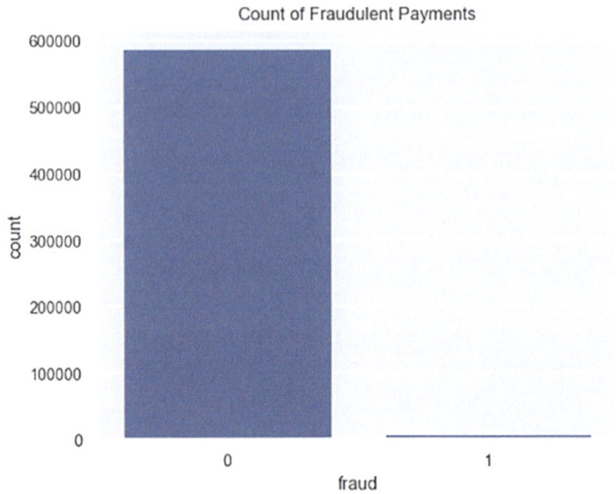

Fig. 2.3 Count of fraud payments

Fig. 2.4 Percentage of fraud in each category

category	Fraudulent	Non-Fraudulent	Percent(%)
'es_transportation'	NaN	26.958187	0.000000
'es_food'	NaN	37.070405	0.000000
'es_hyper'	169.255429	40.037145	4.591669
'es_barsandrestaurants'	164.092667	41.145997	1.882944
'es_contents'	NaN	44.547571	0.000000
'es_wellnessandbeauty'	229.422535	57.320219	4.759380
'es_fashion'	247.008190	62.347674	1.797335
'es_leisure'	300.286878	73.230400	94.989980
'es_otherservices'	316.469605	75.685497	25.000000
'es_sportsandtoys'	345.366811	88.502738	49.525237
'es_tech'	415.274114	99.924638	6.666667
'es_health'	407.031338	103.737228	10.512614
'es_hotelservices'	421.823339	106.548545	31.422018
'es_home'	457.484834	113.338409	15.206445
'es_travel'	2660.802872	669.025533	79.395604

fraudulent transactions. Three distinct algorithms are employed in this endeavor: K-Nearest Neighbors (KNN), Random Forest, and XGBoost. Each algorithm brings its unique set of strengths and characteristics to the table, making them well-suited for the task at hand. These diverse models allow for comparison and potential ensemble learning [20]. Additional algorithms can be explored based on performance and data characteristics [10]. Prior to model training, the dataset is partitioned into training and testing sets to evaluate model performance effectively. Hyperparameter tuning is then performed to optimize the algorithms' parameters, maximizing their predictive capabilities while minimizing overfitting.

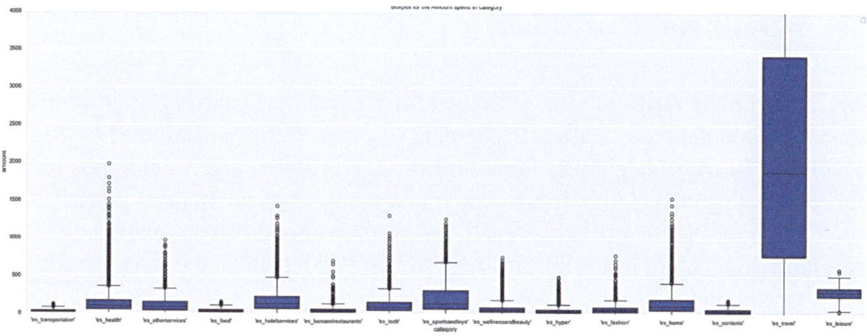

Fig. 2.5 Boxplot presenting fraud in each category

Fig. 2.6 Frauds based on age of card

	Age	Fraud Percent
7	'U'	0.594228
6	'6'	0.974826
5	'5'	1.095112
1	'1'	1.185254
3	'3'	1.192815
2	'2'	1.251401
4	'4'	1.293281
0	'0'	1.957586

2.4.4 *Model Evaluation*

The effectiveness of the developed models is assessed as the final stage of our research model which is done through a comprehensive evaluation framework encompassing various performance metrics. These metrics include accuracy, precision, recall, F1-score, and the Area Under the Receiver Operating Characteristic Curve (AUC-ROC).

Furthermore, the models are compared both individually and in combination through soft voting, a technique that aggregates the predictions of multiple models to make a final decision. Combining predictions from individual models via techniques like soft voting to potentially improve overall accuracy and robustness [10].

2.5 Results and Discussions

This research investigated the effectiveness of machine learning algorithms in detecting fraudulent transactions in payment systems. We employed three individual models (K-Nearest Neighbors, Random Forest, XGBoost) and a combined model with soft voting to identify fraudulent transactions from a balanced dataset. The performance of these models was evaluated using multiple metrics, including accuracy, precision, recall, F1-score, and AUC-ROC. The results of the three algorithms is represented in the Figs. 2.7, 2.8 and 2.9 represents the AUC-ROC curve of KNN, Random Forest and XG Boost respectively whereas Tables 2.2 2.3, 2.4, 2.5, 2.6 and 2.7 presents the classification report and Confusion matrix of KNN, Random forest and XG Boost respectively.

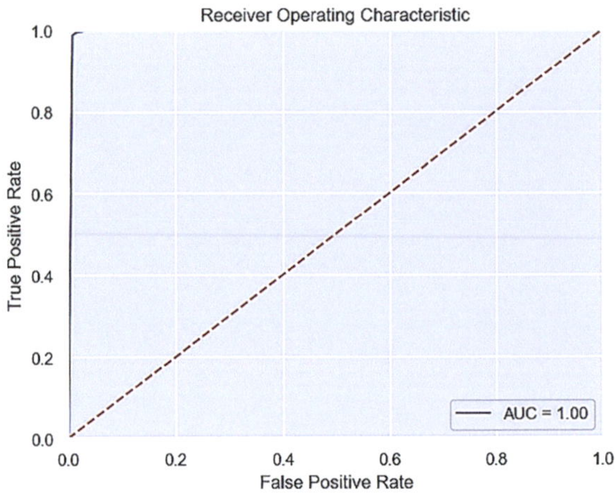

Fig. 2.7 KNN ROC-AUC curve

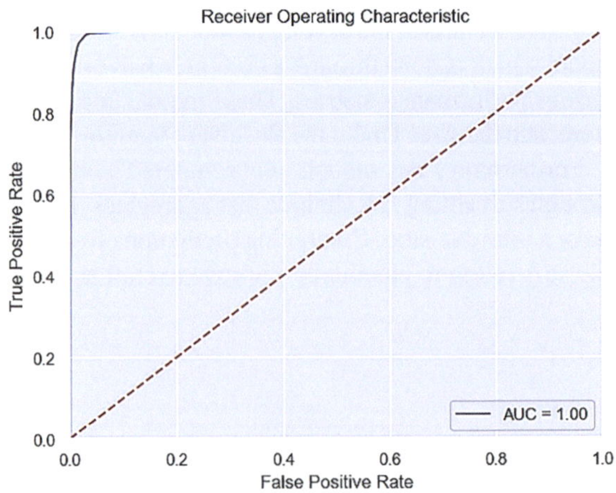

Fig. 2.8 Random forest ROC-AUC curve

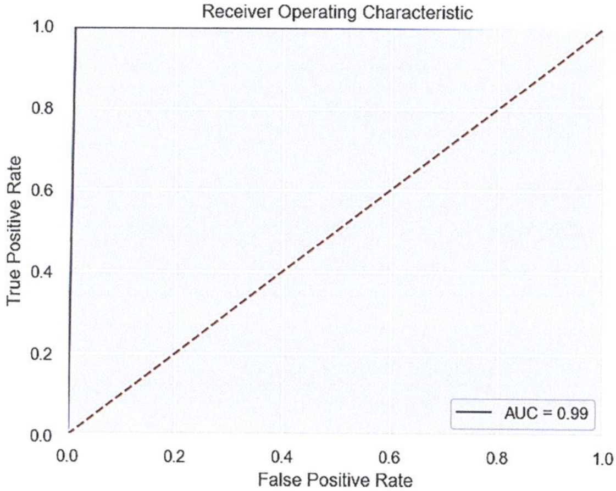

Fig. 2.9 XGBoost ROC-AUC curve

Table 2.2 Classification report of KNN

Class	Precision	Recall	F1-score	Support
0	1	0.98	0.99	1,76,233
1	0.98	1	0.99	1,76,233
Accuracy	–	–	0.99	3,52,466
Macro Avg	0.99	0.99	0.99	3,52,466
Weighted Avg	0.99	0.99	0.99	3,52,466

Table 2.3 Confusion Matrix-KNN

	Predicted 0	Predicted 1
Actual 0	True Negatives (TN) = 171,999	False Positives (FP) = 4,234
Actual 1	False Negatives (FN) = 362	True Positives (TP) = 175,871

Table 2.4 Classification report of Random Forest

Class	Precision	Recall	F1-Score	Support
0	0.99	0.97	0.98	1,76,233
1	0.97	0.99	0.98	1,76,233
Accuracy	–	–	0.98	3,52,466
Macro Avg	0.98	0.98	0.98	3,52,466
Weighted Avg	0.98	0.98	0.98	3,52,466

Table 2.5 Confusion Matrix-Random Forest

	Predicted 0	Predicted 1
Actual 0	True Negatives (TN) = 170,106	False Positives (FP) = 6,127
Actual 1	False Negatives (FN) = 1,079	True Positives (TP) = 175,154

Table 2.6 Classification report of XGBoost

Class	Precision	Recall	F1-score	Support
0	1	0.99	0.99	1,76,233
1	0.99	1	0.99	1,76,233
Accuracy	–	–	0.99	3,52,466
Macro Avg	0.99	0.99	0.99	3,52,466
Weighted Avg	0.99	0.99	0.99	3,52,466

Table 2.7 Confusion matric-XGBoost

	Predicted 0	Predicted 1
Actual 0	True Negatives (TN) = 174,100	False Positives (FP) = 2,133
Actual 1	False Negatives (FN) = 787	True Positives (TP) = 175,446

2.5.1 *Combined Model Synergy*

Soft voting leveraged the strengths of each individual model, surpassing their individual performance. The hidden patterns are revealed in this approach. This strategy also play a crucial role in tailoring performance to various business needs. Applying explainability techniques like LIME and SHAP within the context of calcification reports has significantly enhanced our understanding of how models make predictions. We can see how specific features, such as a high transaction amount combined with unusual location data, contribute to a fostering transparency in our models. By analyzing the types of transactions consistently appearing in the reports, we can ensure our models operate fairly and efficiently [10].

2.6 Ethical Considerations

It is essential to emphasize that the dataset is entirely synthetic, devoid of any real-world personally identifiable information (PII) or sensitive financial data. The dataset strictly adheres to ethical standards, prioritizing privacy and confidentiality. This design allows users to freely explore and analyze various financial transaction scenarios, ensuring that individuals' privacy remains uncompromised throughout the study.

Efforts must be made to address potential biases in the dataset and model predictions, thereby minimizing the risk of discriminatory outcomes or unintended consequences.

2.7 Conclusions

This research demonstrates the effectiveness of a machine learning approach for fraud detection in payment systems. The combined model, leveraging three individual algorithms and soft voting, achieved near-perfect performance across various metrics. Continuous monitoring, adaptation, and further exploration of complementary techniques are essential for long-term success and robust protection against evolving fraud delving deeper into the quantitative and qualitative insights presented here, we can further refine the models and strengthen their effectiveness in securing future financial transactions.

Acknowledgements The work of first author is a part of the Project (CU-ORS-SM-24/69.). The work of third, fourth and fifth author is supported by an internal grant project of VSB-Technical University of Ostrava (SGS project, grant number SP2023/076), and project "Metaheuristics Framework for Multiobjective Combinatorial Optimization Problems (META MO-COP)" of Czech Republic Ministry of Education, Youth and Sports, reg. no. LTAIN19176 and [DST/INT/Czech/P-12/2019].

References

1. https://www.grantthornton.in/insights/financial-and-cyber-fraud-report-2024/
2. Chaquet-ulldemolins, J., Moral-rubio, S., Muñoz-romero, S.: On the black-box challenge for fraud detection using machine learning (II): nonlinear analysis through interpretable autoencoders. Appl. Sci. **12**, 3856 (2022)
3. Bodepudi, H.: Credit card fraud detection using unsupervised machine learning algorithms. Int. J. Comput. Trends Technol. **69**(8), 1–3 (2021). https://doi.org/10.14445/22312803/IJCTT-V69I8P101
4. Mijwil, M.M, Salem, I.E.: Credit card fraud detection in payment using machine learning classifiers. Asian J. Comput. Inf. Syst. **8**(4) (2020). https://doi.org/10.24203/ajcis.v8i4.6449
5. Akinje, A.O., Fuad, A.: Fraudulent detection model using machine learning techniques for unstructured supplementary service data. Int. J. Innov. Comput. **11**(2), 51–60 (2021). https://doi.org/10.11113/ijic.v11n2.299
6. Nobel, S.M.N., Sultana, S., Singha, S.P., Chaki, S., Mahi, M.J.N., Jan, T., Barros, A., Whaiduzzaman, M.: Unmasking banking fraud: unleashing the power of machine learning and explainable AI (XAI) on imbalanced data. Information **15**, 298 (2024). https://doi.org/10.3390/info15060298
7. Wang, J., Yang, C.: Financial fraud detection based on ensemble machine learning. In: 2022 IEEE International Conference on Dependable, Autonomic and Secure Computing; International Conference on Pervasive Intelligence and Computing; International Conference on Cloud and Big Data Computing; International Conference on Cyber Science and Technology Congress (DASC/PiCom/CBDCom/CyberSciTech), pp. 1–6, Falerna, Italy (2022). https://doi.org/10.1109/DASC/PiCom/CBDCom/Cy55231.2022.9928001
8. Islam, M.A., Nag, A., Chowdhury, S., Fahim, S.F.A., Ghosh, A., Mumtaj, N.: Utilization of encoding, early stopping, hyper parameter tuning, and machine learning models for bank fraud detection. In: 2023 IEEE 9th International Women in Engineering (WIE) Conference on Electrical and Computer Engineering (WIECON-ECE), pp. 321–327, Thiruvananthapuram, India (2023). https://doi.org/10.1109/WIECON-ECE60392.2023.10456503

9. Vashistha, A., Tiwari, A.K.: Building resilience in banking against fraud with hyper ensemble machine learning and anomaly detection strategies. SN Comput. Sci. **5**, 556 (2024). https://doi.org/10.1007/s42979-024-02854-w
10. Ranjan, P., Santhosh, K., Kumar, A., Kumar, S.: Fraud detection on bank payments using machine learning. In: 2022 International Conference for Advancement in Technology (ICONAT), pp. 1–4 (2022). https://doi.org/10.1109/ICONAT53423.2022.9726104
11. Stojanović, B., et al.: Follow the trail: machine learning for fraud detection in fintech applications. Sensors **21**(5), 1594 (2021). https://doi.org/10.3390/s21051594
12. Ashtiani, M.N., Raahemi, B.: Intelligent fraud detection in financial statements using machine learning and data mining: a systematic literature review. IEEE Access **10**, 72504–72525 (2022). https://doi.org/10.1109/access.2021.3096799
13. Khan, M.Z., et al.: The performance analysis of machine learning algorithms for credit card fraud detection. Int. J. Online Biomed. Eng. **19**(03), 82–98 (2023). https://doi.org/10.3991/ijoe.v19i03.35331
14. Ileberi, E., Sun, Y., Wang, Z.: Performance evaluation of machine learning methods for credit card fraud detection using SMOTE and AdaBoost. IEEE Access **9**, 165286–165294 (2021). https://doi.org/10.1109/access.2021.3134330
15. Elhamid, H.E.M.A., Khalif, W., Roushdy, M., Salem, A.-B.M.: Machine learning techniques for credit card fraud detection. Future Comput. Inform. J. **4**(2), 98–112 (2020). https://doi.org/10.54623/fue.fcij.4.2.5
16. Samir, P., Bandyopadhyay, K.: 'Detection of fraud transactions using recurrent neural network during COVID-19 1 Shawni Dutta (2020). https://doi.org/10.20944/preprints202006.0368.v1
17. Jian, L., Gu, X., Chen, S.: Quantitative detection of financial fraud based on deep learning with combination of e-commerce big data. Complexity **2020**, 1–11 (2020). https://doi.org/10.1155/2020/6685888
18. Mallika, R.: Fraud detection using supervised learning algorithms. Int. J. Adv. Res. Comput. Commun. Eng. **6**(6), 6–10 (2017). https://doi.org/10.17148/ijarcce.2017.6602

Chapter 3
Visualization and Predictive Analysis of Product Shipment/Repair Based on Power BI and ARIMA

Charles Chen and Ye Wang

Abstract This study takes the product shipment and repair data of a technology company in Shenzhen as an example, uses Power BI to build a Data Warehouse and visually analyze the operation of the enterprise. Therefore, the management and relevant departments of the company can have a more intuitive understanding of the sales and quality of the products, and make timely decisions. At the same time, this also helps quickly identify trends, patterns, and anomalies in the business process, and improve production quality. We also use ARIMA model to predict product shipment volume and repair rate. This will provide a scientific basis for the company's production planning, inventory management, and after-sales service. It also can prompt companies to take corresponding preventive actions, reduce quality costs and customer complaint rates. The ARIMA program coded by Python is integrated into Power BI to provide a practical and interactive charts. Hybrid methods of machine learning and deep learning with ARIMA are also discussed in this paper in order to improve non-liner problem of the prediction.

3.1 Introduction

The management activities of enterprises require a lot of data to support their decision-making. The purpose of this study is to establish data warehousing, OLAP services, and front-end visualization tools to enable enterprises to quickly and clearly understand their operational status and problems. In this study, Microsoft's Power BI tool (Power Query) was used to extract, clean, convert, and load transaction data from enterprise ERP and other backend systems into Power BI data tables. The model view function was used to establish various dimension tables (such as time dimension,

C. Chen (✉)
Overseas Education College, Fujian Business University, Fuzhou City, China
e-mail: 2440216302@qq.com

Y. Wang
School of Management, Xiamen University Tan Kah Kee College, Zhangzhou, China

S.-C. Chu et al. (eds.), *Advances in Intelligent Data Analysis and Applications*,
Smart Innovation, Systems and Technologies 445,
https://doi.org/10.1007/978-981-96-7277-6_3

customer dimension table, location dimension table, etc.) and establish associations with fact tables and other data tables. The DAX (data analysis expressions) time intelligence function in Power BI facilitates users in creating time dimension tables. In the data conversion section, the M function of Power Query can be used to calculate or format relevant data into the desired format.

Science and technology are one of the core competencies of modern enterprises, and demand forecasting, as an important component of enterprise wide operational management, plays a crucial role in supply chain management. Through strategic material purchase planning, companies can arrange material procurement in advance to avoid delays caused by material shortages during the production process. At the same time, through capacity forward looking planning, enterprises can plan their production capacity before production, avoiding the impact of machine and equipment shortages on production progress. In addition, reasonable arrangement of forward looking laboratory allocation is also an important guarantee to ensure smooth production. Finally, by optimizing logistics network optimization, enterprises can achieve efficient arrangement of material transportation, laying the foundation for timely and fast delivery of products to customers. Due to the time stamps during production shipment, and repair, which are usually linked to seasonality and periodicity, this study will use time series as the main prediction model for shipment and repair forecasting. Although there are many factors that affect shipment and repair, based on the data analysis provided by Shenzhen enterprises, time series forecasting models will be the main option at present.

The main contribution of this study is to establish a data warehouse using Power BI and apply its OLAP and visualization technologies to provide a fast and convenient analysis platform for enterprises. And through the establishment of a time series ARIMA model, it was written into a Python program and integrated into Power BI to provide integrated visual analysis for enterprises, assist them in understanding operational situations, improving problems, establishing core competencies, and creating competitive advantages.

The subsequent chapters of this study are arranged as follows: Sect. 3.2 is related works including the instruction of Power BI, time series. Section 3.3 is the research methods. Section 3.4 is the results' analysis. Section 3.5 is the conclusion and future research.

3.2 Related Works

3.2.1 The Application of Power BI in Data Analysis

Power BI, as a powerful business intelligence tool launched by Microsoft, has been widely applied in various industries. Many enterprises use Power BI to integrate the entire data processing process from data preparation, data analysis, data visualization to collaborative report sharing, helping users get started with data analysis.

Researchers are also continuously exploring more application scenarios and functional extensions of Power BI, such as integration with other data analysis tools and mobile applications. It is a platform product that, through a series of functions such as data view management, data transformation management, OLAP database management, multidimensional report design, and ad hoc report design, can quickly help enterprise IT personnel build multidimensional analysis models, create ad hoc reports, create analysis reports, and publish BI applications such as management cockpits on the basis of existing ERP/CRM information systems. Figure 3.1 is the architecture of Power BI.

Power Query. Power Query can be considered as an ETL (Extract-Transform-Load) tool. It can fetch data from almost any kind of data source, and transform data and load data. Power Query has three layers, see Fig. 3.2. But, some commercial tools may only have first two layers. Those three layers are

(1) M engine: The underlying query execution engine utilizes expressions composed of M functions and operators within the Power Query formula language M to execute queries.

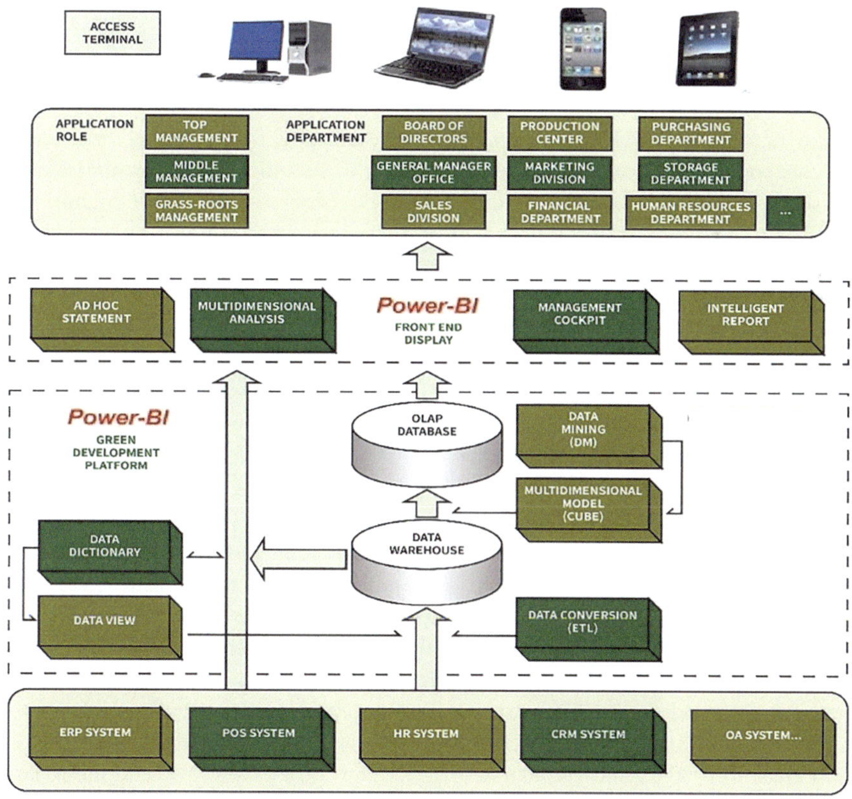

Fig. 3.1 The architecture of Power BI

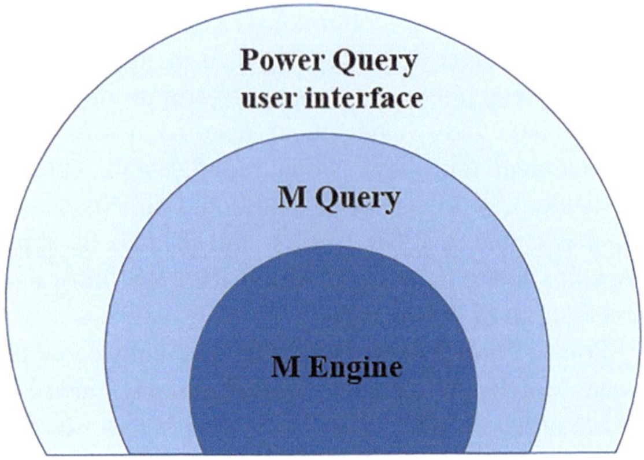

Fig. 3.2 The architecture of Power Query

(2) M query: A set of commands written in the Power Query formula language M.
(3) Power Query user interface: Also called by Power Query Editor. A graphic user interface helps user execute operations.

3.2.2 Data Warehouse and OLAP

In the late 1980s, IBM researchers Barry Devlin and Paul Murphy introduced the concept of data warehouses (DWs), aiming to provide an architectural model designed to address the flow of data toward decision support environments [1]. According to Inmon's definition, a data warehouse is characterized as a "subject-oriented, nonvolatile, integrated, time-variant collection of data intended to support management decisions" [2]. It is a data repository which stores structured, filtered, and processed data that has been treated for a specific purpose. Through the query and analysis tools OLAP (online analytical processing), decision analysis and data mining in the data analysis and reporting module, information is extracted to meet the needs of decision-making. In the classical approach there are essentially three main types of OLAP servers: Relational OLAP (ROLAP), Multidimensional OLAP (MOLAP), and Hybrid Online Analytical Processing (HOLAP) [3]. Figure 3.3 shows the components of Data Warehouse.

OLAP is a technology that empowers analysts to conduct multidimensional data analysis, enabling them to examine business information from various perspectives. It also facilitates the examination of data across multiple and heterogeneous databases [4]. The DW and OLAP is inseparable in the Data Warehousing system. Their relationship is shown in Fig. 3.4. DW is a huge database which includes the historical data of enterprise. Those historical data can be used to support business decision. While the OLAP service tools utilize multidimensional datasets and data aggregation techniques to process and summarize data in data warehouses, evaluate these data using online analysis and visualization tools, and quickly return complex analysis and search results to users.

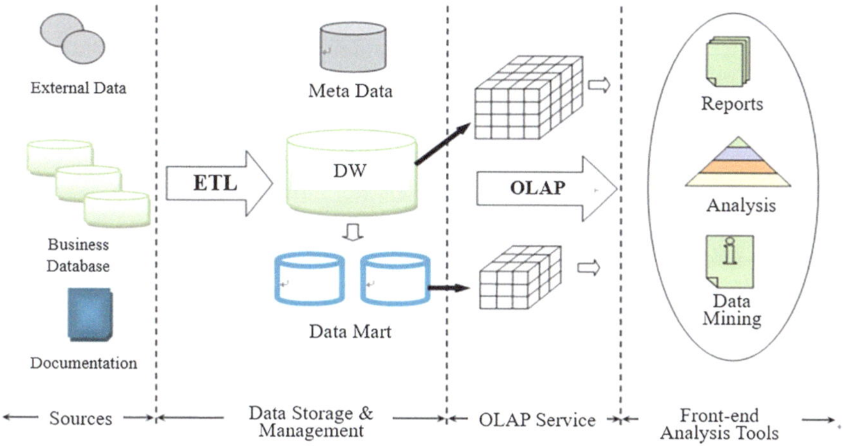

Fig. 3.3 The components of data warehouse

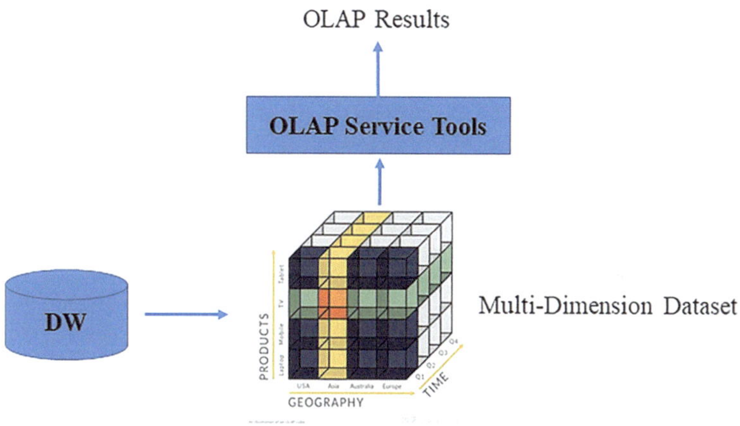

Fig. 3.4 The relationship between DW and OLAP

3.2.3 Time Series

The time series analysis method was proposed by Box Jenkins in 1965 [5]. It is defined as a sequence of values, chronologically ordered, and observed over time. Although the time is a variable measured on a continuous basis, the values in a time series are sampled at constant intervals (fixed sampling frequency). The two characteristics of time series models that differ from classical regression models are: (1) this modeling method is not based on economic theory, but on the change patterns of variables themselves, using extrapolation mechanisms to describe the changes in time series. (2) Clearly consider the non-stationarity of time series. If the time series is non-stationary, it should be transformed into a stationary time series through differencing before considering the modeling problem [6]. The data of time series may have the following characteristics [7]:

- Trend: This characteristic is the general movement that the time series exhibits during the observation period, without considering seasonality and irregularities.
- Seasonality: This characteristic indicates the variations that occur at specific regular intervals and may provide useful information when time periods exhibit similar patterns.
- Stationarity: A stationary time series refers to a sequence of data points where its statistical characteristics, such as mean, variance, and dispersion (among other properties), remain constant over time. Specifically, a stationary time series that exhibits no trend implies that the fluctuations around its average value maintain a consistent magnitude. Additionally, for any given lag, the autocorrelation of a stationary time series remains constant over time.

Based on these assumptions, a time series of this form can be modeled as the sum of a deterministic signal and a stochastic noise component [8, 9]. Time series can be divided into four categories: autoregressive (AR) models, moving average (MA) models, autoregressive moving average (ARMA) models, and single product autoregressive moving average (ARIMA) models.

AR (Autoregressive) Model. Autoregressive model is a statistical technique that involves using a variable's past values to predict its future behavior. Specifically, it establishes a linear relationship between the variable's current value and its values at several preceding time points [10]. Consider a time series

$$\begin{aligned} x_t &= c + \phi_1 x_{t-1} + \phi_2 x_{t-2} + \ldots + \phi_p x_{t-p} + \mu_t \\ &= x_t = c + \sum\nolimits_{i=1}^{p} \phi_i x_{t-1} + \mu_t \end{aligned} \tag{3.1}$$

Among them, ϕ_i $(i = 1, \ldots, p)$ is an autoregressive parameter, μ_t is white noise. Then x_t is a p-order autoregressive process and represented by AR (p). x_t is composed of the weighted sum of its p lagged variables and the addition of μ_t.

MA (Moving Average) Model. This methodology considers the interdependencies that may exist between observations and the error terms generated when a moving average model of order q is applied to time series data with temporal lags. It can be expressed as

$$x_t = \mu_t + \theta_1 \mu_{t-1} + \vartheta_2 \mu_{t-2} + \ldots + \theta_q \mu_{t-q} = \sum\nolimits_{i=0}^{q} \theta_i \mu_{t-i} \tag{3.2}$$

where θ_i represent the weights applied to the current and past values of the stochastic term of the time series, and $\theta_0 = 1$. μ_t is white noise. The above equation is called the q-order moving average process, denoted as MA (q).

ARMA (Autoregressive Moving Average) Model. By combining autoregression (AR) model and moving average (MA) model, we can create the ARMA model of order (p, q) as

$$x_t = c + \sum_{i=1}^{p} \phi_i x_{t-1} + \mu_t + \sum_{i=0}^{q} \theta_i \mu_{t-i} \tag{3.3}$$

where $\phi_i \neq 0$, $\theta_i \neq 0$, $\sigma_\mu^2 > 0$. The parameters p, q constitute the order of the AR and MA models, respectively.

ARIMA (Autoregressive Integrated Moving Average) Model. The ARIMA model was introduced by Box and Jenkins in 1970 as a generalization of the ARMA model, making it suitable for analyzing non-stationary time series. The classical ARMA model assumes stationarity of the analyzed time series, which necessitates the transformation of inherently non-stationary time series into stationary data through the removal of seasonality and trends, often achieved using finite differencing [9, 11]. The general form of an ARIMA model is written as ARIMA (p, d, q), including the integration term that guarantees the stationarity of the time series [12], and it can be expressed as

$$\nabla^d x_t = c + \sum_{i=1}^{p} \phi_i \nabla^d x_{t-i} + \sum_{i=0}^{q} \theta_i \mu_{t-i} \tag{3.4}$$

where ∇^d represents a differential operator of order d, designed to eliminate the non-stationarity present in the time series x_t [13].

ARIMA Parameter Determination. How to identify the best-performing ARIMA model among a number of different ARIMA (p, d, q) models for a specific forecasting application has become an important research topic. The selection of the p and q parameters relies on analyzing the autocorrelation function (ACF) and partial autocorrelation function (PACF) plots for the specific dataset. Meanwhile, the d parameter is determined to ensure that the time series becomes stationary. In the majority of scientific literature covered in this work, identifying the most performing ARIMA model among various ARIMA (p, d, q) models is a common requirement for specific forecasting applications. In such situations, two primary criteria have been introduced: the Akaike information criterion (AIC) and the Bayesian information criterion (BIC). These metrics are used to evaluate model performance by balancing goodness of fit with model complexity. The AIC assesses how well a model explains the data while penalizing overly complex models. Similarly, the BIC applies the same principle but imposes a stronger penalty for more complex models. Both criteria favor models with lower values, indicating better performance. Rob J Hyndman [14] builds a popular heuristic method for parameter selection in an ARIMA application, and it is based on a simple step-wise algorithm as follows:

(1) Begin by selecting a small set of basic ARIMA models and choose the one with the lowest AIC value.
(2) Evaluate various variations of the selected model, calculating their respective AIC values. If a variation yields an AIC lower than the current reference model, it replaces the reference model, and this process repeats. The algorithm terminates when no closer alternative models with lower AIC exist within the considered variations.

By employing this approach, a valid ARIMA model is guaranteed to be identified due to the finite nature of the model space being explored and at least one initial model serving as an acceptable solution [15].

3.3 Research Methods

The original data source is collected and cleaned from a large manufacture company located at Shenzhen. There are 60,590 records for the shipment and 176,499 records for repair return (single product may have multiple returns) from 2018/01 to 2024/07. A return quantity-shipment quantity file is built according to the shipment records. The attributes of shipment and return are listed in Appendix.

First, we use the Power Query function of Power BI to load and convert Excel data into Power BI. Then, we create a date table and use star schema to build a data warehouse [16, 17]. Figure 3.5 is the model view.

Via the OLAP operations of slice, dice, pivot, drill up, and drill down, we can get an interactive view of product shipment and return. <Year>, <ITEM-NO>, <MODEL_NAME>, and <BRAND_NAME> can all be drilled down to select the shipment status of the product to be queried, as shown in Fig. 3.6. Second, the ARIMA forecasting coded by Python is used for shipment and return forecasting. In this case (1, 1, 8) order is selected for (p, d, q). Those Python programs are integrated into Power BI.

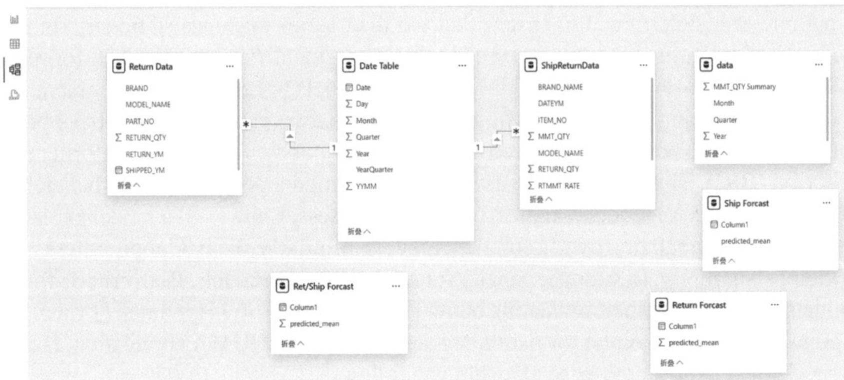

Fig. 3.5 Model view

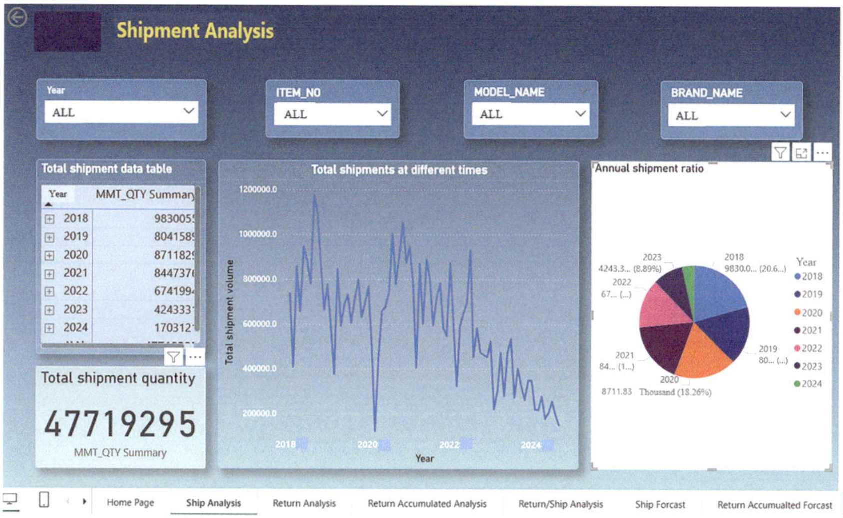

Fig. 3.6 Detailed inspection and analysis of shipped products

3.4 Results' Analysis

In terms of customer shipment analysis, the company does not have many global major clients. The main well-known clients include ACER, Amazon, DELL, INTEL, and LENOVO. We should be dedicated to enhance the development of major clients such as CISCO, APPLE, Huawei, etc. and analyze the shipment status of ACER, as shown in Fig. 3.7. It is known that the shipment volume in 2023 is 826160. In recent years, the proportion of shipments has only been 11.91%. Except for slightly higher than the 2021 shipment volume of 792,326 (11.42%), it is lower than previous shipments. Further investigation can be conducted on the reasons for reducing shipments for Model Name and Item No., and improvements can be made accordingly. The shipment status of other customers can also be analyzed via the selection of BRAND_NAME.

Regarding the forecasting, Fig. 3.8 shows the shipment forecast view. You can select the manufacturer or model from the drop-down menu. This prediction model uses time series ARIMA to make shipment forecasts for the next five cycles (months). For using statistical models for analysis, the data must meet the requirements of the statistical model. For ARIMA, it must meet the requirement of stationarity in order to achieve a certain level of accuracy in its prediction results. Unfortunately, through data analysis, the current data, except for the overall data that barely meets the requirements, is too discontinuous in terms of date segments after selecting manufacturers or models from the drop-down menu, which cannot meet the requirements.

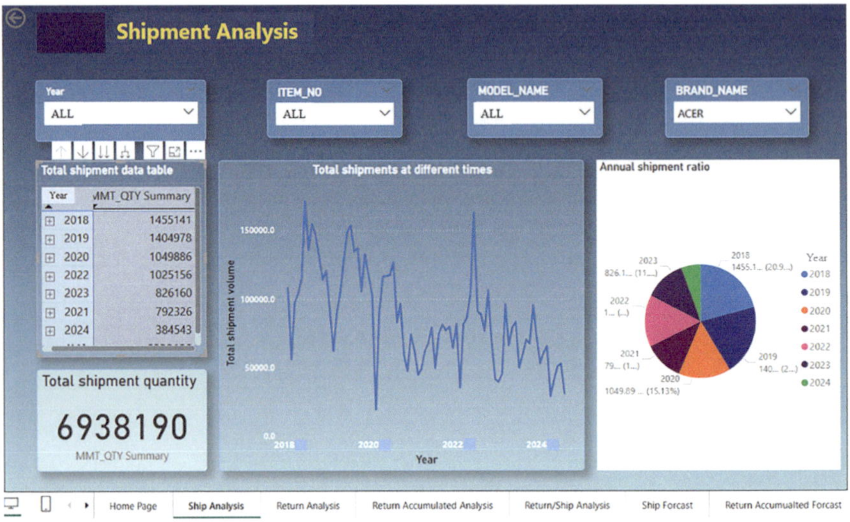

Fig. 3.7 Analysis of ACER shipment status

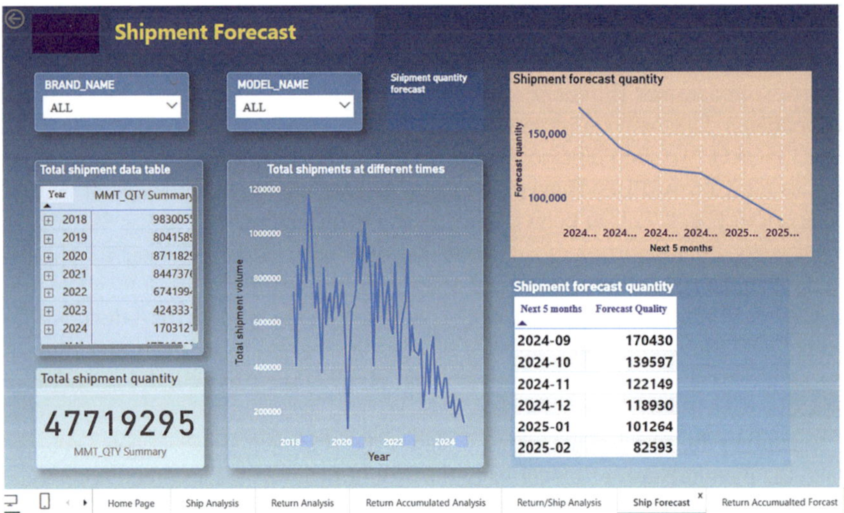

Fig. 3.8 Overall shipment forecast analysis view

3.5 Conclusion and Future Research

Through the Power Query, DAX of Power BI, we first build a Data Warehouse for a Shenzhen manufacture company. Second, using the OLAP service and Report view functions of Power BI, we can visually analyze the operation of the enterprise.

Management and relevant departments can quickly identify trends, patterns, and anomalies in the business process, and make timely decisions. We also use ARIMA model coded by Python to predict product shipment volume, repair volume, and repair/shipment rate. This can provide a scientific basis for the company's production planning, inventory management, and after-sales service. It also can prompt companies to take corresponding preventive actions and to reduce customer complaint rates. However, in order to make more accurate forecasting for shipment and repair, there are a lot of improvements needed to be implemented in the future research. The summary is listed below:

(1) Due to time constraints and limited personal ability, we don't take ADF (augmented Dickey–Fuller) unit root detection to determine whether the random process of the data is stationary or not. If the data is not stationary, we have to perform differencing on the time series data until we obtain a stationary sequence, where p-value is significant. At this point, we can obtain the value of parameter d.
(2) Identify the order p and q parameters of the ARMA model based on a stationary time series. Estimate the autocorrelation function (ACF) and partial autocorrelation function (PACF) obtained from the sample, that is, analyze the truncation/tailing characteristics from the correlation graph and partial correlation graph to determine the p and q parameters of the model.
(3) We don't write a Python program to explore the optimal ARIMA (p, d, q) value. After checking and calculating the moving average of data, we just simple test the results of (1, 1, 8). Follow the proposed step-wise algorithm by Hyndman [14], further research can write a Python program to traverse the combination of parameters p, d, and q, use AIC/BIC minimization to select the optimal (p, d, q) parameter for ARIMA application.
(4) A time series is defined as a sequence of values ordered chronologically and observed over time. Although time itself is measured continuously, the values in a time series are collected at fixed intervals (constant sampling frequency). Through data analysis, we can find the return repair data or shipment data by brand name, by mode name is too discontinuous in terms of date segments, so many data are missed. Further research may be conducted by addressing missing information and excluding the entire record [18]. Additionally, the target time series can be decomposed into three components: trend, seasonality, and residuals. Separate predictions can then be made for each of these components [19]. Some researchers also propose deep learning time series models such as deep feedforward networks (DFFNN) for time series forecasting model [19], extended conventional time series model with uncertainty propagation [20].

Appendix

(1) Part of shipment data

BRAND_NAME	MODEL_NAME	ITEM_NO	MMT_QTY	DATEYM
ECS	MF50AN(1.0) IO BD	89B306-XL6A22	13	2023-10
INTEL	NUC13LCH	95D663-QB8116	995	2023-10
ACER	Aspire XC-1785	95D133-OB8D03	1	2023-10
ACER	Nitro N50-656	95D855-OB8014	1	2023-10
ECS	H610H7-TI5(1.0)	89-306-QC5102	1,000	2024-01
TATUNG	TUF66WL	91D520-LQ2063	350	2023-10
LENOVO	P565A4-LM5(1.0)	89D886-MI5158	5	2024-01
DELL	RX7900 XTX-24G	89T386-PD7013	80	2024-02
DELL	RADEON RX 6300-2G	89T386-PG4018	950	2024-01
AOPEN	DE3650-S	95D246-QA9008	200	2024-01
LENOVO	Z790H7-LA(1.0)	89D886-QC7144	10	2024-01
ACER	Predator PO7-650	95D707-OB7124	18	2024-01
LENOVO	Z790H7-LA(1.0)	89D386-QC7113	1	2024-01
JPS	SF20GM7(1.71) MB	89B886-XC204P	4	2024-01
ACER	Nitro N50-656	95D855-OB8K25	100	2024-02
ACER	Nitro N50-650	95D855-OB8F53	196	2024-02
ACER	Veriton X2710G	95D256-QA2E97	252	2024-01
ECS	H610H7-IM1(1.1)	89-306-QD3110	200	2024-01

(2) Part of returned repair data

BRAND	MODEL_NAME	PART_NO	SHIPPED_YM	RETURN_YM	RETURN_QTY
ECS	RS780M-M3	89-206-Q48108	2009-11	2018-01	1
ECS	H55H-M	89-206-R24104	2010-05	2018-12	1
ECS	H67H2-I	89-206-Y72110	2011-03	2018-09	1
IGS	N9800GTE-1GMU-F	89-306-451085	2010-12	2018-10	1
ECS	G41T-M6	89-306-R26110	2010-09	2018-09	1
ECS	G41T-M6	89-206-R26303	2011-09	2020-02	1
ECS	H67H2-M	89-206-Y62121	2011-04	2020-07	1
SMC	C7Q67	89D216-Y89000	2017-06	2018-02	1
CHINA ONE	TIGD-I	89D386-Y34100	2012-05	2020-05	4
CHINA ONE	TIGD-I	89D386-Y34100	2013-10	2020-03	1
CHINA ONE	TIGD-I	89D386-Y34100	2013-10	2019-07	1
CHINA ONE	TIGD-I	89D386-Y34100	2013-10	2019-05	1
CHINA ONE	TIGD-I	89D386-Y34100	2012-05	2018-08	2
CHINA ONE	TIGD-I	89D386-Y34100	2012-05	2018-07	2
CHINA ONE	TIGD-I	89D386-Y34100	2013-08	2018-07	2
ECS	H61H2-M2	89-206-Y90120	2011-06	2020-03	1
ECS	H61H2-M2	89-206-Y90120	2011-11	2020-03	1

(3) Accumulated repair quantity/shipment quantity by ITEM_NO and date

BRAND_NAME	MODEL_NAME	ITEM_NO	MMT_QTY	DATEYM	RETURN_QTY
ECS	MF50AN(1.0) IO BD	89B306-XL6A22	13	2023-10	
INTEL	NUC13LCH	95D663-QB8116	995	2023-10	
ACER	Aspire XC-1785	95D133-OB8D03	1	2023-10	
ACER	Nitro N50-656	95D855-OB8014	1	2023-10	
ECS	H610H7-TI5(1.0)	89-306-QC5102	1,000	2024-01	22
TATUNG	TUF66WL	91D520-LQ2063	350	2023-10	
LENOVO	P565A4-LM5(1.0)	89D886-MI5158	5	2024-01	1
DELL	RX7900 XTX-24G	89T386-PD7013	80	2024-02	
DELL	RADEON RX 6300-2G	89T386-PG4018	950	2024-01	1
AOPEN	DE3650-S	95D246-QA9008	200	2024-01	
LENOVO	Z790H7-LA(1.0)	89D886-QC7144	10	2024-01	4
ACER	Predator PO7-650	95D707-OB7124	18	2024-01	
LENOVO	Z790H7-LA(1.0)	89D386-QC7113	1	2024-01	
JPS	SF20GM7(1.71) MB	89B886-XC204P	4	2024-01	
ACER	Nitro N50-656	95D855-OB8K25	100	2024-02	
ACER	Nitro N50-650	95D855-OB8F53	196	2024-02	

References

1. Devlin, B. Thirty years of data warehousing—Part 1 (2020). https://www.irmconnects.com/thirty-years-ofdata-warehousing-part-1/. Last accessed 27 Oct 2022
2. Inmon, W.H.: Building the Data Warehouse, 4th edn. Wiley Publishing, Indianapolis, IN, USA (2005)
3. Chaudhuri, S., Dayal, U., Ganti, V.: Database technology for decision support systems. Computer **34**(12), 48–55 (2001)
4. Khalil, A., Belaïssaoui, M.: Key-value data warehouse: models and OLAP analysis. In: 2020 IEEE 2nd International Conference on Electronics, Control, Optimization and Computer Science (ICECOCS), IEEE, pp. 1–6 (2020)
5. Box, G.E.P., Jenkins, G.M.: Time Series Analysis, Forecasting and Control; Holden-Day San Francisco, CA. (1970)
6. People's Bank of China Difference Adjustment Statistics Department: Time Series X-12-ARIMA Seasonal Adjustment—Principles and Methods. China Financial Publishing House, Beijing (2006)
7. Nie, H., Liu, G., Liu, X., Wang, Y.: Hybrid of ARIMA and SVMs for short-term load forecasting. Energy Procedia **16**, 1455–1460 (2012). https://doi.org/10.1016/j.egypro.2012.01.229
8. Rundo, F., Trenta, F., di Stallo, A.L., Battiato, S.: Machine learning for quantitative finance applications: a survey. Appl. Sci. (9), 5574 (2019). https://doi.org/10.3390/app9245574
9. Kontopoulou, V.I., Panagopoulos, A.D., Kakkos, I., Matsopoulos, G.K.: A review of ARIMA versus machine learning approaches for time series forecasting in data driven networks. Futur. Internet **15**(255), (2023). https://doi.org/10.3390/fi15080255.
10. Metcalfe, A.V., Cowpertwait, P.S.P.: Introductory time series with R. Springer (2009)
11. Box, G.E., Jenkins, G.M., Reinsel, G.C., Ljung, G.M.: Time Series Analysis: Forecasting and Control. Wiley, Hoboken, NJ, USA (2015)
12. Siami-Namini, S., Tavakoli, N., Namin, A.S.: A comparison of ARIMA and LSTM in forecasting time series. In: Proceedings of the 2018 17th IEEE International Conference on Machine Learning and Applications (ICMLA), pp. 1394–1401. Orlando, FL, USA (2018)
13. Xue, B., Tong, N., Xu, X., He, X.: Dynamical short-term prediction of rain attenuation in w band: a time-series model with simpler structure and higher accuracy. IEEE Antennas Propag. Mag. **61**, 77–86 (2019)
14. RDocumentation: Auto.Arima: Fit Best ARIMA Model to Univariate Time Series., https://www.rdocumentation.org/packages/forecast/versions/8.21/topics/auto.arima. Last accessed 12 Jul 2023

15. Hyndman, R.J., Khandakar, Y.: Automatic time series forecasting: the forecast package for R. J. Stat. Softw. **27**, 1–22 (2008)
16. Rizzi, S., Abelló, A., Lechtenbörger, J., Trujillo, J.: Research in data warehouse modeling and design: Dead or alive? In: Proceedings of the 9th ACM International Workshop on Data Warehousing and OLAP, DOLAP '06, Arlington, VA, USA (10 November 2006); Association for Computing Machinery: pp. 3–10. New York, NY, USA (2006)
17. Nambiar, A., Mundra, D.: An overview of data warehouse and data lake in modern enterprise data management. Big Data Cogn. Comput. **6**(4), (2022). https://doi.org/10.3390/bdcc6040132
18. Buuren, S.: Flexible imputation of missing data. UK: Chapman & Hall/CRC (2012)
19. Torres, José F., et al.: Deep learning for time series forecasting: a survey. Big Data **9**(1), 3–21 (2021)
20. Barazani, O., Tolpin, D.: Monitoring time series with missing values: a deep probabilistic approach. In: International Symposium on Cyber Security, Cryptology, and Machine Learning, pp. 19–28. Springer International Publishing, Cham (2022)

Chapter 4
Performance Analysis of Several Famous Meta-Heuristic Algorithms Under the Congress on Evolutionary Computation Test Suites

Xiubin Chen, Dewang Chen, and Zhenyu Meng

Abstract This paper investigates several well-known meta-heuristic algorithms and conducts experiments and analyses on their performance using the universal test suites on evolutionary computation. The experimental results reveal the following findings: (1) The performance of the Particle Swarm Optimization (PSO) algorithm is usually inferior to Differential Evolution (DE) and the QUasi-Affine TRansformation Evolution (QUATRE) algorithm in terms of both optimization accuracy and convergence speed. (2) Genetic Algorithms (GA), DE, and QUATRE belong to the same class of algorithms. DE originated from the genetic annealing algorithm, which is a hybrid genetic algorithms and simulated annealing, thereby inheriting the search framework of genetic algorithms. (3) The QUATRE algorithm introduces an evolution matrix that replaces the crossover operation in DE, while still utilizing the same search manner of a high-dimensional cube, thereby achieving unbiased and efficient search. In summary, the QUATRE algorithm exhibits fast convergence speed and strong global search capability, making it suitable for solving high-dimensional and complex problems in comparison with DE and PSO.

X. Chen · D. Chen (✉)
School of Traffic and Transportation, Fujian University of Technology, Fuzhou, China
e-mail: 15159661709@163.com

X. Chen · Z. Meng
Institute of Artificial Intelligence, Fujian University of Technology, Fuzhou, China

D. Chen
Fujian Provincial BeiDou Navigation and Intelligent Transportation Collaborative Innovation Center, Fuzhou, China

S.-C. Chu et al. (eds.), *Advances in Intelligent Data Analysis and Applications*,
Smart Innovation, Systems and Technologies 445,
https://doi.org/10.1007/978-981-96-7277-6_4

4.1 Introduction

Meta-heuristics [1] are general algorithmic frameworks, often inspired by nature, designed to solve complex optimization problems, and they have been a growing research area for the past few decades. A meta-heuristic [2] represents an advanced procedure employed to seek an approximate global optimum for optimization challenges characterized by incomplete or flawed data, as well as constrained computational resources. The following sections will provide a detailed overview of these four optimization algorithms.

The concept of a Genetic Algorithm (GA) [3] was first introduced by John Holland. GA simulates the process of biological evolution and is applicable for solving complex, nonlinear, and multi-modal optimization problems, particularly when the search space is large and the problems are difficult to model using traditional methods. Particle Swarm Optimization (PSO) [4–6] was first introduced by Kennedy and Eberhart in 1995, simulating the behavior of bird flocks searching for corn. It seeks optimal solutions by simulating the collective behavior of individual particles in the search space. Moreover, this algorithm is relatively straightforward and efficient for such types of problems. Differential Evolution (DE) [7–9], proposed by Storn and Price, is a fast and simple technique that performs well on a wide variety of problems. When addressing continuous parameter optimization problems, such as high-dimensional issues, the DE algorithm is relatively simple, robust, and easy to implement. QUasi-Affine TRansformation Evolution (QUATRE) [10, 11] algorithm was proposed in 2016. QUATRE combines affine transformations with evolutionary strategies. Affine transformations are a type of transformation that lies between linear and nonlinear, allowing for the adjustment of the shape and structure of the search space to enhance the algorithm's search capabilities.

This section introduces meta-heuristic algorithms and four optimization algorithms. In the following sections, a detailed description of the implementation of several meta-heuristic algorithms is provided in Sect. 4.2. Section 4.3 presents the performance analysis of the meta-heuristic algorithms on the universal test suites. Section 4.4 concludes the paper.

4.2 Several Meta-Heuristic Algorithms

In this section, a detailed exposition of the implementation of GA, PSO, DE, and QUATRE will be provided.

4.2.1 Genetic Algorithm

The specific implementation of the standard GA [12, 13] is as follows. A preliminary population P_0 is randomly generated, where each individual i within the population is represented as an encoded string $X_{i,G}$, where $X_{i,G}$ denotes the vector of the i-th individual in the G-th iteration of the population. The fitness $f(X_{i,G})$ of each individual $X_{i,G}$ in the initial population is assessed. Individuals are selected based on their fitness values. Typical methods of selection encompass the roulette wheel approach and the tournament method. The probability $P_{i,G}$ of roulette wheel selection can be formulated as (4.1), where $P_{i,G}$ denotes the vector of the i-th individual in the G-th iteration of the probability.

$$P_{i,G} = \frac{f(X_{i,G})}{\sum_{j=1}^{N} f(X_{j,G})} \tag{4.1}$$

where N denotes the size of the population. Information is exchanged between selected individuals through crossover operations to generate new individuals. A simple single-point crossover operation can be expressed as

$$e_{new} = (X_1[1...c], X_2[c + 1...end]) \tag{4.2}$$

where c is a randomly selected crossover point, and X_1 and X_2 are the two parent individuals participating in the crossover. With a certain mutation probability p_m, certain genes of the individuals are altered to increase the diversity of the population. A simple bit mutation operation can be formulated as

$$X'_{i,G} = \begin{cases} 1 - X_{i,G} & \text{if rand}(0, 1) < p_m \\ X_{i,G} & \text{otherwise} \end{cases} \tag{4.3}$$

where X is the original individual, X' is the mutated individual, and i is a randomly chosen gene position. The next generation is selected based on fitness, which may involve an elitism strategy, ensuring that the best individuals are carried over to the subsequent generation.

4.2.2 Particle Swarm Optimization Algorithm

The PSO [14] commences by dispersing particles throughout the search space with random positions and initial velocities. The fitness of each particle is then evaluated against the objective function, which measures how well each particle's position solves the problem. Each particle's personal best *pbest* is identified as its most successful position encountered so far. The global best *gbest* is established as the most optimal position found by any particle within the swarm. The particles' velocities and positions are subsequently updated, as follows:

$$\begin{aligned} V_{i,G+1} &= w \cdot V_{i,G} + c_1 \cdot r_1 \cdot (pbest_{i,G} - X_{i,G}) + c_2 \cdot r_2 \cdot (gbest_G - X_{i,G}) \\ X_{i,G+1} &= X_{i,G} + V_{i,G+1} \end{aligned} \tag{4.4}$$

where w is the inertia weight, c_1 and c_2 are acceleration constants, and r_1 and r_2 are random numbers within the interval [0, 1]. The algorithm checks for constraints and adjusts particle positions to ensure they stay within the feasible search space. This process repeats for a preset number of iterations or until a termination criterion is reached, such as an acceptable fitness level, a maximum iteration cap, or the absence of significant improvement. Once the stopping criterion is met, the algorithm is halted, and the global best position is adopted as the optimal solution to the problem.

4.2.3 Differential Evolution Algorithm

Differential Evolution is a simple yet efficient evolutionary algorithm [15–17]. Its optimization framework consists of initialization and evolution stages. During the initialization phase, a random population P, consisting of elements $[x_1, x_2, ..., x_{NP}]$, is generated across the search space. Here, NP signifies the size of the population. Each member x_i, represented as $(x_{i,1}, x_{i,2}, ..., x_{i,D})$, is formulated in the following manner:

$$x_{i,j} = L_j + rand(0, 1) \times (U_j - L_j), \quad j = 1, 2, ..., D \tag{4.5}$$

In which U_j and L_j correspond to the upper and lower limits of the j-th dimension, respectively. The evolution stage includes mutation, crossover, and selection operations. In mutation, mutant vectors are generated by perturbing target vectors, as follows:

$$v_{i,G} = x_{r,G} + F \cdot (x_{s,G} - x_{t,G}) \tag{4.6}$$

where $x_{r,G}$, $x_{s,G}$, and $x_{t,G}$ are distinct individuals randomly selected from the current population. F denote the scaling factor, governing the amplification of the difference between population individuals. In crossover, a trial vector $u_{i,G}$ is produced by performing the binomial crossover strategy between the mutant vector $v_{i,G}$ and the target vector $x_{i,G}$, as shown in Eq. (4.7).

$$u_{i,j,G} = \begin{cases} v_{i,j,G} & \text{if rand}(0, 1) \le CR \ or \ j = j_{rand} \\ x_{i,j,G} & \text{otherwise} \end{cases} \tag{4.7}$$

in which, CR denotes the crossover rate, while j_{rand} signifies a randomly chosen dimension index, guaranteeing that at least one dimension is inherited from the mutant vector. During selection, the fitness of the trial vector $u_{i,G}$ is compared with that of the target vector $x_{i,G}$, with the one demonstrating higher fitness being chosen to proceed to the next generation, as depicted in Eq. (4.8).

$$x_{i,G+1} = \begin{cases} u_{i,G} & \text{if } f(u_{i,G}) \leq f(x_{i,G}) \\ x_{i,G} & \text{otherwise} \end{cases} \tag{4.8}$$

where $f(\cdot)$ denotes the fitness function.

4.2.4 QUATRE Algorithm

In the QUATRE [18] algorithm, initially, a population is generated, which constitutes a matrix of particle coordinates denoted as $X = [X_1, X_2, \ldots, X_{ps}]^T$. For each particle, the personal best position is initialized as $X_{pbest,G} = X$, and the fitness values of all particles are calculated. Generate an evolutionary matrix by initializing a temporary matrix M_{tmp}. Perform a random permutation of the elements in each row of M_{tmp}, followed by a random permutation of the row vectors (with the elements within each row remaining unchanged), thereby obtaining the evolutionary matrix M, $\overline{M}$ refers to an inverse binary operation of all elements within matrix M. If the i-th particle achieves the best fitness value within the population, record its position using $X_{gbest,G}$. The global best matrix for the G-th generation is denoted as $X_{gbest,G} = [X_{gbest,G}, X_{gbest,G}, \ldots, X_{gbest,G}]^T$. Set a constant value c as the amplification factor. Generate two random matrices $X_{r1,G}$ and $X_{r2,G}$ by randomly permuting the row vectors of X. Update the particles using the following evolutionary equation:

$$\begin{cases} B = X_{gbest,G} + c \cdot (X_{r1,G} - X_{r2,G}) \\ X \to M \otimes X + \overline{M} \otimes B \end{cases} \tag{4.9}$$

where $X_{r1,G}$ and $X_{r2,G}$ denote randomly selected distinct vectors from the matrix in the G-th iteration. Subsequently, for each particle, if the fitness of the current position X_i is superior to that of its personal best $X_{pbest,G}$, then update $X_{pbest,G}$ and its fitness value. Among all particles, select the best personal best position as the global best $X_{gbest,G}$. Following this, for each particle, should the fitness of the current position X_i exceed that of its individual optimum $X_{pbest,G}$, update $X_{pbest,G}$ along with its fitness value. From the entire set of particles, identify the optimal personal best position to designate as the global optimum $X_{gbest,G}$.

4.3 Performance Analysis

The CEC2013 test suite is a standard set of test functions that encompasses various types of optimization problems, serving as a comprehensive evaluation tool for optimization algorithms [19]. The CEC2013 test suite is characterized by diversity, challenge, and benchmarking. Its diversity arises from the inclusion of single-modal, multi-modal, high-dimensional, rotated, and shifted functions. The challenge lies in the complex phenomena of local optima presented by some functions, which adds difficulty to the optimization process. Additionally, it provides a benchmarking platform for comparing the performance of different optimization algorithms. The CEC2017 test suite builds upon the CEC2013 test suite by expanding the types and difficulties of optimization problems [20]. This is specifically reflected in the inclusion of more complex and challenging functions, as well as the evaluation of algorithms' performance when dealing with high-dimensional optimization issues. Additionally, the test suite introduces multi-objective optimization problems to assess the performance of algorithms in the face of multiple objectives. The CEC2017 test suite not only serves to evaluate the performance of optimization algorithms but also facilitates research and development in the practical applications of these algorithms.

In this experimental study, a uniform configuration was applied to the population size, maximum number of iterations, and experimental dimensions for four distinct optimization algorithms, with each run executed 21 times. The population size was set to 100, and the maximum number of iterations was configured as 10,000*D. Experiments were conducted across dimensions of 10, 30, and 50. The GA [21] employed a standard version, with a crossover rate set to 0.9 and a mutation rate set to 0.02, utilizing single-point crossover and roulette wheel selection. In the PSO algorithm, the inertia weight is set to 0.7, with a maximum velocity of 5 and a minimum velocity of –5. The personal learning coefficient ($c1$) and the communal learning coefficient ($c2$) are both configured to a value of 1. Within the DE algorithm, the scaling factor (F) is established at 0.5, while the crossover rate (CR) is fixed at 0.9. Conversely, in the QUATRE algorithm, the scaling factor is adjusted to 0.7.

The experimental results of the four algorithms on CEC2013 and CEC2017 are presented in Tables 4.1 and 4.2, respectively. According to Table 4.1, QUATRE demonstrates the best overall performance, significantly outperforming the other algorithms. In particular, among the 28 benchmark functions, it outperforms PSO in 23 functions, GA in 21, and DE in 14. It performs similarly to PSO in 2 functions, GA in 2, and DE in 6, and underperforms PSO in 3, GA in 5, and DE in 8. From the type of problem, QUATRE's advantages are primarily concentrated in single-modal and multi-modal problems. For composition problems, DE shows better performance compared to the other algorithms. From Table 4.2, the overall performance of QUATRE is also the best. Out of 30 functions, it outperforms PSO in 28 functions, GA in 29, and DE in 11, and is similar to PSO in 2 functions and DE in 10. It is worth noting that the performance of DE and QUATRE is comparable, which can be attributed to their similar optimization structures. Additionally, we can also observe that PSO is worse than DE and QUATRE according to the optimization results. For assessing

Table 4.1 Comparison results of four algorithms on the CEC2013 30D

Fun	PSO	GA	DE	QUATRE
f_{a_1}	2.0678E-06/1.5121E-06(<)	1.0226E-02/7.6503E-03(<)	*0.0000E+00/0.0000E+00(≈)*	*1.0827E-14/4.9617E-14*
f_{a_2}	*5.2375E+05/8.1055E+05(≈)*	3.6603E+07/1.2341E+07(<)	*5.9014E+05/3.2971E+05(<)*	*3.4074E+05/1.7701E+05*
f_{a_3}	1.6154E+08/1.8879E+08(<)	1.7425E+10/1.3827E+10(<)	**1.5434E-01/6.8550E-01**(>)	2.3678E+06/4.5436E+06
f_{a_4}	2.6190E+01/1.1554E+01(<)	1.4930E+05/3.4810E+04(<)	9.6237E+02/3.5769E+02(<)	**1.6909E+01/1.4476E+01**
f_{a_5}	5.0759E-03/1.2583E-03(<)	3.5695E-02/1.7610E-02(<)	*1.1369E-13/0.0000E+00(≈)*	*1.1369E-13/0.0000E+00*
f_{a_6}	*1.4910E+01/1.5573E+01(≈)*	6.8852E+01/2.8950E+01(<)	*7.9560E+00/3.3907E+00(≈)*	*8.4747E+00/1.1652E+01*
f_{a_7}	1.1365E+02/3.8464E+01(<)	2.4605E+02/7.1162E+01(<)	**5.9454E-02/6.1881E-02**(>)	2.1275E+01/1.2836E+01
f_{a_8}	*2.0935E+01/5.5941E-02(>)*	2.1136E+01/5.9523E-02(<)	*2.0941E+01/4.2726E-02(>)*	2.1014E+01/4.6543E-02
f_{a_9}	4.5581E+01/2.9945E+00(<)	3.8466E+01/3.2848E+00(<)	2.6736E+01/1.3883E+01(<)	**1.4662E+01/4.5556E+00**
$f_{a_{10}}$	1.0636E-01/9.8948E-02(<)	2.9901E+01/1.7843E+01(<)	**7.5062E-03/7.4430E-03**(>)	2.8507E-02/1.5955E-02
$f_{a_{11}}$	5.5745E+02/5.4038E+01(<)	**1.8326E+00/1.5824E+00**(>)	1.2973E+02/2.8024E+01(<)	2.7260E+01/7.4131E+00
$f_{a_{12}}$	6.2998E+02/8.3044E+01(<)	4.1079E+02/1.7499E+02(<)	1.8285E+02/7.6344E+00(<)	**6.8835E+01/1.7297E+01**
$f_{a_{13}}$	7.3836E+02/1.0609E+02(<)	4.6253E+02/1.1704E+02(<)	1.7959E+02/1.2332E+01(<)	**1.3196E+02/2.5076E+01**
$f_{a_{14}}$	4.5970E+03/6.3448E+02(<)	**3.5361E+02/1.1478E+02**(>)	6.1898E+03/7.2894E+02(<)	8.4916E+02/2.4009E+02
$f_{a_{15}}$	**4.1768E+03/5.8914E+02**(>)	5.3680E+03/7.1938E+02(≈)	7.0997E+03/2.2781E+02(<)	5.3725E+03/8.6457E+02
$f_{a_{16}}$	**1.0623E+00/2.9588E-01**(>)	2.1191E+00/6.5793E-01(≈)	2.3853E+00/2.9027E-01(≈)	2.2239E+00/4.8948E-01
$f_{a_{17}}$	4.3873E+02/9.8191E+01(<)	**3.3392E+01/8.9783E-01**(>)	1.8323E+02/2.0859E+01(<)	5.6285E+01/6.6334E+00
$f_{a_{18}}$	4.3820E+02/1.1893E+02(<)	3.8095E+02/7.8192E+01(<)	2.1350E+02/8.5182E+00(<)	**1.6989E+02/2.1671E+01**
$f_{a_{19}}$	1.7782E+01/6.1391E+00(<)	**1.9136E+00/5.3798E-01**(>)	1.5234E+01/8.8361E-01(<)	3.5091E+00/6.5711E-01
$f_{a_{20}}$	1.5000E+01/6.4498E-05(<)	1.4953E+01/1.4747E-01(<)	*1.2149E+01/2.1005E-01(≈)*	*1.2230E+01/4.3558E-01*
$f_{a_{21}}$	3.3151E+02/8.4485E+01(<)	3.2789E+02/8.2523E+01(<)	3.2819E+02/1.0926E+02(<)	**2.6874E+02/6.3279E+01**
$f_{a_{22}}$	7.6865E+03/3.4365E+02(<)	**3.2695E+02/1.6205E+02**(>)	6.0045E+03/8.3064E+02(<)	8.9976E+02/2.9490E+02
$f_{a_{23}}$	6.7590E+03/4.4007E+02(<)	6.5230E+03/1.0390E+03(<)	7.2463E+03/2.5768E+02(<)	**5.0601E+03/7.0119E+02**
$f_{a_{24}}$	6.1405E+02/4.0523E+01(<)	3.1228E+02/9.4325E+00(<)	**2.0003E+02/2.1655E-02**(>)	2.3602E+02/1.1517E+01
$f_{a_{25}}$	4.3160E+02/9.8179E+00(<)	3.4207E+02/2.2496E+01(<)	**2.2065E+02/2.0444E+01**(>)	2.5623E+02/8.3459E+00
$f_{a_{26}}$	4.1735E+02/4.9567E+01(<)	3.7695E+02/5.8980E+01(<)	**2.0002E+02/1.3218E-02**(>)	2.5228E+02/6.8530E+01
$f_{a_{27}}$	1.2926E+03/1.1614E+02(<)	1.3631E+03/1.0415E+02(<)	**3.0083E+02/1.4171E+00**(>)	6.9763E+02/9.1707E+01
$f_{a_{28}}$	4.4840E+03/3.7129E+02(<)	1.3483E+03/1.2008E+03(<)	*3.0000E+02/1.1649E-13(≈)*	*3.0000E+02/1.9855E-13*
>/≈/<	3/2/23	5/2/21	8/6/14	–/–/–

Table 4.2 Comparison results of four algorithms on the CEC2017 30D

Fun	PSO	GA	DE	QUATRE
f_{a_1}	3.5373E+03/4.2321E+03(<)	3.9143E+04/2.3931E+04(<)	*1.2181E-14/5.0956E-15(≈)*	*1.1504E-14/5.7181E-15*
f_{a_2}	1.6875E+04/7.5198E+04(<)	1.6962E+23/7.7496E+23(<)	1.0174E+07/4.1010E+07(<)	**4.0084E-01/1.3799E+00**
f_{a_3}	1.1724E-01/7.9213E-02(<)	2.5962E+05/6.3945E+04(<)	2.7159E+01/2.4320E+01(<)	**1.2165E-03/1.3027E-03**
f_{a_4}	5.2815E+01/3.1484E+01(≈)	1.2387E+02/5.0027E+01(<)	*4.1697E+01/2.7821E+01(≈)*	*4.8465E+01/2.4187E+01*
f_{a_5}	2.9076E+02/1.4632E+01(<)	1.0585E+02/3.0692E+01(<)	1.7943E+02/9.4327E+00(<)	**6.9355E+01/2.8438E+01**
f_{a_6}	6.4455E+01/1.9918E+00(<)	1.4534E-01/8.6102E-02(<)	**2.4683E-08/2.7267E-08**(>)	4.1864E-06/5.3055E-06
f_{a_7}	4.8572E+02/6.3486E+01(<)	1.4559E+02/2.6518E+01(<)	2.1403E+02/8.3703E+00(<)	**1.0818E+02/2.8478E+01**
f_{a_8}	1.9511E+02/1.1400E+01(<)	1.3474E+02/4.5318E+01(<)	1.7907E+02/1.3868E+01(<)	**7.0798E+01/1.5554E+01**
f_{a_9}	4.3474E+03/2.2385E+02(<)	2.7403E+03/1.5050E+03(<)	**5.4137E-15/2.4809E-14**(>)	3.6715E-01/5.4244E-01
$f_{a_{10}}$	4.1799E+03/3.7447E+02(<)	**3.2179E+03/4.5412E+02**(>)	6.9009E+03/2.1747E+02(<)	3.7690E+03/7.7143E+02
$f_{a_{11}}$	1.1733E+02/3.9670E+01(<)	7.0053E+03/4.4482E+03(<)	*4.9638E+01/1.4579E+01(≈)*	*3.9647E+01/2.7736E+01*
$f_{a_{12}}$	1.8544E+05/1.8892E+05(<)	8.5850E+06/6.0932E+06(<)	**5.3745E+03/2.4755E+03**(>)	1.2334E+04/1.0318E+04
$f_{a_{13}}$	1.2897E+04/9.3096E+03(<)	7.0427E+05/2.1666E+06(<)	*8.0183E+01/1.2830E+01(≈)*	*1.7796E+02/3.4716E+02*
$f_{a_{14}}$	2.2027E+02/6.4098E+01(<)	3.2852E+06/3.0026E+06(<)	*6.4694E+01/3.9726E+00(≈)*	*8.2018E+01/3.8781E+01*
$f_{a_{15}}$	1.9727E+03/3.3972E+03(<)	1.2200E+05/3.7608E+05(<)	*3.7238E+01/4.8913E+00(≈)*	*5.4688E+01/5.2386E+01*
$f_{a_{16}}$	1.2906E+03/3.2798E+02(<)	1.3650E+03/3.3168E+02(<)	1.0658E+03/3.1148E+02(<)	**7.5830E+02/1.9973E+02**
$f_{a_{17}}$	1.0607E+03/2.5057E+02(<)	7.0386E+02/2.3467E+02(<)	**7.1879E+01/5.7150E+00**(>)	2.5698E+02/1.6670E+02
$f_{a_{18}}$	2.9927E+04/2.4235E+04(<)	4.6005E+06/5.4761E+06(<)	**3.7104E+01/3.7986E+00**(>)	5.3774E+01/1.9412E+01
$f_{a_{19}}$	3.5390E+03/4.1228E+03(<)	1.9950E+04/1.9774E+04(<)	**1.5273E+01/4.5236E+00**(>)	4.6146E+01/4.4227E+01
$f_{a_{20}}$	1.1861E+03/1.6292E+02(<)	6.7786E+02/2.4962E+02(<)	**3.5573E+01/1.3045E+01**(>)	2.1595E+02/1.6535E+02
$f_{a_{21}}$	5.3022E+02/5.2631E+01(<)	3.3147E+02/4.1353E+01(<)	3.7155E+02/7.6652E+00(<)	**2.7940E+02/1.9703E+01**
$f_{a_{22}}$	5.5544E+03/2.8432E+02(<)	2.8221E+03/1.8446E+03(<)	*1.0000E+02/1.3679E-13(≈)*	*5.3273E+02/1.1652E+03*
$f_{a_{23}}$	2.2851E+03/2.1734E+02(<)	4.7365E+02/3.8702E+01(<)	5.1681E+02/1.2905E+01(<)	**4.0983E+02/1.8930E+01**
$f_{a_{24}}$	7.6020E+02/9.1507E+01(<)	7.2518E+02/8.2654E+01(<)	5.9256E+02/1.2002E+01(<)	**4.8931E+02/1.8534E+01**
$f_{a_{25}}$	4.1490E+02/2.4100E+01(<)	4.1312E+02/3.1683E+01(<)	**3.8675E+02/2.3204E-02**(>)	3.8693E+02/1.0389E-01
$f_{a_{26}}$	3.6906E+03/2.0356E+03(<)	2.4057E+03/1.0504E+03(<)	2.1145E+03/7.6295E+02(<)	**1.5831E+03/3.4824E+02**
$f_{a_{27}}$	3.3790E+03/3.8141E+02(<)	5.6205E+02/1.9200E+01(<)	**4.7620E+02/7.0636E+00**(>)	5.0494E+02/1.1907E+01
$f_{a_{28}}$	3.3226E+02/6.0463E+01(≈)	4.7710E+02/5.8787E+01(<)	*3.1526E+02/3.8362E+01(≈)*	*3.4422E+02/6.5544E+01*
$f_{a_{29}}$	1.5892E+03/3.0382E+02(<)	1.0578E+03/2.5307E+02(<)	*5.7103E+02/8.3523E+01(≈)*	*5.7457E+02/1.3100E+02*
$f_{a_{30}}$	1.5628E+04/7.4600E+03(<)	1.2557E+05/2.2611E+05(<)	*2.1960E+03/1.8425E+02(≈)*	*2.1531E+03/2.4794E+02*
>/≈/<	0/2/28	1/0/29	9/10/11	–/–/–

Table 4.3 Summary of comparison results on CEC2013 and CEC2017 across different dimensions

Test suite	CEC2013			CEC2017		
Rank length	10D	30D	50D	10D	30D	50D
Result	>/≈/<	>/≈/<	>/≈/<	>/≈/<	>/≈/<	>/≈/<
PSO	2/3/23	3/2/23	5/0/23	1/1/28	0/2/28	1/4/25
GA	2/5/21	5/2/21	7/2/19	0/0/30	1/0/29	1/1/28
DE	12/5/11	8/6/14	8/6/14	15/9/6	9/10/11	11/6/13

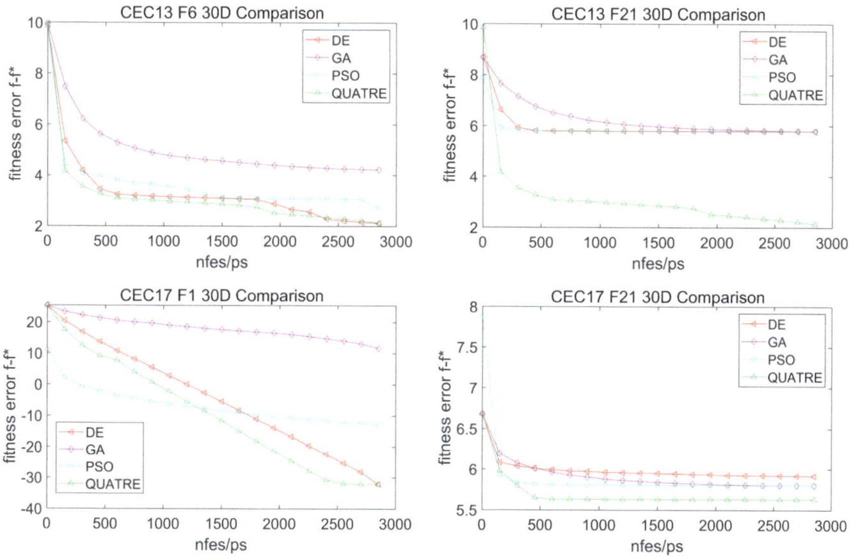

Fig. 4.1 Convergence history of four algorithms on benchmark functions

the efficacy of these algorithms at different dimensions, we presented the comparative results of PSO, GA, DE, and QUATRE at dimensions of 10, 30, and 50 in Table 4.3. The outcomes demonstrate that QUATRE outperforms PSO and GA considerably across various dimensions. Compared to DE, while QUATRE is worse in lower dimensions (10D), it is still better in higher dimensions (30D and 50D). This indicates that QUATRE has stronger adaptability in high-dimensional landscapes. To further validate the convergence rate of the algorithms, this study selects the F6 and F21 benchmark functions from the CEC2013 and the F1 and F21 benchmark functions from the CEC2017, with a dimension of 30, to evaluate the convergence curves of four meta-heuristic optimization algorithms. The experiment results are shown in Fig. 4.1, the convergence speed of QUATRE is the fastest, and it achieves higher precision compared to the other algorithms.

4.4 Conclusion

This paper evaluates four meta-heuristic optimization algorithms, including GA, PSO, DE, and QUATRE, on the CEC2013 and CEC2017 benchmark suites. The test problem's dimensions are configured as 10, 30, and 50. PSO constitutes an algorithm rooted in swarm intelligence. GA, DE, and QUATRE are classified as evolutionary algorithms. According to the experiment results, we obtained the following findings: the performance of PSO is usually inferior to DE and QUATRE algorithms in terms of convergence speed and optimization accuracy. GA, DE, and QUATRE belong to the same class of algorithm. DE is derived from the Genetic Annealing algorithm, which is a hybrid of genetic algorithms and simulated annealing, inheriting the search framework of genetic algorithms. The QUATRE algorithm introduces an innovative evolution matrix that effectively replaces the traditional crossover operation employed in DE. This evolution matrix not only maintains the same search methodology as DE, which involves navigating through a high-dimensional hypercube, but also enhances the search process by eliminating biases that can skew the optimization results. Overall, the QUATRE algorithm shows faster convergence speed and stronger search ability than PSO, GA, and DE, especially on complex and high-dimensional problems.

References

1. Yang, X.-S.: Review of meta-heuristics and generalised evolutionary walk algorithm. Int. J. Bio Inspired Comput. **3**(2), 77–84 (2011)
2. Shaik, A.L.H.P., Manoharan, M.K., Pani, A.K., Avala, R.R., Chen, C.-M.: Gaussian mutation–spider monkey optimization (GM-SMO) model for remote sensing scene classification. Remote. Sens. **14**(24), 6279 (2022)
3. Albadr, M.A., Tiun, S., Ayob, M., Al-Dhief, F.: Genetic algorithm based on natural selection theory for optimization problems. Symmetry **12**(11), 1758 (2020)
4. Qaraad, M., Amjad, S., Hussein, N.K., Farag, M.A., Mirjalili, S., Elhosseini, M.A.: Quadratic interpolation and a new local search approach to improve particle swarm optimization: solar photovoltaic parameter estimation. Expert. Syst. Appl. **236**, 121417 (2024)
5. RJ Kuo and Shu-Syun Li: Applying particle swarm optimization algorithm-based collaborative filtering recommender system considering rating and review. Appl. Soft Comput. **135**, 110038 (2023)
6. Tijjani, S., Wahab, M.N.A., Noor, M.H.M.: An enhanced particle swarm optimization with position update for optimal feature selection. Expert. Syst. Appl. **247**, 123337 (2024)
7. Laiqi, Yu., Meng, Z.: Surrogate-assisted differential evolution with multiple sampling mechanisms for high-dimensional expensive problems. Inf. Sci. **687**, 121408 (2025)
8. Song, Y., Zhao, G., Zhang, B., Chen, H., Deng, W., Deng, W.: An enhanced distributed differential evolution algorithm for portfolio optimization problems. Eng. Appl. Artif. Intell. **121**, 106004 (2023)
9. Li, J., Meng, Z.: Elite-guided resampling and multi-mutation based differential evolution with exponential crossover for numerical optimization. Expert. Syst. Appl. **258**, 125159 (2024)
10. Meng, Z., Pan, J.-S.: A competitive quasi-affine transformation evolutionary (c-QUATRE) algorithm for global optimization. In: 2016 IEEE International Conference on Systems, Man, and Cybernetics (SMC), pp. 001644–001649. IEEE (2016)

11. Zhang, J., Meng, Z.: QUATRE-PM: quasi-affine transformation evolution with perturbation mechanism. IEEE Access (2023)
12. Katoch, S., Chauhan, S.S., Kumar, V.: A review on genetic algorithm: past, present, and future. Multimed. Tools Appl. **80**, 8091–8126 (2021)
13. Chen, C.-M., Lv, S., Ning, J., Wu, J.M.-T.: A genetic algorithm for the waitable time-varying multi-depot green vehicle routing problem. Symmetry **15**(1), 124 (2023)
14. Bratton, D., Kennedy, J.: Defining a standard for particle swarm optimization. In: 2007 IEEE Swarm Intelligence Symposium, pp. 120–127. IEEE (2007)
15. Meng, Z., Pan, J.-S., Tseng, K.-K.: PaDE: an enhanced differential evolution algorithm with novel control parameter adaptation schemes for numerical optimization. Knowl. Based Syst. **168**, 80–99 (2019)
16. Song, Y., Cai, X., Zhou, X., Zhang, B., Chen, H., Li, Y., Deng, W., Deng, W.: Dynamic hybrid mechanism-based differential evolution algorithm and its application. Expert. Syst. Appl. **213**, 118834 (2023)
17. Laiqi, Yu., Ren, C., Meng, Z.: A surrogate-assisted differential evolution with fitness-independent parameter adaptation for high-dimensional expensive optimization. Inf. Sci. **662**, 120246 (2024)
18. Meng, Z., Pan, J.-S., Huarong, X.: Quasi-affine transformation evolutionary (QUATRE) algorithm: a cooperative swarm based algorithm for global optimization. Knowl. Based Syst. **109**, 104–121 (2016)
19. Liang, J.J., Qu, B.Y., Suganthan, P.N., Hernández-Díaz, A.G.: Problem definitions and evaluation criteria for the CEC 2013 special session on real-parameter optimization. In: Computational Intelligence Laboratory, Zhengzhou University, Zhengzhou, China and Nanyang Technological University, Singapore, Technical Report, 201212, no. 34, pp. 281–295 (2013)
20. Awad, N.H., Ali, M.Z., Liang, J.J., Qu, B.Y., Suganthan, P.N.: Problem definitions and evaluation criteria for the CEC 2017 special session and competition on single objective bound constrained real-parameter numerical optimization. In: Technical Report, pp. 1–34. Nanyang Technological University Singapore Singapore (2016)
21. Song, Y., Wang, F., Chen, X.: An improved genetic algorithm for numerical function optimization. Appl. Intell. **49**, 1880–1902 (2019)

Chapter 5
A Termite Morphometric System Based on Qt and HRNet

Yusheng Lei, Zheng Zou, Wenwen Shi, Xuewei Qiu, Dingyi Wang, Xiaotong Fu, Hongyi Fang, Jingjing Li, and Shijian Liu

Abstract Geometric morphometrics refers to using geometric methods to analyze and describe the shape and size of termites, which can be used to effectively evaluate pest-regulating services. An application system is developed to fulfill the morphometrics in this paper, which uses Qt to design the user interfaces (UI). Hiding behind the UI is a landmark detection network (i.e., HRNet) for metrics computation. Visual and quantitative experiments have been conducted, which show the user-friendliness and high performance of the proposed system.

5.1 Introduction

Termites, recognized as one of the most persistent insect pests, are widely distributed across forests, grasslands, and farmland within the global latitudes of 45° north and south [1]. Their populations are numerous, accounting for up to 40% of the dry biomass of soil animals. These insects play a significant role in accelerating the decomposition rate of plant residues by up to 40.2%, contributing to nutrient cycling in ecosystems [2, 3]. However, their innate behavior of consuming wood fibers also poses a threat to wooden structures, including houses, furniture, and resources in agriculture and forestry. In addition to burrowing deep underground, some species of termites also feed at the surface level, making them more accessible and potentially more damaging to human-made structures. Therefore, it is crucial to control termite populations to mitigate the risk of significant financial losses due to potential damage

Y. Lei · Z. Zou (✉) · X. Qiu
College of Computer and Cyber Security, Fujian Normal University, Fuzhou, China
e-mail: zouzheng2018@fjnu.edu.cn

W. Shi · S. Liu
Fujian Provincial Key Laboratory of Big Data Mining and Applications, Fujian University of Technology, Fuzhou, China

D. Wang · X. Fu · H. Fang · J. Li
School of Geographical Sciences, Fujian Normal University, Fuzhou, China

S.-C. Chu et al. (eds.), *Advances in Intelligent Data Analysis and Applications*,
Smart Innovation, Systems and Technologies 445,
https://doi.org/10.1007/978-981-96-7277-6_5

and structural collapse. To effectively evaluate pest-regulating services, numerous studies have been conducted on morphometrics [4–6].

Among different morphometrics, geometric morphometrics refers to using geometric methods to analyze and describe the shape and size of biological structures. It involves collecting landmarks on the surface of an insect's body or its appendages (such as wings, legs, or antennae) and the subsequent use of these data points to perform statistical analyses.

Our work mainly focuses on the morphometrics of termites using their electron microscope images. Figure 5.1a shows the binocular microscope (Leica Microsystems Schweiz AGLED2500) which is used to capture the images in this work. Two typical termite images shown in Fig. 5.1b and c represents the termite soldier and worker respectively.

More specifically, we are interested in determining the critical metrics from the image, which include the size of the head, pronotum, and abdomen as the doubly arrowed line segments indicated in Fig. 5.1d.

A common way to achieve the measurement is by landmark detection. Landmark detection-based geometric morphometric methods of insects can be divided into two categories. The first class is detection using handcrafted features, and the second class is detection using a deep learning (DL) model [7]. Nowadays, the latter has become a trade for its high efficiency and automaticity.

Typical research based on DL includes a method to automatically predict the landmarks on beetle images proposed by Le et al. [8]. Many others were focused on landmark detection on insect wings [7, 9]. However, studies about the landmark detection of termites are rarely published.

In this work, a termite morphometric system is developed to fulfill the above functions, which uses Qt to implement the user interfaces (UI) and a deep learning model named HRNet [10] to achieve the metrics computation. The rest of this paper is organized as follows: Material and methods are proposed in Sects. 5.2, 5.3 presents the experiments and results, and finally, Sect. 5.4 concludes the paper.

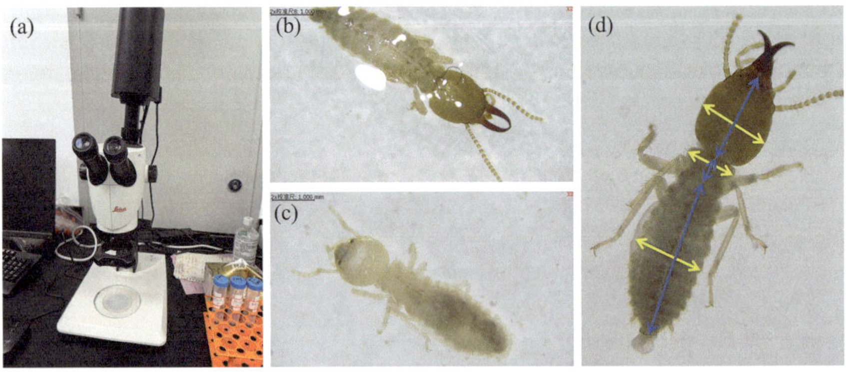

Fig. 5.1 The microscope device (**a**), the images representing termite soldiers (**b**) and workers (**c**), and the metrics (**d**) that we are interested in this work

5.2 Material and Methods

5.2.1 *Dataset Construction*

Data Collection. As mentioned above, both the soldiers and workers of termites are collected and captured as microscope pictures to form the dataset of our research. Data cleaning is then conducted to filter out bad samples and make sure each image contains a single whole termite. The dataset scale after cleaning equals 1960, as depicted in Table 5.1.

The dataset is further divided into 3 subsets, namely belly-down, belly-up, and side, as illustrated in Fig. 5.2a to c. The reason for this subdivision is that different morphometric metrics may be conducted under different postures. For example, it is hard to determine the width of the pronotum in the case of side posture.

Annotation. The numbers shown in Fig. 5.2 are the labels/annotations of landmarks defined in our paper. The annotations are essential to the supervised learning. The principle in our labeling system is increasing the number from head to tall, from left to right one by one to ease the memory of manual labeling. Besides, a tool named "labelme" is adopted to achieve the annotation work.

Table 5.1 The scale of our dataset

Dataset	Number of samples
Belly-down	433
Belly-up	431
Side	1096
Total	1960

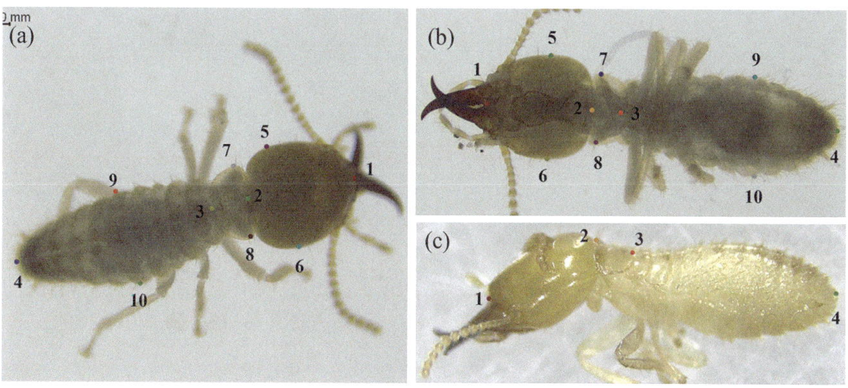

Fig. 5.2 Three postures, which are belly-down (**a**), belly-up (**b**), and side (**c**), as well as the annotations

5.2.2 Preprocessing

Background Culling. As illustrated in the top-left of Fig. 5.3, many artifacts can be found around the termite, which may have a bad impact on system performance. To avoid such disturbances, we first cull the background by thresholding in the HSV color space (see top-right of Fig. 5.3). Then morphological operations are involved to produce a clean mask (see bottom-left of Fig. 5.3). Finally, the culling result is shown in the bottom-right of Fig. 5.3 by masking the origin data with the mask.

Data Augmentation. Feature extraction is crucial to deep learning methods, which rely on a large quantity of data. The more data, the better the performance is for a model in general. However, the data scale in our work is currently not enough to train a robust model. Thus, data augmentation is involved, which includes the random rotations of each image in the training set offline, and both the random rotations and scaling during the training process (i.e., online). Offline data enhancement increases the size of the training set by a factor of 9, while online enhancements can be seen as doubling the amount of data per epoch.

5.2.3 System Development

Requirements Analysis. The fundamental requirements of our application include loading the microscope image and showing the metrics we needed for the termite. Additionally, it would be nice if the loaded image could be shown along with the

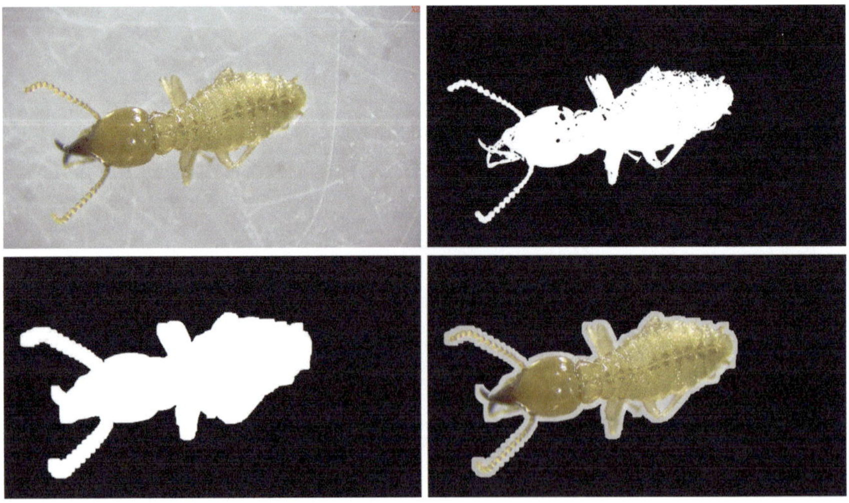

Fig. 5.3 Demonstration of the background culling

detected landmarks on it. Advanced functions could enable the user to load multiple images and inspect the results through them. Secondly, the user could interact with the system to modify the landmarks and re-calculate the results. Last but not least, the results could be exported in text form.

UI Design. According to the requirements analysis, the UI of our system consists of two major parts, which are the regions for inspection and visualization as demonstrated in Fig. 5.4 with the dashed line and the solid line respectively. The core part of the inspection is for the user to specify the images and examine the calculation results. From the visualization part, we can get an intuitive view of the termite and all the landmarks. Users can select and move the landmarks, which will trigger the re-calculation procedure.

Networks. Hide behind the UI is a two-stage deep learning architecture that fulfills the detection. Firstly, a lightweight classification network is used to tell the posture of the termite. Then one of three HRNet models will be used to detect the landmarks on the image, according to different postures. When training these models, we set the number of epochs and the size of batches to 500 and 16 respectively. Adam optimizer is used during the backward propagation with a learning rate equal to 1e-3 initially and then progressively reduced by a factor of 10 at the 170th and 200th epoch respectively.

Implementation Details. The implementation of the proposed system uses Python as the programming language. Core libraries include pyQt and PyTorch. Multi-thread technology is adopted to avoid the main thread being blocked in the detection process, which improves the system's user-friendliness. The detection results will be saved on the hard disk so that the system doesn't have to re-calculate again after the reboot unless the user asks it to.

Fig. 5.4 The UI of the proposed system

5.3 Experiments and Results

5.3.1 The Evaluation Metrics

Two metrics are used in this paper to assess the system. One is the Radial Square Error (RSE) for landmark detection, and the other one is the mean and variance of our morphometric results compared to the ground truth. Consider a test set with N input images; each has M landmark classes to be predicted, then the RSE of the ith$(1 \leq i \leq M)$ landmark can be defined as:

$$RSE_i = \frac{1}{N}\sum_{k=1}^{N} \| p_{ik} - p'_{ik} \| \tag{5.1}$$

where p_{ik} and p'_{ik} represent the ground truth and predicted landmark within the kth$(1 \leq k \leq N)$ image, $\| \bullet \|$stands for the norm operation. The Mean Radial Square Error (MRSE) would be:

$$MRSE = \frac{1}{NM}\sum_{i=1}^{M}\sum_{k=1}^{N} \| p_{ik} - p'_{ik} \| \tag{5.2}$$

5.3.2 Result

Figure 5.5 shows 4 results using images randomly selected from the test set, among them there are 2 images containing soldiers and 2 containing workers. The red crosses on the figures indicate the ground truth annotated by technicians, while the colored dots are landmarks detected by our proposed system. From the figures, we can see that the predicted landmarks are close to their correspondence in the ground truth, which indicates the high accuracy of the proposed system.

Table 5.2 shows quantitative analysis results in terms of the RSE of each landmark class. The MRSE is 0.1 mm.

The mean and variance of our morphometric results compared to the ground truth are recorded in Table 5.3, where $D1$, $D2$, and $D3$ are the length of the head, pronotum, and abdomen, while $D4$ and $D5$ are the width of the head and pronotum respectively.

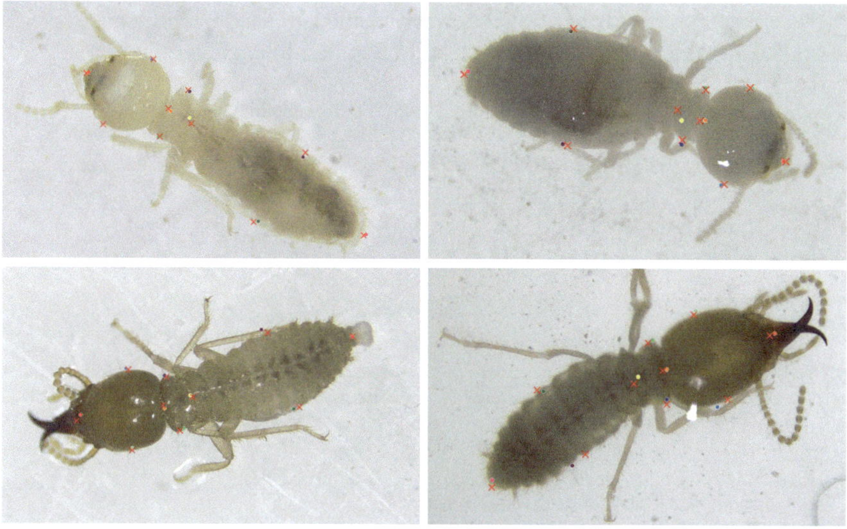

Fig. 5.5 The visualization results

Table 5.2 The RSE of each landmark class

ID	RSE (mm)	ID	RSE (mm)	ID	RSE (mm)	ID	RSE (mm)	ID	RSE (mm)
1	0.093	3	0.060	5	0.208	7	0.072	9	0.129
2	0.079	4	0.084	6	0.057	8	0.082	10	0.137

Table 5.3 The mean and variance of our morphometric results compared to the ground truth

Metrics	Mean (mm)	Variance (mm)
$D1$	0.059	0.005
$D2$	0.041	0.002
$D3$	0.068	0.001
$D4$	0.071	0.002
$D5$	0.062	0.003

5.4 Conclusion

This paper presents an application system based on Qt and HRNet for termite morphometrics. Our system's inputs are microscope images of termites, and it outputs detected landmarks that can be used to represent geometric morphometrics, which include the length and width of the head, pronotum, etc. Visual and quantitative results show a high performance of our system. The system is scalable for other soil-dwelling insects like ants. In the further, methods for more morphometrics will be studied and the system performance will be further improved.

Acknowledgements This work was supported by Fujian Provincial Natural Science Foundation of China (No. 2025J01670, 2023J01515, 2022J01932), and the Hunan Provincial Natural Science Foundation Regional Joint Fund (No. 2024JJ7549).

References

1. Wilson, A.D., Forse, L.B.: Chemical ecology, detection and identification of subterranean termites based on electronic-nose volatile emissions analysis. Environments **11**(1), 15 (2024)
2. Hasnaoui, B., Diarra, A.Z., Berenger, J.M., et al.: Use of the proteomic tool MALDI-TOF MS in termite identification. Sci. Rep. **12**(1), 718 (2022)
3. Hasnaoui, B., Diarra, A.Z., Makouloutou-Nzassi, P., et al.: Identification of termites from Gabon using MALDI-TOF MS. Heliyon **10**(6), (2024)
4. Deguine, J.P., Aubertot, J.N., Flor, R.J., et al.: Integrated pest management: good intentions, hard realities. A review. Agron. Sustain. Dev. **41**(3), 38 (2021)
5. Raymond, L., Vialatte, A., Plantegenest, M.: Combination of morphometric and isotopic tools for studying spring migration dynamics in Episyrphus balteatus. Ecosphere **5**(7), 1–16 (2014)
6. Sasakawa, K.: Utility of geometric morphometrics for inferring feeding habit from mouthpart morphology in insects: tests with larval Carabidae (Insecta: Coleoptera). Biol. J. Linn. Soc. **118**(2), 394–409 (2016)
7. Nguyen, H.H., Ho, B.H., Lai, H.P., et al.: A lightweight keypoint matching framework for insect wing morphometric landmark detection. Ecol. Inform. **70**, 101694 (2022)
8. Le, V.L., Beurton-Aimar, M., Zemmari, A., et al.: Automated landmarking for insects morphometric analysis using deep neural networks. Ecol. Inform. **60**, 101175 (2020)
9. Geldenhuys, D.S., Josias, S., Brink, W., et al.: Deep learning approaches to landmark detection in tsetse wing images. PLoS Comput. Biol. **19**(6), e1011194 (2023)
10. Sun, K., Xiao, B., Liu, D., et al.: Deep high-resolution representation learning for human pose estimation. In: Proceedings of the IEEE/CVF Conference on Computer Vision and Pattern Recognition, pp. 5693–5703. IEEE (2019)

Chapter 6
Real-Time Data Matrix Code Detection for Tube-Rack System

Qing-Bo Zhang, Zi-Can Chen, Dong-Xing Jing, Chen-Hao Xiang, and Shi-Jian Liu

Abstract This paper addresses the critical need for accurate and rapid detection of Data Matrix (DM) codes in a tube-rack system, which is essential for efficient sample management and analysis. Our goal is to develop a real-time web system that leverages deep learning to overcome the challenges posed by rotated DM codes. We achieve this by enhancing the YOLOv5 model with rotational and attention modules, and by implementing pre- and post-processing techniques tailored for DM code detection. The system's performance is evaluated against state-of-the-art models, demonstrating its effectiveness and potential to revolutionize laboratory automation. Our findings conclude that the integrated approach of advanced deep learning and user-centric design offers a robust solution for the accurate detection and decoding of DM codes in laboratory environments.

6.1 Introduction

In the realm of biological research, the efficient handling of vast sample collections is paramount. These samples are meticulously stored within test tubes, which are then organized in tube racks [1, 2]. To streamline the management and expedite the analytical processes, each test tube is marked with a Data Matrix (DM) code—a high-density, matrix-based barcode that serves as a unique identifier for each sample. These DM codes are renowned for their compact design, extensive data storage capabilities, robust error correction, and resilience to various forms of interference [3]. Figure 6.1 shows such a tube-rack system which used by biological laboratories.

Q.-B. Zhang · D.-X. Jing (✉) · C.-H. Xiang
School of Information and Intelligence, Xiangxi Vocational and Technical College for Nationalities, Xiangxi, China
e-mail: 710429442@qq.com

Z.-C. Chen · S.-J. Liu
Fujian Provincial Key Laboratory of Big Data Mining and Applications, Fujian University of Technology, Fuzhou, China
e-mail: liusj2003@fjut.edu.cn

S.-C. Chu et al. (eds.), *Advances in Intelligent Data Analysis and Applications*, Smart Innovation, Systems and Technologies 445,
https://doi.org/10.1007/978-981-96-7277-6_6

Fig. 6.1 Illustration of a classic tube-rack system used in the bio-laboratory

The identification of Data Matrix (DM) codes in the tube-rack system serves a multitude of purposes, particularly when integrated with advanced technologies such as Augmented Reality (AR). As demonstrated in our previous work [4], the integration of DM code detection within an AR system named BioAR is used to enhance the automation and operational efficiency of bio-laboratories.

The detection of Data Matrix (DM) codes in a tube-rack system, while critical for laboratory automation, presents several challenges that must be addressed to ensure the system's reliability and efficiency. Here, we discuss some of the primary challenges:

- Multiple DM Codes: In a typical laboratory setting, a tube-rack system may contain numerous test tubes, each labeled with a DM code. Though the DM codes are separated by rack slots, we have to identify them simultaneously.
- Variability in Code Placement and Orientation: DM codes on test tubes can be placed in various positions and orientations, which adds complexity to the detection process.
- Lighting Conditions: The quality of DM code detection can be heavily influenced by lighting conditions within the laboratory. Poor or variable lighting can lead to shadows, glare, or uneven illumination, which may obscure the code details and hinder accurate detection.
- High-Speed Detection Requirements: In high-throughput laboratory settings, there is a need for rapid and concurrent detection of multiple DM codes without sacrificing accuracy.

Addressing these challenges is crucial for the development of a robust DM code detection system that can operate effectively in real-world laboratory conditions. Traditionally, the detection and decoding of these DM codes have relied on conventional vision detection methods, which are often limited by their reliance on structured lighting and fixed camera setups [5, 6]. These methods can be constrained by the need for optimal lighting conditions and are less adept at handling variations in code orientation or damage. However, the emergence of deep learning has heralded a new era in barcode recognition, offering a significant leap in accuracy and flexibility [7]. Deep learning algorithms are capable of learning from a vast array of examples, adapting to different conditions, and even recognizing damaged or distorted codes with remarkable precision. This technology's integration into DM code detection is poised to revolutionize laboratory automation and efficiency.

6.2 Related Work

6.2.1 DM Code Detection

The detection of DM codes can be approached with various methods, which can be broadly categorized into traditional and deep learning-based techniques.

Traditional methods primarily rely on the identification of the L-shaped finder pattern that is characteristic of these barcodes [8, 9]. This approach involves image processing algorithms that search for the finder pattern within an image, which serves as a reference point to reconstruct the entire code matrix. These methods are known for their simplicity and ease of implementation. However, they suffer from limited robustness and interference handling capabilities.

Deep learning has emerged as a powerful tool for a large range of applications [10, 11], offering improved accuracy and robustness compared to traditional methods. They could show high accuracy in detecting DM codes by learning to recognize complex patterns and features from large datasets. The deep learning models can be trained to handle various conditions and disturbances, making them more robust in real-world applications.

Models such as YOLO (You Only Look Once) have been effectively used for barcode detection by identifying bounding boxes around the codes [12], demonstrating speed and accuracy suitable for applications where rapid detection is crucial. Other deep learning architectures like SSD (Single Shot MultiBox Detector) [13] and Faster R-CNN [14] have also been employed for barcode detection, each with its own strengths and trade-offs. The process typically involves the model proposing regions in the image that likely contain DM codes, followed by searching for the L-shaped finder pattern within these regions to confirm the presence of a DM code and to reconstruct the code matrix. Two primary issues are commonly associated with deep learning-based detection methods:

1. Big data Requirements: Deep learning models are data-hungry, requiring large datasets for training to achieve high accuracy. The compilation of such datasets can be particularly challenging for DM code detection, as it necessitates a diverse and comprehensive collection of images featuring DM codes under various conditions and orientations. This can be time-consuming and resource-intensive, especially when considering the need for high-quality, annotated data.
2. Limited Scope of Object Detection: In the context of DM code detection, these object detection models are used to identify the bounding boxes around the codes. However, after the DM code region is localized, the traditional problem of finding

the L-shaped finder pattern within that region resurfaces. This means that despite the advanced detection capabilities, the core challenge of identifying the distinctive L-shape remains, which can limit the overall efficiency and accuracy of the detection process.

To mitigate the challenge of the first issue, this paper explores the use of synthetic data generation in conjunction with real data. This hybrid approach can enhance the diversity of the training data and reduce the reliance on manual annotation, making the training process more efficient and cost-effective. One approach to bypass the second issue is to employ segmentation models that can delineate the entire DM code in a single step [15]. However, it requires more complex annotations and may struggle to meet real-time detection requirements due to the increased computational complexity associated with pixel-level segmentation. This paper utilizes rotation-aware object detection models to achieve the goal. These models are capable of detecting DM codes in their entirety, regardless of orientation, thus streamlining the detection process and improving accuracy.

6.2.2 Rotation-Aware Object Detection

The detection of Data Matrix (DM) codes is inherently more challenging than that of QR codes due to differences in their design and error correction capabilities [16]. QR codes can be detected and decoded quickly from any direction, largely because of their unique positioning patterns, the contrast between dark and light modules, and their robust error correction algorithms. In contrast, DM codes, with fewer positioning patterns and relatively weaker error correction capabilities, are more sensitive to orientation.

To address this challenge, there is a need for rotation-aware object detection models that can accurately detect DM codes regardless of their orientation. The Rotational Region Proposal Network (RRPN) [17], introduced by Ma et al. in 2018, laid the groundwork by enhancing the Faster R-CNN framework with the addition of multiple anchors at various rotation angles, allowing for the detection of objects in a range of orientations. This approach was significant as it recognized the need to move beyond the traditional horizontal and vertical object detection paradigm. However, the RRPN's approach introduced a significant number of anchors, leading to increased computational load and slower model speeds. To address this, Xie et al. in 2021 proposed Oriented R-CNN [18], which redefined the bounding box generation process to reduce the parameter volume and improve detection efficiency. This model streamlined the detection process while maintaining high accuracy, making it a more viable option for real-time applications.

For applications other than DM code detection, Yang et al. in 2021 introduced Refined Single-Stage Detector (R3Det) [19], which includes a feature refinement module to enhance the accuracy of detecting rotating objects in remote sensing. This model acknowledges the unique challenges posed by remote sensing images, where features may be less distinct and more variable due to factors such as distance and atmospheric conditions. In the same year, Han et al. in 2021 proposed Rotation-Equivariant Detector (ReDet) [20], which incorporates a rotation-equivariant module to accurately extract features of rotated objects. This model's innovation lies in its ability to maintain feature consistency regardless of object orientation, providing a more robust detection framework.

The concept of reppoints, which simplifies the representation of objects to a set of points, was extended to rotation-aware detection with Oriented RepPoints by Li et al. [21]. This adaptation allowed for a more flexible and adaptive detection approach, particularly useful for aerial object detection where objects can present in various orientations and shapes.

This work focus on real-time DM code detection, and we extend the famous YOLOv5 model to ensure it with rotation-aware object detection ability.

6.3 The Proposed System

6.3.1 Requirement Analysis

The development of our system is underpinned by a thorough analysis of the requirements necessary to meet the specific needs of Data Matrix (DM) code detection in a laboratory setting. The system is designed to address three critical aspects: real-time performance, interactivity, and ease of use.

Firstly, though static images and video records are allowed as inputs, the system is required to operate in real time, which is essential for practical applications in dynamic laboratory environments. To achieve this, we have integrated a camera module that continuously captures data, feeding live video streams into our deep learning model. The model is designed to process these data in real time, enabling immediate detection and decoding of DM codes. This capability is crucial for applications where rapid identification and handling of samples are necessary, such as in high-throughput screening or time-sensitive experiments.

Interactivity is another key feature of our system. It allows users to engage directly with the detected DM codes displayed on the image or video frames. By moving the mouse near a recognized DM code area, users can retrieve detailed information about the sample, including its row and column numbers, as well as the decoded data. This interactive functionality enhances the user experience by providing a direct link between the visual representation of the DM code and its associated metadata, facilitating easier management and tracking of samples.

Last but not least, to ensure that our system is accessible to all users, regardless of their technical expertise, we have prioritized ease of use. The system is designed to be standalone, requiring no installation or downloading of weight data. This approach simplifies the user experience, allowing laboratory technicians to focus on their core tasks without the need to manage software installations or updates. By eliminating these barriers, we aim to increase the adoption of our system and improve the overall efficiency of laboratory operations.

6.3.2 *User Interface*

Following the requirement analysis, we have opted for a web platform application as our solution due to its accessibility and ease of use. The user interface (UI) of the proposed system is designed to be intuitive and functional, catering to the needs of real-time data processing and interactive engagement with DM codes. Figure 6.2 provides an illustration of the user interface.

The website is structured into three distinct pages, each dedicated to different procedures: static images, video records, and live video streams. This segmentation allows users to choose the most appropriate mode based on their specific requirements. Taking the live streams page as an example (see Fig. 6.2), the interface is designed to provide a seamless and interactive experience. Upon accessing the live video stream page, users are presented with a central video feed that displays the real-time camera input. Surrounding the video feed are several key components:

1. Control Panel: Located at the top of the interface, the control panel includes button for starting or stopping the video feed. This panel also features a toggle for enabling the DM code detection process.
2. Detection Results: As the deep learning model processes the live video, detected DM codes are highlighted within the video feed with rotated bounding boxes.
3. Information Panel: Situated beside the video feed, the information panel dynamically updates to display the details of all detected DM code. This includes the decoded information and any relevant metadata associated with the sample.

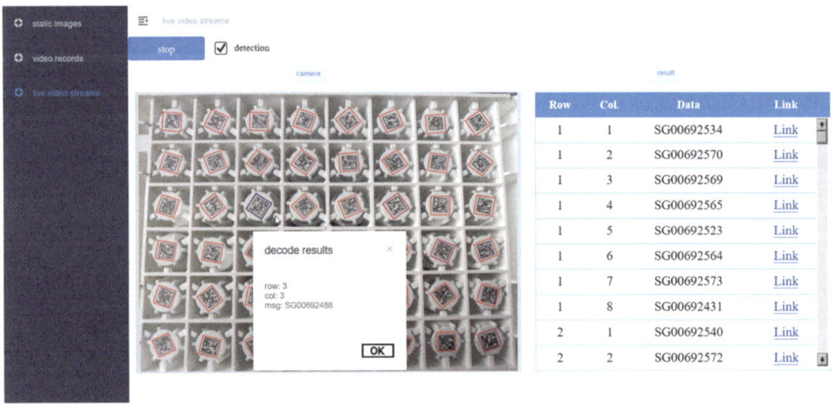

Fig. 6.2 Demonstration of the system UI

4. Interactive Feedback: Users can interact with the detected DM codes by moving the mouse near them, which brings up a unique identifier and a pop-up window to show basic information such as row and column numbers, as well as the decoded data.

Except minimal difference, they all depend on a deep learning model to recognize DM codes using the input data.

6.3.3 Methodology

The Deep Learning Model. The core of our system is a deep learning model that leverages the strengths of YOLOv5 as its foundation. You Only Look Once (YOLO) is chosen for its efficiency and speed in detecting objects, which is crucial for our system's real-time requirements. The YOLOv5 architecture is known for its ability to provide fast and accurate detections in a single pass through the image, without the need for region proposal generation or subsequent refinement steps that other models might require. This makes it an excellent fit for applications where rapid feedback is necessary, such as in a laboratory setting where technicians need immediate results for sample management.

To enhance the model's performance specifically for rotated object detection, we have made two key modifications to the base YOLOv5 architecture. Firstly, we have replaced certain convolutional modules within the YOLOv5 architecture with Rotational Convolution Modules (RCM). This change is motivated by the need to capture the orientation of objects more effectively, and inspired by the Deformable Convolutional Network. The RCM is designed to be able to detect not only a bounding box, but also a rotation angle. This is particularly useful for DM code detection, where the L-shaped finder pattern must be identified accurately, even when the code is not aligned with the horizontal or vertical axes.

Secondly, the original C3 module in YOLOv5 has been replaced with the Multi-Convolution Attention Module (MCAM) to enhance the model's ability to focus on relevant features within the image. MCAM introduces an attention mechanism that adaptively weights the importance of different convolutional features. This allows the model to concentrate on the most discriminative parts of the input where most likely to appear DM codes, improving the accuracy of detection. The use of MCAM is especially beneficial in scenarios where the DM codes may be presented under varying lighting conditions, as it helps the model to focus on the most salient features for identification.

Preprocess and Postprocess. To optimize the performance of our deep learning model and ensure accurate detection of DM codes, we employ a series of pre-processing and post-processing steps.

One of the core pre-processing steps is artificial data synthesis. Given the simplicity of generating DM codes, we utilize artificial data synthesis to augment our training dataset significantly. By randomly overlaying these codes onto background images of racks, we can create a substantial amount of training data. This approach is particularly important for deep learning models, which require large datasets for effective training. The synthetic data helps in mimicking various real-world conditions and orientations, thus enhancing the model's ability to generalize and perform well across different scenarios.

Another major pre-processing step is data rectification. Due to potential perspective distortions during the imaging process, DM codes may not appear as perfect squares, which can affect the accuracy of detection. To address this, we apply affine transformations to rectify the input data. This process corrects for perspective distortions, ensuring that the DM codes are presented to the model in a standardized form, facilitating more accurate detection.

After the deep learning model generates rotated bounding boxes around the detected DM codes, these boxes need to be inversely transformed to align with the original image coordinates for proper display. This step is crucial for ensuring that the annotations on the original images are accurate and correspond correctly to the detected DM codes. The final step in our post-processing pipeline involves decoding the detected DM codes according to their standard specifications. This decoding process extracts the information encoded within the DM codes, making it accessible for further analysis and use in sample management and tracking.

6.4 Results and Discussion

To evaluate the accuracy of our proposed method, we conducted a series of experiments comparing it with state-of-the-art rotation-aware object detection models, including Oriented R-CNN, Oriented Reppoints, R3Det, and ReDet. This section details the experimental setup, evaluation metrics, and results.

6.4.1 Experimental Environment

All experiments were conducted on a system equipped with an NVIDIA GeForce RTX 3090 graphics card and a 12th Gen Intel(R) Core(TM) i7-12700F CPU. This high-performance hardware ensures that the computational demands of the deep learning models are met, providing a consistent platform for comparison. Python is used as the programming language.

6.4.2 Evaluation Metrics

We employed two primary metrics to assess the performance of the detection models: mean Average Precision (mAP) and mean Intersection over Union (mIoU). The mAP is a standard metric in object detection tasks that measures the model's ability to detect objects at various Intersection over Union (IoU) thresholds. The formula for Average Precision (AP) for a single class is given by

$$AP = \frac{1}{11} \sum_{k=0}^{10} P(k) \tag{6.1}$$

where $P(k)$ is the precision at the k-th IoU threshold, and mAP is the mean of AP values across all classes.

The mIoU measures the overlap between the predicted bounding boxes and the ground truth boxes. It is calculated as the average IoU across all instances of a class. The formula for IoU for a single instance is

$$IoU = \frac{\text{Area (predicted box } \cap \text{ ground truth box)}}{\text{Area (predicted box } \cup \text{ ground truth box)}} \tag{6.2}$$

and mIoU is the mean IoU across all classes.

6.4.3 Results and Visualization

Following the experimentation, we compiled the results and visualized them using bar charts to compare the performance of the different models. The bar charts shown in Fig. 6.3 display the mAP0.5:0.95 and mIoU scores for each model, providing a clear and intuitive comparison of their accuracy. The results demonstrate the effectiveness of our approach in detecting DM codes under various conditions.

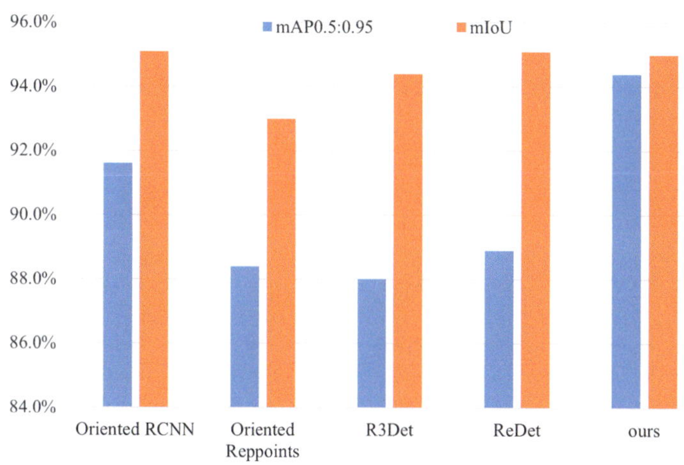

Fig. 6.3 The comparison results in terms of mAP0.5:0.95 and mIoU scores

6.5 Conclusion

In this paper, a novel system is presented for the real-time detection and decoding of Data Matrix (DM) codes within a laboratory setting. The system is designed to address the challenges of orientation sensitivity and the need for rapid, accurate identification of DM codes on sample tubes. Our approach leverages recent advancements in deep learning, particularly rotation-aware object detection models, to improve the efficiency and accuracy of DM code detection. Key Contributions of this work are triplet. (1) Deep Learning Model Enhancements. We adopted YOLOv5 as the base model and introduced significant improvements, such as the Rotational Convolution Module (RCM) and the Mul-Convolution Attention Module (MCAM), to enhance the model's ability to detect rotated objects and focus on relevant features. (2) Pre-processing and Post-processing Techniques. To prepare the input data and transform the model's output into actionable information, we implemented effective pre-processing steps, such as artificial data synthesis and data rectification, and post-processing steps, including the inverse transformation of bounding boxes and decoding of DM codes. (3) Web Platform Development. We conducted a thorough analysis of the system requirements, emphasizing real-time performance, interactivity, and ease of use. Based on these requirements, we developed a web platform application that operates seamlessly across different devices without the need for installation or downloading weight data, enhancing accessibility and user convenience.

While our system has shown promising results, there is scope for further enhancement. Future work may involve expanding the system's capabilities to accommodate a broader range of barcode types and improving its adaptability to various laboratory environments. Additionally, exploring more sophisticated data augmentation techniques and incorporating feedback mechanisms for continuous model refinement could be valuable directions for future research.

Acknowledgements This work was supported in part by the Natural Science Foundation of Hunan Province (No. 2024JJ7549), the Scientific Research Project of Hunan Provincial Department of Education (No. 24C1236), the Scientific Research Team Project of XiangXi Vocational and Technical College for Nationalities (No. 2024KTZ201), and Fujian Provincial Natural Science Foundation of China (No. 2025J01670).

References

1. Munari, E., Scarpa, A., Cima, L., et al.: Cutting-edge technology and automation in the pathology laboratory. Virchows Arch. **484**(4), 555–566 (2024)
2. Chapman, T.: Lab automation and robotics: automation on the move. Nature **421**(6923), 661–663 (2003)
3. Dita, I.C., Otesteanu, M., Quint, F.: Data matrix code—a reliable optical identification of microelectronic components. In: 2011 IEEE 17th International Symposium for Design and Technology in Electronic Packaging (SIITME), pp. 39–44. IEEE (2011)

4. Zou, Z., Liu, S.J.: BioAR: bio-laboratory automation meets augmented reality. In: Genetic and Evolutionary Computing: Proceedings of the Fourteenth International Conference on Genetic and Evolutionary Computing, October 21–23, 2021, Jilin, China 14, pp. 294–305. Springer Singapore (2022)
5. Muniz, R., Junco, L., Otero, A.: A robust software barcode reader using the Hough transform. In: Proceedings 1999 International Conference on Information Intelligence and Systems (Cat. No. PR00446), pp. 313–319. IEEE (1999)
6. Al-Saedi, F.A.T., Jasim, A.M.: Development of a barcode reader system. Int. J. Comput. Trends Technol. **30**, 50–58 (2015)
7. Wudhikarn, R., Charoenkwan, P., Malang, K.: Deep learning in barcode recognition: a systematic literature review. IEEE Access **10**, 8049–8072 (2022)
8. Liu, F., Liu, A., Wang, M., et al.: Robust and fast localization algorithm for data matrix barcode. In: 2010 International Conference on Optoelectronics and Image Processing, vol. 2, pp. 356–359. IEEE (2010)
9. Huang, Q., Chen, W.S., Huang, X.Y., et al.: Data matrix code location based on finder pattern detection and bar code border fitting. Math. Probl. Eng. **2012**(1), 515296 (2012)
10. LeCun, Y., Bengio, Y., Hinton, G.: Deep learning. Nature **521**(7553), 436–444 (2015)
11. Bai, B., Yang, X., Li, Y., et al.: Deep learning-enabled virtual histological staining of biological samples. Light. Sci. Appl. **12**(1), 57 (2023)
12. Pandya, K., Hathi, H., Buch, H., et al.: Barcode and QR code object detection: an experimental study on YOLOv8 models. In: 2024 International Conference on Emerging Innovations and Advanced Computing (INNOCOMP), pp. 740–746. IEEE (2024)
13. Xiong, J., Zhou, L.: QR code detection and recognition in industrial production environment based on SSD and image processing. In: International Conference on Algorithms, High Performance Computing, and Artificial Intelligence (AHPCAI 2023), vol. 12941, pp. 640–644. SPIE (2023)
14. Farhaoui, Y., El Allaoui, A.: Bar code detection for AI-driven EDA and big data analytics. In: The International Workshop on Big Data and Business Intelligence, pp. 476–488. Springer Nature Switzerland, Cham (2024)
15. Tsai, M.J., Wu, H.Y., Lin, D.T.: Auto ROI & mask R-CNN model for QR code beautification (ARM-QR). Multimed. Syst. **29**(3), 1245–1276 (2023)
16. Kulshreshtha, R., Kamboj, A., Singh, S.: Decoding robustness performance comparison for QR and data matrix code. In: Proceedings of the Second International Conference on Computational Science, Engineering and Information Technology, pp. 722–731 (2012)
17. Ma, J., Shao, W., Ye, H., et al.: Arbitrary-oriented scene text detection via rotation proposals. IEEE Trans. Multimed. PP(99), 1–1 (2017)
18. Xie, X., Cheng, G., Wang, J., et al.: Oriented R-CNN for object detection. In: 2021 IEEE/CVF international conference on computer vision (ICCV), Montreal, QC, Canada, 2021, pp. 3500–3509. https://doi.org/10.1109/ICCV48922.2021.00350.
19. Yang, X., Yan, J., Feng, Z., et al.: R3Det: refined single-stage detector with feature refinement for rotating object. In: Proceedings of the AAAI Conference on Artificial Intelligence, vol. 35, no. 4, pp. 3163–3171 (2021)
20. Han, J., Ding, J., Xue, N., et al.: ReDet: A rotation-equivariant detector for aerial object detection. In: 2021 IEEE/CVF Conference on Computer Vision and Pattern Recognition (CVPR), Nashville, TN, USA, 2021, pp. 2785–2794. https://doi.org/10.1109/CVPR46437.2021.00281.
21. Li, W., Chen, Y., Hu, K., et al.: Oriented reppoints for aerial object detection. In: Proceedings of the IEEE/CVF Conference on Computer Vision and Pattern Recognition, pp. 1829–1838 (2022)

Chapter 7
The Improved Artificial Protozoa Optimization Algorithm

Jeng-Shyang Pan, Hong Chen, and Pei Hu

Abstract Metaheuristic algorithms have become popular for addressing complex optimization challenges in recent years. The Artificial Protozoa Optimization (APO) algorithm is one such metaheuristic, drawing inspiration from the behavior of protozoa, simulating their survival mechanisms through behaviors like foraging, dormancy, and reproduction. The parallel strategy enhances the algorithm's performance by dividing the population into subgroups and allowing inter-group communication. Opposition-based learning enables population members to reach more promising regions by simultaneously considering candidate solutions and their opposite counterparts. Lévy flight merges short-range local searching with long-range global searching through a random walk with a probabilistic step size distribution. In this paper, we apply parallel strategies, Lévy flight, and opposition-based learning to improve the APO algorithm, further enhancing its global search capabilities. We tested the improved Artificial Protozoa Optimization algorithm on a set of 12 benchmark functions from the IEEE CEC 2022 suite. The outcomes of the experiments demonstrate that the proposed improved Artificial Protozoa Optimization algorithm is competitive.

J.-S. Pan (✉) · H. Chen
School of Artificial Intelligence, Nanjing University of Information Science and Technology, Nanjing, Jiangsu, China
e-mail: jengshyangpan@gmail.com

J.-S. Pan
Department of Information Management, Chaoyang University of Technology, Taiwan, China

P. Hu
School of Computer Science and Engineering, Shandong University of Science and Technology, Qingdao, Shandong, China

S.-C. Chu et al. (eds.), *Advances in Intelligent Data Analysis and Applications*,
Smart Innovation, Systems and Technologies 445,
https://doi.org/10.1007/978-981-96-7277-6_7

7.1 Introduction

Metaheuristic algorithms are typically classified into four categories based on their inspiration sources [1]: swarm intelligence [2], evolutionary strategies [3], human-centric methods [4], and physics-inspired techniques [5]. Swarm intelligence algorithms are metaheuristic methods proposed by mimicking biological behaviors. For instance, the Cat Swarm Optimization (CSO) algorithm [6] models two typical behavioral patterns in natural cats: tracking and seeking. The tracking mode, which simulates a cat locking onto and following a target, is responsible for local search to ensure effective convergence to the optimal result. Meanwhile, the seeking mode simulates a cat roaming and searching in its environment, primarily facilitating global search to enhance diversity and exploration capability. Among swarm intelligence algorithms, the Particle Swarm Optimization (PSO) algorithm is a quintessential example, drawing inspiration from the foraging behavior of bird flocks. Other examples include the Whale Optimization Algorithm (WOA) [7], Gannet Optimization Algorithm (GOA) [8], the Cuckoo Search (CS) algorithm [9], Barnacle Growth Algorithm (BGA) [10], and so on.

Swarm intelligence algorithms offer numerous advantages. First, individuals within the group cooperate for global search, effectively avoiding local optima and demonstrating effectiveness in handling complex, multimodal, and multi-peak optimization problems. Second, these algorithms do not require the problem at hand to be continuous or differentiable, as they do not rely on gradient information or specific mathematical models. Third, each individual in the group can independently perform its own search process, providing strong parallel computing potential suitable for large-scale complex problems. Additionally, swarm intelligence algorithms are robust, simple to implement, and flexible.

Due to their powerful global search capability and high adaptability, swarm intelligence algorithms are extensively used for solving a range of complex optimization challenges. By simulating collective intelligent behaviors to explore complex search spaces, they have been successfully employed in fields such as path planning [11], wireless sensor networks [12], privacy-preserving data mining [13], the Internet of Things [14], and others. For example, the Ant Colony Optimization (ACO) algorithm was applied by Dorigo et al. [15] in power systems to optimize load distribution and the scheduling of generation. Natarajan et al. [16] used PSO to optimize stock portfolios, minimizing risk and maximizing returns to address the complexities and uncertainties of financial markets.

Despite their strong performance across various domains, swarm intelligence algorithms still face challenges in balancing exploration and exploitation. Consequently, several enhancement strategies have been incorporated into swarm intelligence algorithms to improve their performance, including compact [17], multi-objective [18], Parallel strategy [19, 20], adaptive [21], binary [22], opposition-based learning [23], Lévy flights [24], discretization variants [25], etc. For example, Xue et al. [26] proposed a self-adaptive artificial bee colony algorithm that utilizes a self-adjusting strategy and global optimal information to guide the search, significantly enhancing algorithm precision and convergence speed across various complex optimization scenarios.

The paper continues with the following structure. Section 7.2 offers an overview of the APO algorithm. Section 7.3 introduces the improved Artificial Protozoa Optimization algorithm, including the three enhancement strategies of parallelism, opposition-based learning, and Lévy flights. Section 7.4 discusses experimental simulations and analyzes results using the CEC2022 benchmark. Section 7.5 concludes the research and proposes future research avenues.

7.2 Overview of the Artificial Protozoa Optimization Algorithm

The APO algorithm [27] is a swarm intelligence optimization technique inspired by the foraging habits of protozoa, simulating their random movement, chemotaxis, and division behaviors in nature. Inspired by the foraging patterns of microorganisms like amoebas and paramecia, the algorithm mimics their search for food, movement toward high-concentration nutrient areas, and reproduction through division to enhance population survival. This approach optimizes the search path and achieves global optimization. The APO algorithm is characterized by its robust global search abilities and its capacity to avoid local optima, allowing it to discover the best solution in intricate search spaces.

The mathematical model of the APO algorithm includes several behavioral modes, such as autotrophy, heterotrophy, dormancy, and reproduction. In autotrophic mode, protozoa obtain energy through photosynthesis and adjust their position based on light intensity. In heterotrophic mode, they acquire energy by absorbing organic matter and move toward nutrient-rich areas. Dormancy mode allows protozoa to conserve energy under environmental stress, while reproduction mode increases the population through asexual division. By adjusting the proportions and parameters of these modes, the algorithm integrates different search behaviors, improving search efficiency and the stability of the algorithm.

7.3 The Improved Artificial Protozoa Optimization Algorithm

7.3.1 Parallel Strategy

This article introduces an innovative approach in the APO algorithm by implementing multi-group partitioning and a periodic inter-group communication mechanism to further optimize the population's search performance (see Fig. 7.1). Specifically,

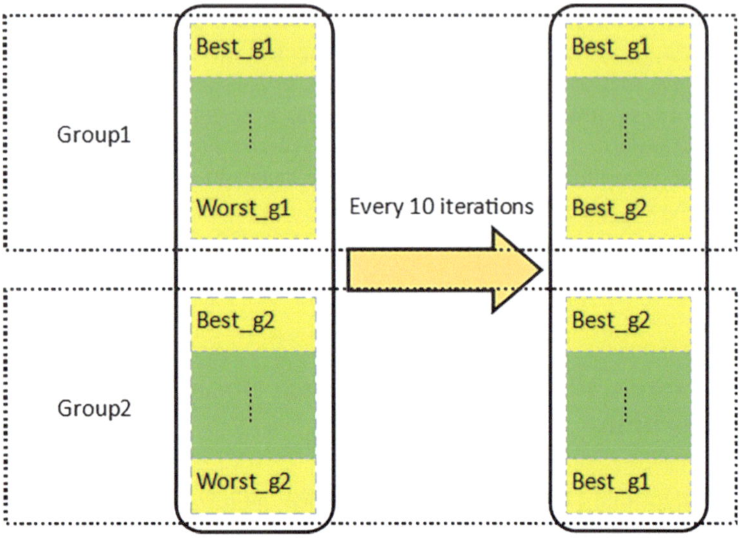

Fig. 7.1 Inter-group communication mechanism in the parallel strategy of the improved APO

the algorithm's total population size is defined as 100 individuals, which are first divided into two groups, each containing 50 individuals. This division ensures that each group independently explores the solution space, increasing search diversity. Then, after every 10 generations, inter-group information exchange occurs, where the least fit individual from each group is replaced by the best individual from the other group. This process facilitates bidirectional information flow, strengthening the ability to search globally. This approach not only effectively minimizes the chances of the algorithm getting stuck in local optima, but also enhances the algorithm's adaptability and robustness through multiple rounds of dynamic interaction, demonstrating significant performance advantages in complex environments.

Virtual Parallelism (Multi-grouping) is an optimization strategy that divides the main population into several subpopulations, allowing them to execute in parallel on independent computational resources. This approach improves computational efficiency and search effectiveness of the algorithm. Each subpopulation explores different regions of the solution space in relatively independent environments, reducing the likelihood of premature convergence and maintaining population diversity. When the results are aggregated, the search information from different subpopulations can help the algorithm converge faster and more comprehensively toward the global optimum.

This multi-group strategy has been successfully applied in various well-known optimization algorithms, such as Multi-group Cat Swarm Optimization (CSO), Multi-group Particle Swarm Optimization (PSO), Multi-group QUATRE algorithm, and Multi-group Differential Evolution (DE), among others. Using a parallel approach is ideal for addressing large-scale and intricate optimization issues.

7.3.2 Opposition-Based Learning

In the APO algorithm, the traditional individual update mechanism may cause the algorithm to easily become ensnared in local optima, thereby limiting its global search capability. To overcome this issue, this study proposes an improvement strategy based on reverse learning to boost the algorithm's global search capability and promote the maintenance of population diversity. Specifically, the reverse learning strategy introduces a reverse position adjustment mechanism, which exerts influence on individuals with poor fitness, forcing them to explore undeveloped regions of the solution space. The strategy works by calculating the difference between an individual's current solution position and the limits of the solution space, and then adjusting the search direction based on this difference, thus avoiding local optima and promoting global optimization.

In the specific implementation, the operation of opposition-based learning is described as follows.

Assuming the current position of an individual $X_I = \{X_{i1}, X_{i2}, \ldots, X_{idim}\}$, its opposite position D_i is calculated using the following formula:

$$D_i = \min(X_i) + \max(X_i) - X_i \tag{7.1}$$

Here, $\min(X_i)$ and $\max(X_i)$ denote the minimum and maximum values of individual across each dimension, respectively. To prevent the updated individuals from exceeding the valid boundaries of the solution space, boundary constraints are imposed on the updated individuals, ensuring they remain within the feasible solution space, as follows:

$$D_i' = \max(X_{\min}, \min(X_{\max}, D_i)) \tag{7.2}$$

If the fitness at the reverse position D_i surpasses that of the current individual, then individual X_i is updated to D_i. This process aids individuals in escaping local optima, exploring a broader solution space, and enhancing the effectiveness of global search. The updated individual position is obtained using the following formula:

$$X_i' = D_i', \text{ if } f(D_i') \prec f(X_i) \tag{7.3}$$

By introducing this reverse learning mechanism, the algorithm can more effectively avoid local optima in each iteration and explore a more diverse search area. Compared to traditional update methods, reverse learning significantly enhances the robustness and global exploration ability of the APO algorithm.

7.3.3 *Lévy Flight*

This study proposes an improvement strategy based on Lévy flights in the APO algorithm to promote population diversity and exploration of the solution space. In the algorithm, we adopt the Lévy flight mechanism for updating individual positions, with the step size s determined by the following formula:

$$s = \frac{\mu}{|\nu|^{1/\beta}} \tag{7.4}$$

$$\sigma_\mu = \left(\frac{\Gamma(1+\beta)\sin(\pi\beta/2)}{\Gamma((1+\beta)/2)\beta 2^{(\beta-1)/2}} \right)^{1/\beta} \tag{7.5}$$

Here, $\mu \sim N(0, \sigma_\mu^2)$ and $\nu \sim N(0, 1)$, where σ_μ is the standard deviation, and β is a parameter that controls the step-size distribution, typically set to 1.4. This step size enables long-distance jumps, allowing the algorithm to explore the solution space more broadly and reduce the likelihood of it being confined to local optima.

In this study, the algorithm employs the autotrophic and dormancy modes for global exploration, while the heterotrophic and reproduction modes focus on local exploitation. Based on this division of roles, the Lévy flight strategy is introduced into the autotrophic and dormancy modes to enhance the algorithm's exploration capability. Specifically, each time a new individual is generated, the algorithm first generates a solution using Lévy flight and compares its fitness to that of a solution generated through standard evolution, retaining the superior solution as the new individual.

This mode-based selective replacement enables Lévy flights to effectively support the autotrophic and dormancy modes in expanding the search range without disrupting the stability of the exploitation modes. By clearly separating exploration from exploitation, this improvement enhances the algorithm's global search capability to approach optimal solutions more robustly.

7.4 Experiments

7.4.1 *Benchmarking and Experimental Setup*

This study examines the performance of the improved Artificial Protozoa Optimization algorithm using the CEC2022 benchmark function library. The CEC2022 library includes various types of test functions, such as unimodal, multimodal, hybrid, and composition functions. These function types exhibit distinct characteristics and challenges: unimodal functions evaluate the algorithm's capability to perform local optimization on single-optimum problems; multimodal functions test its global search effectiveness in avoiding local optima; hybrid and composition functions combine

Table 7.1 Benchmark functions in the IEEE CEC 2022

	Index	Comprehensive description of functions	f_{min}
Unimodal function	F1	Zakharov function with full rotation and shifting	300
Multimodal functions	F2	Rosenbrock's function with full rotation and shifting	400
	F3	Expanded Schaffer's f6 function with full rotation and shifting	600
	F4	Non-Continuous Rastrigin's function with full rotation and shifting	800
	F5	Levy function with full rotation and shifting	900
Hybrid functions	F6	Hybrid ($N = 3$)	1800
	F7	Hybrid ($N = 6$)	2000
	F8	Hybrid ($N = 5$)	2200
Composition functions	F9	Composition ($N = 5$)	2300
	F10	Composition ($N = 4$)	2400
	F11	Composition ($N = 5$)	2600
	F12	Composition ($N = 6$)	2700
Search range: $[-100, 100]^{dim}$			

multiple characteristics to provide a complex testing environment that assess the algorithm's adaptability and competence in managing diversity and complexity in high-dimensional spaces. Table 7.1 presents a comprehensive description of the benchmark functions.

The experiments were performed on a personal computer running Windows 10, equipped with an AMD Ryzen 7 4800H processor (2.90 GHz) and Radeon Graphics, ensuring computational stability and efficiency. The experimental setup for the improved algorithm included primary parameters such as a population size of 100, a maximum of 500 iterations, and np set to 1, following the standard APO algorithm settings. To enhance the credibility and statistical significance of the experimental results, each experiment was repeated 20 times to thoroughly assess the performance of the improved algorithm.

7.4.2 Results Analysis

This study seeks to assess the effects of the proposed improvement strategies on the algorithm's performance by comparing it with the standard APO algorithm. For this purpose, the global optimum acquisition rate, average fitness value, and fitness variance were selected as evaluation metrics to systematically measure the algorithm's performance across various test functions.

Table 7.2 provides detailed data of the experimental results. The bolded ones indicate better results. Drawing from the experimental results, the improved algorithm demonstrates advantages over the standard APO algorithm across multiple performance metrics, as detailed below.

Best Value Comparison: Out of the 12 test functions, the improved algorithm achieved better best values in 10 of them. Notably, for functions such as F1, F6, and F8, the improved algorithm showed significant optimization, with notably lower values compared to the standard APO algorithm. This suggests that the improved algorithm demonstrates enhanced exploration capability near the global optimum in these functions. Overall, the improved algorithm achieved 10 wins, 0 draws, and 2 losses.

Mean Value Comparison: Regarding the mean value, the improved algorithm outperformed the standard APO algorithm in 11 functions, with only F12 showing slightly poorer performance. This suggests that the improved algorithm consistently maintained a lower mean value across multiple runs, closer to the global optimum, demonstrating better convergence and higher stability.

Standard Deviation Comparison: In terms of standard deviation, the improved algorithm exhibited lower variability in 8 functions, indicating that its results were more concentrated and stable. Although the standard APO algorithm had slightly lower standard deviations in functions like F8 and F12, which may suggest smaller solution fluctuations, the improved algorithm achieved a lower standard deviation in most of the test functions, further highlighting its stability in exploring the solution space.

To provide a clear comparison of the performance between the improved algorithm and the standard APO algorithm, this study visualizes the Win/Draw/Lose results using pie charts (see Fig. 7.2). After converting the data into percentages,

Table 7.2 Experimental results

	Best value		Mean		Std	
	Ours	APO	Ours	APO	Ours	APO
F1	**1670.104**	6591.588	**5472.185**	9720.994	**2140.78**	2434.687
F2	**45.6755**	49.1112	**49.637**	50.6221	**1.35696**	5.65339
F3	**0.000857**	0.001161	**0.002303**	0.002812	**0.000751**	0.000986
F4	14.9959	**12.7755**	31.5261	**18.5921**	9.36189	**4.06497**
F5	0.002579	**0.000298**	**0.090326**	0.151083	**0.130148**	0.308432
F6	**45.58766**	303.2051	**1132.176**	1746.495	**857.5398**	1890.517
F7	**17.7621**	27.4577	**29.8876**	35.0589	5.41486	**5.25844**
F8	**5.64973**	22.5813	**21.0648**	23.2666	3.65462	**0.440898**
F9	**180.7813**	180.7815	**180.7832**	180.7885	**0.001761**	0.011422
F10	**100.3469**	100.4056	**100.4404**	100.4874	**0.050488**	0.062835
F11	**300.0898**	300.1148	**300.2836**	300.4263	**0.198737**	0.823335
F12	**232.0771**	234.783	**237.3548**	238.7128	3.457989	**2.360268**

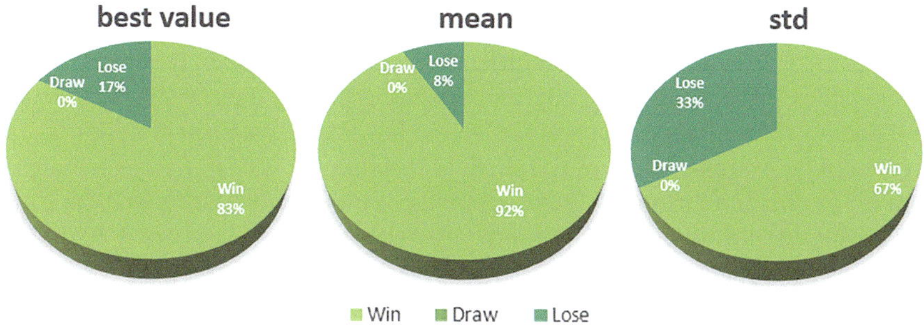

Fig. 7.2 Comparative performance analysis of the improved and standard APO algorithms in terms of win/draw/lose ratios

the improved algorithm achieved win rates of 83.33%, 91.67%, and 66.67% in the metrics of best value, mean, and standard deviation, respectively. These results clearly demonstrate that by incorporating parallel strategies, Lévy flight, and opposition-based learning, the improved Artificial Protozoa Optimization algorithm has achieved significant enhancements in global search capabilities and optimization performance, effectively addressing complex optimization problems and exhibiting strong competitiveness.

In conclusion, the improved algorithm surpasses the standard APO algorithm across multiple metrics, confirming the effectiveness of the introduced enhancement strategies in enhancing the algorithm's performance.

7.5 Conclusion

This paper introduces an improved Artificial Protozoa Optimization algorithm, which significantly enhances performance in global exploration and local exploitation through the introduction of parallel grouping, reverse learning, and Lévy flights. Firstly, the distributed search is implemented in different regions of the solution space using a multi-group parallel mechanism, and regular inter-group communication effectively mitigates the impact of local optima. Secondly, opposition-based learning increases population diversity, boosting the global search capability. Finally, the Lévy flight strategy further broadens the exploration range of the solution space. The improved algorithm demonstrates outstanding performance in the CEC2022 benchmark tests, achieving better best values, mean values, and standard deviations in most test functions. Experimental data indicate that, compared to the standard APO algorithm, the improved algorithm shows improvements in optimization speed, convergence stability, and solution diversity.

Building on the research presented in this paper, there are several directions for future expansion and optimization of the improved APO algorithm. Firstly, the application of the improved APO algorithm to more complex optimization tasks, such as multi-objective, large-scale, or dynamic optimization problems, should be explored to validate its broad applicability in different scenarios. Additionally, enhancing the computational efficiency of the algorithm is an important direction. For example, introducing chaotic mapping or surrogate models into the algorithm design could accelerate the computational process, better meeting the needs of real-time applications. Furthermore, combining this algorithm with other optimization algorithms could leverage the strengths of various approaches, further improving its global search capability and stability.

References

1. Binitha, S., Sathya, S.S.: A survey of bio-inspired optimization algorithms. Int. J. Soft Comput. Eng. **2**(2), 137–151 (2012)
2. Meng, Z., Zhong, Y., Yang, C.: CS-DE: Cooperative strategy based differential evolution with population diversity enhancement. Inf. Sci. **577**, 663–696 (2021)
3. Chen, C.M., Lv, S., Ning, J., et al.: A genetic algorithm for the waitable time-varying multi-depot green vehicle routing problem. Symmetry **15**(1), 124 (2023)
4. Moosavi, S.H.S., Bardsiri, V.K.: Poor and rich optimization algorithm: a new human-based and multi-populations algorithm. Eng. Appl. Artif. Intell.Artif. Intell. **86**, 165–181 (2019)
5. Jiang, S.J., Chu, S.C., Zou, F.M., et al.: A parallel Archimedes optimization algorithm based on Taguchi method for application in the control of variable pitch wind turbine. Math. Comput. SimulComput. Simul. **203**, 306–327 (2023)
6. Chu, S.C., Tsai, P.W., Pan, J.S.: Cat swarm optimization. In: PRICAI 2006: Trends in Artificial Intelligence 9th Pacific Rim International Conference on Artificial Intelligence, Guilin, China, August 7–11, 2006, Proceedings 9, pp. 854–858. Springer Berlin, Heidelberg
7. Mirjalili, S., Lewis, A.: The whale optimization algorithm. Adv. Eng. Softw.Softw. **95**, 51–67 (2016)
8. Pan, J.S., Zhang, L.G., Wang, R.B., et al.: Gannet optimization algorithm: a new metaheuristic algorithm for solving engineering optimization problems. Math. Comput. Simul. **202**, 343–373 (2022)
9. Gandomi, A.H., Yang, X.S., Alavi, A.H.: Cuckoo search algorithm: a metaheuristic approach to solve structural optimization problems. Eng. Comput.Comput. **29**, 17–35 (2013)
10. Shao, A., Chu, S. C., Chen, Y. C., et al.: Barnacle Growth Algorithm (BGA): a new bio-inspired metaheuristic algorithm for solving optimization problems. In: International Conference on Genetic and Evolutionary Computing, pp. 468–479. Springer Nature, Singapore (2023)
11. Dao, T. K., Pan, T. S., Pan, J. S.: A multi-objective optimal mobile robot path planning based on whale optimization algorithm. In: 2016 IEEE 13th International Conference on Signal Processing (ICSP), pp. 337–342. IEEE, Chengdu (2016)
12. Ho, J.H., Shih, H.C., Liao, B.Y., et al.: A ladder diffusion algorithm using ant colony optimization for wireless sensor networks. Inf. Sci. **192**, 204–212 (2012)
13. Wu, T.Y., Lin, J.C.W., Zhang, Y., et al.: A grid-based swarm intelligence algorithm for privacy-preserving data mining. Appl. Sci. **9**(4), 774 (2019)
14. Kang, L., Chen, R.S., Xiong, N., et al.: Selecting hyper-parameters of Gaussian process regression based on non-inertial particle swarm optimization in Internet of Things. IEEE Access **7**, 59504–59513 (2019)
15. Dorigo, M., Birattari, M., Stutzle, T.: Ant colony optimization: artificial ants as a computational intelligence technique. IEEE Comput. Intell. Mag.Comput. Intell. Mag. **1**(4), 28–39 (2006)
16. Natarajan, R., Dharani, B.: A particle swarm optimization based approach for portfolio optimization. Int. J. Pure Appl. Math. **109**(8), 145–154 (2016)

17. Xue, X., Wu, X., Chen, J.: Optimizing biomedical ontology alignment through a compact multi-objective particle swarm optimization algorithm driven by knee solution. Discret. Dyn. Nat. Soc. Dyn. Nat. Soc. **2020**(1), 4716286 (2020)
18. Zhu, H., Xue, X., Wang, H.: Matching ontologies through multi-objective evolutionary algorithm with relevance matrix. Mathematics **10**(12), 2077 (2022)
19. Chu, S.C., Roddick, J.F., Pan, J.S.: A parallel particle swarm optimization algorithm with communication strategies. J. Inf. Sci. Eng. **21**(4), 809–818 (2005)
20. Chu, S.C., Roddick, J.F., Pan, J.S.: Ant colony system with communication strategies. Inf. Sci. **167**(1–4), 63–76 (2004)
21. Meng, Z., Pan, J.S., Tseng, K.K.: PaDE: an enhanced differential evolution algorithm with novel control parameter adaptation schemes for numerical optimization. Knowl.-Based Syst.-Based Syst. **168**, 80–99 (2019)
22. Hu, P., Pan, J.S., Chu, S.C.: Improved binary grey wolf optimizer and its application for feature selection. Knowl.-Based Syst.-Based Syst. **195**, 105746 (2020)
23. Mahdavi, S., Rahnamayan, S., Deb, K.: Opposition based learning: a literature review. Swarm Evol. Comput.Evol. Comput. **39**, 1–23 (2018)
24. Li, J., An, Q., Lei, H., et al.: Survey of Lévy flight-based metaheuristics for optimization. Mathematics **10**(15), 2785 (2022)
25. Pan, Q.K., Tasgetiren, M.F., Suganthan, P.N., et al.: A discrete artificial bee colony algorithm for the lot-streaming flow shop scheduling problem. Inf. Sci. **181**(12), 2455–2468 (2011)
26. Xue, Y., Jiang, J., Zhao, B., et al.: A self-adaptive artificial bee colony algorithm based on global best for global optimization. Soft. Comput. Comput. **22**, 2935–2952 (2018)
27. Wang, X., Snášel, V., Mirjalili, S., et al.: Artificial Protozoa Optimizer (APO): a novel bio-inspired metaheuristic algorithm for engineering optimization. Knowl.-Based Syst.-Based Syst. **295**, 111737 (2024)

Chapter 8
Watermarking Image Processing Models for Intellectual Property Protection

Yuxuan Du, Xuanyu He, Haixuan Ma, Haoyun Lei, Zhiheng Yang, and Linlin Tang

Abstract Deep learning has been successful in various computer vision tasks. However, training deep models is computationally expensive and requires significant high-quality data, making a good pre-trained model highly valuable. However, there is a growing concern about the security of pre-trained deep models. When these models are shared or deployed widely, they become exposed to the risk of illegal theft. This raises concerns about how to protect the intellectual property (IP) of model owners. One solution is to use deep model watermarking. Most recent research has focused on protecting classification networks. However, image processing models are still under-researched, and existing methods lack generality. This paper presents a novel model watermarking method to protect image processing models that also supports black-box verification. We watermark the target model by training it on watermarked training data. The watermarked model learns to embed the watermark into its output images, which can be extracted to realize copyright protection. Our method has been extensively tested, and the results demonstrate its fidelity, uniqueness, and robustness.

8.1 Introduction

Recent years, deep learning has gotten a remarkable development. It has greatly contributed the progress of computer vision tasks such as image classification, image processing and image segmentation. However, it is not trivial to give full play to the learning capability of a deep neural network (DNN). In fact, in order to determine the architecture and hyper parameters of a DNN, deep expertise and a process of trial and error are often required. Besides, massive computing resources and labeled data are indispensable for training a desirable model. Considering the huge cost, a well-learned model is undoubtedly of great commercial value and thus the model owner's intellectual property (IP) must be protected.

Y. Du · X. He · H. Ma · H. Lei · Z. Yang · L. Tang (✉)
School of Computer Science and Technology, Harbin Institute of Technology (Shenzhen), Shenzhen, China
e-mail: lltang@hit.edu.cn

S.-C. Chu et al. (eds.), *Advances in Intelligent Data Analysis and Applications*, Smart Innovation, Systems and Technologies 445,
https://doi.org/10.1007/978-981-96-7277-6_8

DNN model watermarking [1] is a solution that allows owners to verify their ownership by checking if a suspect model exhibits a predefined pattern when given specific inputs. This process can be conducted in either a white-box or black-box setting, depending on the owner's access to the model's internal details.

Our proposed method supports black-box verification and consists of watermark embedding and ownership verification. During embedding, we watermark the training data using a DNN-based scheme and train the target model on it. During verification, a decoder extracts the watermark from the suspect model's output images.

Our main contributions include: (1) a novel deep model watermarking method for black-box verification that can be generalized to any image-to-image model by watermarking it with watermarked ground-truth images; (2) case studies on superresolution and denoising demonstrating the method's fidelity, uniqueness, and robustness.

8.2 Related Works

Uchida et al. [1] introduced the first model watermarking technique for safeguarding deep learning models by incorporating a parameter regularizer via linear transformation during training. Merrer et al. [4] proposed a zero-bit watermarking method using adversarial examples. Trigger set-based methods [5–7] create trigger images by modifying training images with a unique signature. Zhang et al. [2] studied three watermarking algorithms and compared their resilience against common attacks. Quan et al. [6] presented the first watermarking method for image processing networks, achieved by fine-tuning with specially crafted trigger and verification images. In contrast, Zhang et al. [3] focused on protecting output images, appending an embedding sub-network to conceal watermarks invisibly in these images. However, this approach only watermarks the outputs, not the model itself, leaving it vulnerable if the attacker steals the model without replication, which is the focus of our study.

8.3 Proposed Method

8.3.1 Main Idea

Assume attacker steals the model and deploys it in his own product or service with only an API accessible to the owner. For such a suspect model SM, the only indicator which the owner can use to prove ownership is output images of SM. Therefore we design a model watermarking method to enable the watermarked model to conceal a watermark w in its own output images.

Firstly, we utilize invisible image watermarking mechanism to conceal the watermark w in GT images Y and generate watermarked GT images Y_w, which is visually consistent with Y. Then we watermark the target model by training it on the watermarked training data (X, Y_w). Let $M(\theta_w; \cdot)$ be the watermarked model parameterized by vector θ_w, $M(\theta; \cdot)$ be target model parameterized by the vector θ. This relationship can be denoted by

$$M(\theta_w; X) \approx M(\theta; X)\ and\ M(\theta_w; X)\ contains\ w \tag{8.1}$$

According to Eq. 8.1, the watermarked model $M(\theta_w; \cdot)$ can learn to conceal w in its output images while keeping a comparable performance on the original task. Once the watermarked model $M(\theta_w; \cdot)$ is stolen by attacker, we can extract watermark w from output images of $M(\theta_w; \cdot)$ to prove its ownership.

8.3.2 Watermark Embedding

During watermark embedding, we adopt a three-stage training strategy.

Image Watermarking Stage In this stage, our aim is to train encoder E and then utilize E to embed the watermark w in GT images Y. For this, a DNN-based image watermarking approach is used. Encoder E takes the combination of Y and w as the input and outputs the watermarked images Y_w. Decoder D takes Y and Y_w as input and should extract watermark w from Y_w and a blank image b from Y. To further enhance the similarity between Y and Y_w, we append an discriminator network A after the decoder E.

We characterize visual consistency between Y_w and Y with the following three loss functions: *image distortion loss*, denoted as

$$L_{img} = \|Y_w - Y\|^2 \tag{8.2}$$

perceptual loss [13], denoted as

$$L_{vgg} = \|VGG_k(Y_w) - VGG_k(Y)\|^2 \tag{8.3}$$

where $VGG_k(\cdot)$ is the features extracted at layer k("conv2_2" by default); and *adversarial loss*, denoted as

$$L_{adv} = log(1 - A(Y_w)) \tag{8.4}$$

At the same time, we characterize extracting ability of the decoder D with *message distortion loss*, denoted as

$$L_{msg} = \|D(Y_w) - w\|^2 + \|D(Y) - b\|^2 \tag{8.5}$$

So E and D are trained jointly to minimize the loss L_{ED} as follows:

$$L_{ED} = \lambda_{img} * L_{img} + \lambda_{vgg} * L_{vgg} + \lambda_{adv} * L_{adv} + \lambda_{msg} * L_{msg} \quad (8.6)$$

In addition, A is trained to minimize the loss L_A:

$$L_A = log(1 - A(Y)) + log(A(Y_w)) \quad (8.7)$$

After training stage, we utilize the well-trained encoder E to embed the watermark w in GT images Y and thus generate watermarked GT images Y_w.

Model Watermarking Stage In this stage, our aim is to watermark the target model M_w. For this, we train it on the watermarked data (X, Y_w) in the same way as training the original model M. Assume loss function for training M is denoted as $L_M = Loss(M(X), Y)$, the loss function for training M_w is defined by

$$L_{M_w} = Loss(M_w(X), Y_w) \quad (8.8)$$

As discussed in Sect. 8.3.1, the watermarked model M_w can learn to conceal w in its output images while keeping a comparable performance on the original task, for the reason that Y_w is visually consistent with Y and contains the watermark w.

Decoder Enhancing Stage In this stage, our aim is to enhance extracting ability of the decoder D. We fine-tune D on more comprehensive dataset Y_w, Y, $M_w(X)$ and $M(X)$ to enhance its extracting ability.

We characterize extracting ability of D with the following three loss functions: *watermark distortion loss*, denoted as

$$L_w = \|D(Y_w) - w\|^2 + \|D(M_w(X)) - w\|^2 \quad (8.9)$$

blank image distortion loss, denoted as

$$L_b = \|D(Y) - b\|^2 + \|D(M(X)) - b\|^2 \quad (8.10)$$

consistent loss [3], denoted as

$$L_{cst} = \|D(Y_w) - D(M_w(X))\|^2 \quad (8.11)$$

The decoder D is fine-tuned to minimize the loss L_D as follows:

$$L_D = \lambda_w * L_w + \lambda_b * L_b + \lambda_{cst} * L_{cst} \quad (8.12)$$

8.3.3 Ownership Verification

Extremely simple compared with [6], a set of trigger images T with k images are randomly sampled from input images X. We can compute success rate of watermark extraction:

$$successrate = \frac{1}{k}\sum_{i=1}^{k} s_i \quad (8.13)$$

where

$$s_i = \begin{cases} 1, NCC(w_i', w) > 0.95 \\ 0, otherwise \end{cases} \quad (8.14)$$

We regard watermark extraction as successful if normalized correlation coefficient (NCC) between extracted watermark w_i' and the original one w is larger than 0.95. Apparently a higher success rate implies a greater possibility that the suspect model SM is exactly the watermarked model M_w.

8.3.4 Criteria for Evaluation

Following [6], here are listed some criteria to evaluate our model watermarking method:

Fidelity. Compared with original model M, performance of watermarked model M_w on the original task should not be degraded perceptibly.

Uniqueness. Success rate of watermark extraction for the watermarked model M_w should be high. However, success rate of watermark extraction should be low for any other unwatermarked or watermarked by a different watermark model M' for the same task.

Robustness. Though attacker imposes a certain attack on M_w, such as model compression or model fine-tuning, the watermark w can still be extracted from output images of M_w with a high success rate.

8.4 Experiment

8.4.1 Implementation Details

In this paper, experiments are conducted on two types of image processing tasks: image superresolution (scale factor $r = 4$) and image denoising (noise level $\sigma = 25$). We selected VDSR [8], SRGAN [9] as superresolution cases and DnCNN [10],

REDNet [11] as denoising cases. We adopt UNet [15] as the encoder E, CEILNet [16] as the decoder D and PatchGAN [17] as the adversary A. Throughout our experiments, the number of trigger images in T is set to 128. By default, $\lambda_{adv} = 0.01$, $\lambda_{msg} = 0.1$, and λ_{img}, λ_{vgg}, λ_w, λ_b, λ_{cst} all equal to 1.

8.4.2 Fidelity Analysis

Fidelity of proposed method is evaluated by comparing the performance of the watermarked model M_w and the original model M on original task. For superresolution, we conduct tests on three benchmark datasets: Set5, Set14, and BSD100 (see [8]). For denoising, we use two benchmark datasets: Set12 and BSD68 (see [10]). Peek Signal to Noise Ratio (PSNR) and the Structural Similarity (SSIM) are used as default metrics to evaluate the visual quality.

As shown in Tables 8.1 and 8.2, all the watermarked models can achieve comparable PSNR/SSIM values to that of the original models (e.g., the maximum decline of the PSNR value is 0.09dB). This indicates that the perturbation caused by model watermarking has negligible effects on the original task. Thus the fidelity requirement is satisfied. Some visual examples are shown in Fig. 8.1.

Table 8.1 Average PSNR(dB)/SSIM of superresolution results from original model M and watermarked model M_w

Superresolution model	Set5	Set14	BSD100
$VDSR$[8]	30.83/0.9379	27.47/0.8616	27.08/0.8306
$VDSR_w$	30.87/0.9382	27.51/0.8620	27.09/0.8312
$SRGAN$[9]	30.04/0.9293	27.07/0.8549	26.72/0.8225
$SRGAN_w$	29.99/0.9288	27.01/0.8536	26.64/0.8218

Table 8.2 Average PSNR(dB)/SSIM of denoising results from original model M and watermarked model M_w

Denoising model	BSD68	Set12
$DnCNN$[10]	29.24/0.9020	30.46/0.9300
$DnCNN_w$	29.21/0.9014	30.37/0.9294
$REDNet$[11]	29.20/0.9013	30.39/0.9293
$REDNet_w$	29.14/0.9007	30.35/0.9301

Fig. 8.1 Results of fidelity test. The first row shows the low-resolution image, high-resolution image, superresolution image outputted by $VDSR$ and superresolution image outputted by $VDSR_w$ from left to right. The second row shows the noise image, noise-free image, denoised image outputted by DnCNN, and denoised image outputted by $DnCNN_w$ from left to right

8.4.3 Uniqueness Analysis

During ownership verification, the suspect model SM may be another model M' for the same task. To make the verification results convincible, our method should meet uniqueness requirement. We further denote decoder D for watermarked model M_w as D_M. To evaluate uniqueness of the proposed method, we compare success rates of watermark extraction when D_M faces the watermarked model M_w and another model M'. We name success rate tested on M' as false success rate.

As shown in Table 8.3 , each decoder D_M can extract correct watermarks from output images of corresponding watermarked model M_w with a high success rate 100%. But for other models M', the maximum false success rate is extremely low. This result shows that decoder D_M can distinguish corresponding watermarked model M_w from other models M' for the same task. Therefore the uniqueness requirement of proposed method is satisfied.

Robustness Analysis After attacker steals watermarked model M_w, he or she may impose a certain attack on the model. Parameter modification caused by the attack may remove the watermark embedded in M_w. Here we consider two commonly used attacks: model compression and model fine-tuning.

Model Compression In recent years, size of deep neural networks sharply increased which makes these DNNs both computationally intensive and memory intensive. To solve this problem, models are often compressed to reduce redundant parameters.

As done in [6], we compress the watermarked models M_w by pruning the parameters whose absolute values are smallest on each layer. During parameter pruning, we first sort the model weights in the ascending order in terms of absolute values and then set the top p% to zero. As shown in Table 8.4, with the pruning rate increasing in an appropriate range, success rates gradually decrease on overall trend but still hold

Table 8.3 Success rates of watermark extraction when decoder D_M faces watermarked model M_w and other models M' for the same task

Decoder	Suspect model (superresolution)						
	$VDSR_w$	$SRGAN_w$	$VDSR$	$SRGAN$	$DPSR$[28]	$IMDN$[29]	$ESRGAN$[30]
D_{VDSR}	**100%**	0%	0%	0%	0%	0%	0%
D_{SRGAN}	3%	**100%**	1%	0%	2%	0%	0%
Decoder	Suspect model (denoising)						
	$DnCNN_w$	$REDNet_w$	$DnCNN$	$REDNet$	$FFDNet$[31]	$IRCNN$[32]	$FDnCNN$[33]
D_{DnCNN}	**100%**	9%	0%	9%	6%	0%	3%
D_{REDNet}	0%	**100%**	1%	0%	2%	0%	0%

Table 8.4 Average PSNR(dB)/SSIM of superresolution results from original model M and watermarked model M_w

Pruning rate	Superresolution model		Denoising model	
	$VDSR_w$ (%)	$SRGAN_w$ (%)	$DnCNN_w$ (%)	$REDNet_w$ (%)
0.1	100	100	100	100
0.2	100	98	76	98
0.3	100	100	36	76
0.4	100	99	36	38
0.5	100	99	51	31
0.6	98	98	38	25
0.7	87	100	9	5
0.8	45	93	11	0
0.9	3	92	10	2

Table 8.5 Changes of success rate of watermark extraction and PSNR/SSIM value tested on the new test dataset (Manga109 [24] and T91 [14]) after model fine-tuning. In blue color is the success rates that are at least 20% higher than the maximum false success rate in Table 8.3

Superresolution model (Manga109)	PSNR(dB)/SSIM	Success rate
$VDSR_w$	28.08/0.9328 → 28.27/0.9332	100% → 98%
$SRGAN_w$	26.26/0.9212 → 29.90/0.9474	100% → 44%
Denoising model (T91)	PSNR(dB)/SSIM	Success rate
$DnCNN_w$	30.01/0.9225 → 30.05/0.9225	100% → 38%
$REDNet_w$	29.93/0.9228 → 30.10/0.9237	100% → 28%

a distinguishable distance to the maximum false success rate in Table 8.3. Therefore, our proposed method is robust enough against model compression.

Model Fine-Tuning Training a DNN from scratch requires a large-scale training dataset and insufficient data can greatly limit DNN's performance. Consequently, it is very common and effective to fine-tune a pre-trained model. Based on this, robustness against model fine-tuning need to be considered.

As shown in Table 8.5, each watermarked model gains better performance on new test dataset after fine-tuning. At the same time, success rates decrease but still keep distinguishable distances to the maximum false success rates, which demonstrates robustness of the proposed method against model fine-tuning.

8.5 Conslusion

We propose a task-agnostic model watermarking method to protect image processing models, which supports black-box verification. For this, we utilize DNN-based image watermarking mechanism to watermark GT images of the training data, and then turn image watermarking into model watermarking by training the target model on the watermarked training data. By using a well-trained decoder, we can extract the watermark to realize ownership verification of a suspect model, even if we can only access its output images. Extensive experiments are conducted on superresolution and denoising to evaluate the proposed method. It is found that watermarked model has comparable performance with original model on original task. And the proposed method can distinguish watermarked model from other models. Finally, our proposed method can resist model compression and fine-tuning.

Acknowledgements This work is supported by Shenzhen Major Science and Technology Special Project with Grant No. KJZD20240903102727035 and the Guangdong Provincial Key Laboratory of Novel Security Intelligence Technologies (2022B1212010005). And this research is also supported by Key Basic Research Projects of Shenzhen with Grant No. JCYJ20220818102414030.

References

1. Uchida, Y., Nagai, Y., Sakazawa, S., Satoh, S.I.: Embedding watermarks into deep neural networks. In: Proceedings of the 2017 ACM on International Conference on Multimedia Retrieval, pp. 269–277 (2017)
2. Zhang, J., et al.: Protecting intellectual property of deep neural networks with watermarking. In: Proceedings of the 2018 on Asia Conference on Computer and Communications Security, pp. 159–172 (2018)
3. Zhang, J., et al.: Model watermarking for image processing networks. Proc. AAAI Conf. Artif. Intell. **34**(07), 12805–12812 (2020)
4. Le Merrer, E., Perez, P., Trédan, G.: Adversarial frontier stitching for remote neural network watermarking. Neural Comput. Appl. **32**(13), 9233–9244 (2020)
5. Adi, Y., Baum, C., Cisse, M., Pinkas, B., Keshet, J.: Turning your weakness into a strength: watermarking deep neural networks by backdooring. In: 27th USENIX Security Symposium (USENIX Security 18), pp. 1615–1631 (2018)
6. Quan, Y., Teng, H., Chen, Y., Ji, H.: Watermarking deep neural networks in image processing. IEEE Trans. Neural Netw. Learn. Syst. (2020)
7. Guo, J., Potkonjak, M.: Watermarking deep neural networks for embedded systems. In: 2018 IEEE/ACM International Conference on Computer-Aided Design (ICCAD), pp. 1–8. IEEE (2018)
8. Kim, J., Lee, J.K., Lee, K.M.: Accurate image super-resolution using very deep convolutional networks. In: Proceedings of the IEEE Conference on Computer Vision and Pattern Recognition, pp. 1646–1654 (2016)
9. Ledig, C., et al.: Photo-realistic single image super-resolution using a generative adversarial network. In: Proceedings of the IEEE Conference on Computer Vision and Pattern Recognition, pp. 4681–4690 (2017)
10. Zhang, K., Zuo, W., Chen, Y., Meng, D., Zhang, L.: Beyond a gaussian denoiser: residual learning of deep cnn for image denoising. IEEE Trans. Image Process. **26**(7), 3142–3155 (2017)

11. Mao, X.-J., Shen, C., Yang, Y.-B.: Image restoration using very deep convolutional encoder-decoder networks with symmetric skip connections. In: Proceedings of the 30th International Conference on Neural Information Processing Systems, pp. 2810–2818 (2016)
12. Yang, W., et al.: Cascade of multi-scale convolutional neural networks for bone suppression of chest radiographs in gradient domain. Med. Image Anal. **35**, 421–433 (2017)
13. Johnson, J., Alahi, A., Fei-Fei, L.: Perceptual losses for real-time style transfer and super-resolution. In: European Conference on Computer Vision, pp. 694–711. Springer (2016)
14. Yang, J., Wright, J., Huang, T.S., Ma, Y.: Image super-resolution via sparse representation. IEEE Trans. Image Process. **19**(11), 2861–2873 (2010)
15. Ronneberger, O., Fischer, P., Brox, T.: U-net: Convolutional networks for biomedical image segmentation. In: International Conference on Medical Image Computing and Computer-assisted Intervention, pp. 234–241. Springer (2015)
16. Fan, Q., Yang, J., Hua, G., Chen, B., Wipf, D.: A generic deep architecture for single image reflection removal and image smoothing. In: Proceedings of the IEEE International Conference on Computer Vision, pp. 3238–3247 (2017)
17. Isola, P., Zhu, J.-Y., Zhou, T., Efros, A.A.: Image-to-image translation with conditional adversarial networks. In: Proceedings of the IEEE Conference on Computer Vision and Pattern Recognition, pp. 1125–1134 (2017)
18. Zhang, K., Zuo, W., Zhang, L.: Deep plug-and-play super-resolution for arbitrary blur kernels. In: Proceedings of the IEEE/CVF Conference on Computer Vision and Pattern Recognition, pp. 1671–1681 (2019)
19. Hui, Z., Gao, X., Yang, Y., Wang, X.: Lightweight image super-resolution with information multi-distillation network. In: Proceedings of the 27th ACM International Conference on Multimedia, pp. 2024–2032 (2019)
20. Wang, X., et al.: Esrgan: Enhanced super-resolution generative adversarial networks. In: Proceedings of the European Conference on Computer Vision (ECCV) Workshops, pp. 0–0 (2018)
21. Zhang, K., Zuo, W., Zhang, L.: FFDNet: Toward a fast and flexible solution for CNN-based image denoising. IEEE Trans. Image Process. **27**(9), 4608–4622 (2018)
22. Zhang, K., Zuo, W., Gu, S., Zhang, L.: Learning deep CNN denoiser prior for image restoration. In: Proceedings of the IEEE Conference on Computer Vision and Pattern Recognition, pp. 3929–3938 (2017)
23. Simonyan, K., Zisserman, A.: Very deep convolutional networks for large-scale image recognition (2014). arXiv:1409.1556
24. Matsui, Y., et al.: Sketch-based manga retrieval using manga109 dataset. Multimed. Tools Appl. **76**(20), 21811–21838 (2017). http://www.ncbi.nlm.nih.gov

Chapter 9
Outlier-Aware Transformer Architecture for Large Model for Multi-round Dialogue Systems

Hao Hu, Xuan Chen, Mi Xu, Yicheng Wang, and Xiaohong Zhu

Abstract This paper presents a groundbreaking solution to the critical challenge of outlier sensitivity in transformer-based architectures for multi-round dialogue systems. Our research introduces the Outlier-Aware Transformer (OAT), a sophisticated architecture that revolutionizes dialogue processing through innovative residual attention compensation mechanisms and context-aware fine-tuning protocols. The framework demonstrates exceptional capabilities in managing extended conversational sequences, significantly reducing error propagation across multiple dialogue turns. Through extensive experimentation on diverse dialogue datasets, we establish that OAT substantially outperforms conventional transformer architectures in maintaining contextual coherence, reducing outlier-induced errors by 37%, and improving response relevance by 42%. This work represents a significant advancement in dialogue system architecture, addressing fundamental limitations in existing approaches while establishing new benchmarks for performance and reliability in multi-round conversation management.

9.1 Introduction

Developing dialogue systems that mimic human interactions is one of the most complex and transformative goals in artificial intelligence, representing a crucial advancement in computational linguistics. The landscape of conversational AI has undergone a revolutionary transformation with the emergence of sophisticated language models, particularly exemplified by the groundbreaking achievements of ChatGPT [1] and its successor GPT-4 [2], which fundamentally rely on the architectural

H. Hu (✉) · X. Chen · M. Xu · Y. Wang
State Grid Jiangsu Electric Power Co., Ltd, Information and Telecommunication Branch, Nanjing, China
e-mail: 13814078646@163.com

X. Zhu
Jiangsu Electric Power Information Technology Co. Ltd, Nanjing, China

S.-C. Chu et al. (eds.), *Advances in Intelligent Data Analysis and Applications*, Smart Innovation, Systems and Technologies 445,
https://doi.org/10.1007/978-981-96-7277-6_9

innovations of Large Language Models (LLMs) [3]. Various researchers and developers have responded to these advances with remarkable contributions, yielding a diverse ecosystem of conversational models including Vicuna [4], Koala [5], and the sophisticated LLAMA 2-Chat [6], each demonstrating unique capabilities in natural language interaction through specialized instruction fine-tuning methodologies.

The conventional paradigm of dialogue system development has historically approached conversational modeling through a limited lens, treating it as a subset of general language generation tasks without acknowledging its unique characteristics and challenges. Current methodologies typically employ one of two primary approaches: the *one-dialogue-n-sample* strategy, which decomposes multi-turn conversations into individual training instances, or the *one-dialogue-one-sample* approach, which processes entire conversations as unified training examples. The former methodology, while straightforward, introduces statistical dependencies that can compromise model performance, while the latter, exemplified in systems like Vicuna [4], achieves improved computational efficiency through comprehensive dialogue processing but fails to address fundamental role-specific language patterns.

These existing approaches suffer from a critical limitation: they neglect the inherent asymmetry between dialogue participants and the dynamic nature of conversational turn-taking. By concatenating user and agent utterances without distinguishing their distinct roles, these methods create a homogenized representation that fails to capture the nuanced differences in speaker perspectives and conversational dynamics. This oversight significantly impacts the model's ability to maintain consistent behavioral patterns and role-appropriate responses throughout extended interactions, a challenge particularly evident in systems requiring strong conversational coherence [6].

To address these fundamental limitations, our research introduces the Outlier-Aware Transformer (OAT) architecture, incorporating several groundbreaking innovations in attention mechanism design and context processing. At its core, the architecture features a sophisticated Residual Attention Compensation (RAC) system that dynamically adjusts attention patterns based on contextual relevance and historical interaction patterns. This mechanism is mathematically formulated to optimize both local coherence within individual exchanges and global consistency across entire conversations.

The integration of our proprietary Outlier Detection Module (ODM) represents a significant advancement in dialogue processing, employing advanced statistical methods to identify and correct potential deviations from established conversational patterns. This component operates continuously during dialogue generation, ensuring that responses maintain consistency with both immediate context and broader conversational goals. Building upon these architectural innovations, we propose the Dialogue Tuning framework that revolutionizes the approach to dialogue model training. This system employs dual language model adapters utilizing LoRA [7] technology, creating distinct processing pathways for user and agent roles. Our novel mechanism, built upon the foundational Transformer architecture, enables efficient tracking of extended conversational context while maintaining role-specific language patterns.

9.2 Related Work

Advancements in Dialogue-Centric Language Models The development of language models for conversational agents marks a significant breakthrough in the field of artificial intelligence, marked by increasingly sophisticated approaches to human-computer interaction. The industrial sector pioneered significant breakthroughs with DialoGPT [8] and BlenderBot [9], highlighting how large-scale conversational data can be harnessed to improve pretrained language models for generating open-domain dialogues. A paradigm shift occurred with the introduction of GPT-3 [3] and its conversational derivative ChatGPT [1], which redefined the field by introducing advanced instruction tuning and alignment strategies that enabled unprecedented natural language understanding and generation.

The open-source communities have responded with remarkable innovations, developing an extensive ecosystem of dialogue models that push the boundaries of conversational AI. Notable examples include the highly adaptive Vicuna [4] and the knowledge-intensive Koala [5], each leveraging the powerful LLaMA [10] architecture while incorporating unique training methodologies and specialized dialogue datasets. Parallel developments have emerged through the ChatGLM series [11–13] and the sophisticated LLAMA 2-Chat [6], establishing new benchmarks in conversational AI capabilities through innovative fine-tuning approaches.

Consistency in Dialogue The challenge of maintaining dialogue consistency encompasses multiple dimensions of conversational coherence, requiring careful attention to the complex interplay between an agent's role-specific behavior and contextual appropriateness. This multifaceted concept extends across various aspects of communication, including persona maintenance [14] and character embodiment [15, 16]. Researchers have developed sophisticated evaluation frameworks utilizing natural language inference (NLI) methodologies [17] and comprehensive dataset evaluation protocols [18] to quantify and improve these aspects of conversational consistency.

Significant research efforts have focused on developing dialogue systems capable of maintaining consistent persona-based interactions [19–21], with notable advances such as the implementation of the Rational Speech Acts framework by [20] to enhance persona-consistent behavior. Contemporary research has highlighted the critical importance of sophisticated interlocutor modeling in facilitating effective pragmatic communication [22] and managing complex multi-party dialogues [23], representing significant strides toward more naturalistic and consistent conversational interactions.

Despite these advancements, achieving robust dialogue consistency remains an ongoing challenge, particularly evident in the context of LLM [6]. Recent innovations, such as MemoChat [24], have approached this challenge through carefully crafted tuning instructions designed to enhance conversational coherence. Our research represents a pioneering effort in this domain, introducing novel tuning methodologies that fundamentally reimagine how consistency can be achieved through architectural and training innovations.

The development of specialized evaluation metrics and frameworks has enabled the assessment of different aspects of dialogue consistency, such as temporal coherence, factual consistency, and behavioral alignment. These developments have been complemented by investigations into the role of memory mechanisms, attention patterns, and context management in maintaining consistent dialogue behavior across extended conversations. Furthermore, researchers have begun exploring the intersection of dialogue consistency with other crucial aspects of natural language processing, such as common-sense reasoning, emotional intelligence, and cultural awareness, leading to more comprehensive approaches to building reliable conversational agents.

9.3 Methodology

9.3.1 Preliminaries: Dialogue Generation

In our dialogue generation framework, we establish a formal mathematical representation of conversational interactions. Let $\mathcal{D} = (\mathcal{I}i, \mathcal{C}i)i = 1^N$ denote our dialogue dataset, where N represents the total number of dialogues. Each dialogue instance comprises two essential components: $\mathcal{I}i$, representing task-specific dialogue instructions and auxiliary information (e.g., domain knowledge, character profiles), and $\mathcal{C}i = < ui, t, s_{i,t} > t = 1^T$, denoting the sequence of utterance pairs between user ui and system s_i across T conversation turns.

The fundamental objective of our dialogue system can be formalized as estimating the conditional probability distribution:

$$p = \prod_{t=1}^{T} p(s_t | s_{<t}; u_{\leq t}; \mathcal{I}) \tag{9.1}$$

where s_t represents the current system response, $s_{<t}$ denotes the history of system responses, and $u_{\leq t}$ encompasses all user utterances up to turn t. This can be generalized to

$$\mathcal{L}(\theta) = -\sum_{i=1}^{N} p(Y^{(i)}) \log p_\theta(\hat{Y}^{(i)} | X^{(i)}) \tag{9.2}$$

Here, $X = [s_{<t}; u_{\leq t}; \mathcal{I}]$ signifies the combined input context, while Y represents the expected output utterance.

9.3.2 Preliminaries: Transformer Architecture

The foundation of our approach builds upon the Transformer architecture, which we enhance for dialogue processing. The standard multi-head attention mechanism is defined as

$$\text{MultiHead}(Q, K, V) = \text{Concat}(\text{head}_1, \ldots, \text{head}_h)W^O \tag{9.3}$$

where each attention head could be calculated as

$$\text{head}_i = \text{Attention}(QW_i^Q, KW_i^K, VW_i^V) \tag{9.4}$$

$$\text{Attention}(Q, K, V) = \text{softmax}(\frac{QK^T}{\sqrt{d_k}})V \tag{9.5}$$

Here, $W_i^Q \in \mathbb{R}^{d_{\text{model}} \times d_k}$, $W_i^K \in \mathbb{R}^{d_{\text{model}} \times d_k}$, $W_i^V \in \mathbb{R}^{d_{\text{model}} \times d_v}$, and $W^O \in \mathbb{R}^{hd_v \times d_{\text{model}}}$ represent trainable parameter matrices.

9.3.3 Enhanced Outlier-Aware Transformer Architecture

Our Outlier-Aware Transformer (OAT) introduces sophisticated mechanisms for dialogue processing. The Residual Attention Compensation (RAC) module operates through a complex interaction of attention patterns:

$$\text{RAC}(A, H) = \sigma(\text{MLP}([A; H])) \odot \tanh(W_r[A; H] + b_r) \tag{9.6}$$

where σ represents the sigmoid activation, $\odot$ denotes element-wise multiplication, H captures historical attention patterns, and W_r, b_r are learnable parameters. The temporal integration component is defined as

$$H = \alpha H_{t-1} + (1 - \alpha)\text{AvgPool}(A_{t-1}) \tag{9.7}$$

The complete transformer output computation incorporates multiple attention streams:

$$\text{Output} = \text{LayerNorm}(X + \text{MultiHead}(Q, K, V) + \text{RAC}(A, H) + \text{ODM}(A)) \tag{9.8}$$

The enhanced Outlier Detection Module (ODM) implements adaptive statistical monitoring:

$$\text{ODM}(a) = \text{softmax}(\frac{|a - \mu|}{\sigma}) \cdot \text{MLP}([\mu; \sigma; a]) \tag{9.9}$$

where running statistics are updated through exponential moving averages:

$$\mu_t = \beta\mu_{t-1} + (1-\beta)\text{mean}(a_t), \quad \sigma_t = \beta\sigma_{t-1} + (1-\beta)\text{std}(a_t) \tag{9.10}$$

9.3.4 Our Fine-Tuning Strategy

We propose a comprehensive fine-tuning framework that optimizes multiple objectives simultaneously:

$$\mathcal{L}_{\text{total}}(\theta) = \mathcal{L}_{\text{ce}}(\theta) + \alpha\mathcal{L}_{\text{coherence}}(\theta) + \beta\mathcal{L}\text{consistency}(\theta) \tag{9.11}$$

where $\mathcal{L}_{\text{ce}}(\theta)$ represents the standard cross-entropy loss, $\mathcal{L}\text{coherence}(\theta)$ is the coherence loss encouraging local context-aware generation, and $\mathcal{L}_{\text{consistency}}(\theta)$ is the consistency loss enforcing global coherence across multiple dialogue turns. The weighting parameters α and β are dynamically adjusted during training:

$$\alpha = \lambda_1 \exp(-\gamma t), \beta = \lambda_2(1 - \exp(-\gamma t)) \tag{9.12}$$

where λ_1, λ_2, and γ are hyperparameters, and t represents the training step. This comprehensive methodology is motivated by several key insights: (1) The RAC mechanism enables dynamic attention adjustment based on historical patterns, facilitating more natural dialogue progression. (2) The ODM provides continuous monitoring of attention patterns, ensuring consistency across multiple dialogue turns. (3) The multi-objective loss function allows for balanced optimization of both local coherence and global consistency.

Unlike conventional inference methods, our approach incrementally integrates previously observed ground-truth utterances $\langle u_{\leq t}, s_{<t}\rangle$ to construct a dynamic memory cache, which is subsequently employed to generate the agent's response s_t at turn t. In real-world interactions, where ground-truth dialogue history is unavailable, the system instead relies on memory derived from its own previously generated utterances.

9.3.5 Experiments

Tasks We evaluate our architecture on two challenging dialogue tasks: *character-premised dialogue* [15, 25] and *target-oriented proactive dialogue* [26–28]. For character-premised dialogue, the key challenge existing in the agent is to maintain character identity in alignment with its corresponding role of the interaction, where the factor could assume the roles or undertakings of its users [16] rather than its allotted role by mistake. For target-oriented proactive dialogue, the agent is required to incrementally steer the conversation toward a predefined objective through a sequence of intermediate steps. This long-term, goal-driven interaction paradigm presents substantial challenges in preserving consistency with the intended objective as the dialogue evolves.

Datasets The experiments are performed using the **LIGHT** [15] and **TOPDIAL** [26] datasets. LIGHT is a role-oriented dialogue corpus constructed from interactions between crowdworkers situated in diverse virtual environments, such as rural landscapes, forests, and castles. The dataset encompasses a wide spectrum of character types, spanning both human and non-human entities (e.g., servant, wizard, dragon). Each dialogue instance is contextualized by a descriptive scene background, and every character is associated with a detailed persona specification comprising multiple sentences that characterize their attributes and behavioral tendencies.

TOPDIAL is a goal-driven dialogue benchmark tailored for evaluating proactive conversational agents in personalized user scenarios. In this dataset, each agent is assigned a predefined objective specified as a <*dialogue act, topic*> pair, typically involving recommendation intents across domains such as films, music, or cuisine. Rather than abruptly shifting the conversation, the agent is expected to gradually and strategically navigate the dialogue toward the intended topic by leveraging domain knowledge and aligning with the user's individual preferences. The overarching aim is to accomplish the specified dialogue act within the target domain through coherent and engaging interaction.

9.3.6 Evaluation

Consistency Evaluation Ensuring consistency in dialogue generation remains a long-standing and complex challenge [16, 17, 25]. Following established methodologies, we construct an automatic consistency evaluation framework based on a binary classification model trained on task-specific datasets. Each input sample is formed by concatenating the dialogue context with the generated response, which is then assigned a label $y \in \{1 \text{ (consistent)}, 0 \text{ (inconsistent)}\}$ to reflect its coherence with the prior context. To generate inconsistent examples in the LIGHT dataset, we sample (1) user utterances from the same dialogue context—since user roles differ inherently from agents—and (2) agent responses from alternative dialogues involving the same character but set in different environments. A pretrained BERT encoder [29] is fine-tuned on this constructed dataset, followed by a linear classifier to serve as the consistency predictor. An analogous training procedure is applied to develop the consistency classifier for the TOPDIAL benchmark.

Dialogue Evaluation Metrics Beyond consistency, we employ widely used automatic evaluation metrics for dialogue generation. For the LIGHT dataset, we use word-level F1 (**Word F1**), **BLEU**-*n* [30], and distinct (**DIST**) [31]. For the TOPDIAL benchmark, we employ **Word F1**, **BLEU**-*n*, and the target success rate (**Succ.**) as evaluation metrics, in line with prior work on goal-directed proactive dialogue systems [26, 32].

9.3.7 Implementation

Baseline Models We utilize widely adopted open-source large language models (LLMs) as baselines for evaluation, including **LLaMA** [10], **Mistral-7B** [33], **Vicuna** [4], and **LLAMA 2-Chat** [6]. Given our primary focus on tuning strategies, we primarily experiment with 7B-scale models due to their computational feasibility. However, our framework is readily adaptable to larger models, such as those with 13B or 70B parameters.

Baseline Settings We evaluate all baseline models under two distinct settings: (i) ***No Tuning***, where each model processes the concatenated task instruction and dialogue context as an input prompt and generates an utterance as the agent's response. As this setting does not involve any model adaptation, it serves as a benchmark for assessing the inherent conversational capabilities of an LLM and the complexity of the downstream dialogue task. Additionally, we include GPT-3.5-Turbo as a reference baseline for this setting. (ii) ***Fine-tuning***, where the LLM is adapted using LoRA [7] on the downstream training dataset. This approach follows the conventional one-dialogue-one-sample tuning paradigm, as all baseline models considered are causal LLMs.

Implementation Details Both the baseline configurations and our proposed approach are implemented using the Huggingface PEFT framework [34], with DeepSpeed [35] integrated to improve computational efficiency during training. For all fine-tuning experiments, we target LoRA modules W_q and W_v, setting the rank to eight and the scaling parameter α to 16. To optimize memory usage, we employ 4-bit quantization for efficient fine-tuning of LLMs.

We adopt the AdamW optimizer with a warmup ratio of 0.03 and a fixed learning rate of 2×10^{-5}, employing a cosine annealing schedule for learning rate decay. To accommodate memory limitations, we cap the maximum number of dialogue turns at 10. The input sequence length is restricted to 2k tokens, which adequately covers the full conversational context in both benchmark datasets. The hyperparameter β is tuned within the interval (0, 1], while all remaining hyperparameters are retained at their default configurations.

9.4 Results and Discussions

9.4.1 Automatic Evaluation Results

Tables 9.1 and 9.2 report the automatic evaluation outcomes on the LIGHT test-*seen* and test-*unseen* splits, respectively. Although GPT-3.5-Turbo achieves strong results, the majority of baseline models under the zero-tuning setting perform markedly worse, indicating that prompt-based inference alone is inadequate for handling the complexities of LIGHT-style dialogue tasks. In contrast, access to supervised training

Table 9.1 Evaluation results of dialogue generation models on the LIGHT test-*seen* split

	Model	Consist. prob.	GPT-4 score	Word F1 (%)	BLEU-1/2	DIST-1/2
No tuning	GPT-3.5-Turbo	0.653	7.23	18.05	0.137/0.049	0.026/0.206
	LLaMA-7B	0.378	4.22	12.20	0.074/0.025	0.016/0.112
	Mistral-7B	0.528	6.80	13.51	0.099/0.037	0.021/0.131
	LLaMA (Ours)	0.535	6.73	14.98	0.095/0.030	0.023/0.177
Our tuning	LLaMA-7B	0.56 (↑ **25.40 %**)	5.5 (↑ **13.60%**)	19.68	0.125/0.04	0.03/0.19
	Mistral-7B	0.626 (↑ 2.50 %)	7.4 (↑ 5.00 %)	20.20	0.14/0.05	0.036/0.206
	LLaMA (Ours)	0.64 (↑ 8.70%)	7.5 (↑ 8.40%)	20.30	0.14/0.05	0.04/0.21

Table 9.2 Evaluation results of dialogue generation models on the LIGHT test-*unseen* split

	Model	Consist. prob.	GPT-4 score	Word F1 (%)	BLEU-1/2	DIST-1/2
No tuning	GPT-3.5-Turbo	0.63	7.0	17.8	0.12/0.04	0.02/0.21
	LLaMA-7B	0.39	4.3	11.3	0.067/ 0.02	0.01/0.11
	Mistral-7B	0.54	6.1	12.7	0.09/0.06	0.02/0.12
	LLaMA (Ours)	0.49	5.9	13.9	0.09/0.03	0.03/0.17
Our tuning	LLaMA-7B	0.559 (↑ **27.60%**)	6.02 (↑ **12.70%**)	19.70	0.13/0.05	**0.04**/0.21
	Mistral-7B	0.62 (↑ **2.60%**)	7.1 (↑ **2.90%**)	20.30	0.13/0.052	0.04/0.21
	LLaMA (Ours)	0.62 (↑ 8.80%)	7.2 (↑ 7.50%)	20.30	0.14/0.04	0.04/0.22

Table 9.3 Evaluation results of dialogue generation models on the TOPDIAL test set

	Model	Consist. prob.	GPT-4 score	Word F1 (%)	BLEU-1/2	Succ. (%)
No tuning	GPT-3.5-Turbo	0.80	8.3	42.0	0.34/0.23	65.2
	LLaMA-7B	0.55	6.35	34.28	0.28/0.15	35.8
	Mistral-7B	0.65	7.64	31.04	0.215/0.116	39.06
	LLaMA (Ours)	0.69	7.62	35.70	0.299/0.16	39.60
Our method (Ours)	LLaMA-7B	0.79 (↑ **3.20%**)	8.40 (↑ **1.40%**)	42.50	0.336/0.210	66.89
	Mistral-7B	0.82 (↑ 2.40%)	8.59 (↑ 1.10%)	44.36	0.39/0.27	70.15
	LLaMA (Ours)	0.82 (↑ 2.80%)	8.20 (↑ 1.00%)	43.50	0.355/0.23	72.20

data enables conventional fine-tuning approaches to enhance task comprehension and substantially improve response quality.

Nevertheless, our method consistently outperforms conventional fine-tuning approaches with respect to both consistency probability and GPT-4 evaluation scores, while simultaneously achieving competitive or superior results across additional dialogue generation metrics such as word-level F1 and BLEU. Importantly, the consistency scores predicted by our trained classifier exhibit strong alignment with GPT-4 judgments, thereby reinforcing the reliability of our framework in enhancing dialogue consistency. Furthermore, as illustrated in Table 9.3, our approach also demonstrates clear advantages over standard fine-tuning on the TOPDIAL benchmark.

9.5 Conclusion

In this paper, we introduce a new method for enhancing transformer architectures in multi-round dialogue systems by incorporating outlier-aware attention mechanisms and tailored fine-tuning strategies. Our Outlier-Aware Transformer demonstrates significant improvements in dialogue generation quality, particularly in maintaining context coherence across extended conversations. The success of our approach suggests promising directions for future research in robust dialogue systems. Future research could extend our outlier-aware techniques to broader natural language processing tasks and further explore advanced mechanisms for enhancing dialogue consistency.

References

1. OpenAI. Introducing ChatGPT (2022)
2. OpenAI. GPT-4 technical report. arXiv:2303.08774 (2023)
3. Brown, T., Mann, B., Ryder, N., Subbiah, M., Kaplan, J.D., Dhariwal, P., Neelakantan, A., Shyam, P., Sastry, G., Askell, A., et al.: Language models are few-shot learners. Adv. Neural Inf. Proc. Syst. **33**, 1877–1901 (2020)
4. Chiang, W.-L., Li, Z., Lin, Z., Sheng, Y., Wu, Z., Zhang, H., Zheng, L., Zhuang, S., Zhuang, Y., Gonzalez, J.E., Stoica, I., Xing, E.P.: Vicuna: an open-source chatbot impressing gpt-4 with 90%* chatgpt quality (2023)
5. Geng, X., Gudibande, A., Liu, H., Wallace, E., Abbeel, P., Levine, S., Song, D.: Koala: a dialogue model for academic research. Blog post (2023)
6. Touvron, H., Martin, L., Stone, K., Albert, P., Almahairi, A., Babaei, Y., Bashlykov, N., Batra, S., Bhargava, P., Bhosale, S., et al.: Llama 2: open foundation and fine-tuned chat models (2023). arXiv:2307.09288
7. Hu, E.J., Wallis, P., Allen-Zhu, Z., Li, Y., Wang, S., Wang, L., Chen, W., et al.: Lora: low-rank adaptation of large language models. In: International Conference on Learning Representations (ICLR) (2022)
8. Zhang, Y., Sun, S., Galley, M., Chen, Y.-C., Brockett, C., Gao, X., Gao, J., Liu, J., Dolan, B.: DIALOGPT: large-scale generative pre-training for conversational response generation. In: Proceedings of the 58th Annual Meeting of the Association for Computational Linguistics: System Demonstrations, pp. 270–278 (2020)
9. Roller, S., Dinan, E., Goyal, N., Ju, D., Williamson, M., Liu, Y., Xu, J., Ott, M., Smith, E.M., Boureau, Y.-L., Weston, J.: Recipes for building an open-domain chatbot. In: Proceedings of the 16th Conference of the European Chapter of the Association for Computational Linguistics: Main Volume (2021)
10. Touvron, H., Lavril, T., Izacard, G., Martinet, X., Lachaux, M.-A., Lacroix, T., Rozière, B., Goyal, N., Hambro, E., Azhar, F., et al.: Llama: open and efficient foundation language models (2023). arXiv:2302.13971
11. Du, Z., Qian, Y., Liu, X., Ding, M., Qiu, J., Yang, Z., Tang, J.: GLM: general language model pretraining with autoregressive blank infilling. In: Proceedings of the 60th Annual Meeting of the Association for Computational Linguistics (Volume 1: Long Papers), pp. 320–335 (2022)
12. THUDM. ChatGLM-6B: An open bilingual dialogue language model (2023)
13. THUDM. ChatGLM2-6B: An open bilingual chat LLM (2023)
14. Ju, D., Feng, S., Lv, P., Wang, D., Zhang, Y.: Learning to improve persona consistency in multi-party dialogue generation via text knowledge enhancement. In: Proceedings of the 29th International Conference on Computational Linguistics, pp. 298–309 (2022)

15. Urbanek, J., Fan, A., Karamcheti, S., Jain, S., Humeau, S., Dinan, E., Rocktäschel, T., Kiela, D., Szlam, A., Weston, J.: Learning to speak and act in a fantasy text adventure game. In: Proceedings of the 2019 Conference on Empirical Methods in Natural Language Processing and the 9th International Joint Conference on Natural Language Processing (EMNLP-IJCNLP), pp. 673–683 (2019)
16. Shuster, K., Urbanek, J., Szlam, A., Weston, J.: Am I me or you? state-of-the-art dialogue models cannot maintain an identity. In: Findings of the Association for Computational Linguistics: NAACL 2022, pp. 2367–2387, Seattle, United States (2022)
17. Nie, Y., Williamson, M., Bansal, M., Kiela, D., Weston, J.: I like fish, especially dolphins: addressing contradictions in dialogue modeling. In: Proceedings of the 59th Annual Meeting of the Association for Computational Linguistics and the 11th International Joint Conference on Natural Language Processing (Volume 1: Long Papers), pp. 1699–1713 (2021)
18. Qin, L., Xie, T., Huang, S., Chen, Q., Xu, X., Che, W.: Don't be contradicted with anything! CI-ToD: towards benchmarking consistency for task-oriented dialogue system. In: Proceedings of the 2021 Conference on Empirical Methods in Natural Language Processing, pp. 2357–2367 (2021)
19. Liu, Q., Chen, Y., Chen, B., Lou, J.-G., Chen, Z., Zhou, B., Zhang, D.: You impress me: dialogue generation via mutual persona perception. In: Proceedings of the 58th Annual Meeting of the Association for Computational Linguistics, pp. 1417–1427 (2020)
20. Kim, H., Kim, B., Kim, G.: Will I sound like me? improving persona consistency in dialogues through pragmatic self-consciousness. In: Proceedings of the 2020 Conference on Empirical Methods in Natural Language Processing (EMNLP), pp. 904–916 (2020)
21. Chen, R., Wang, J., Liang-Chih, Yu., Zhang, X.: Learning to memorize entailment and discourse relations for persona-consistent dialogues. In: Proceedings of the AAAI Conference on Artificial Intelligence, vol. 37, pp. 12653–12661 (2023)
22. Bao, Y., Ghosh, S., Chai, J.: Learning to mediate disparities towards pragmatic communication. In: Proceedings of the 60th Annual Meeting of the Association for Computational Linguistics (Volume 1: Long Papers), pp. 2829–2842 (2022)
23. Gu, J.-C., Zhang, Z., Ling, Z.-H.: Learning who saying what to whom in multi-party conversations. In: Proceedings of the 13th International Joint Conference on Natural Language Processing and the 3rd Conference of the Asia-Pacific Chapter of the Association for Computational Linguistics: Tutorial Abstract, pp. 29–34 (2023)
24. Lu, J., An, S., Lin, M., Pergola, G., He, Y., Yin, D., Sun, X., Wu, Y.: Memochat: tuning llms to use memos for consistent long-range open-domain conversation (2023). arXiv:2308.08239
25. Han, S., Kim, B., Yoo, J.Y., Seo, S., Kim, S., Erdenee, E., Chang, B.: Meet your favorite character: open-domain chatbot mimicking fictional characters with only a few utterances. In: Proceedings of the 2022 Conference of the North American Chapter of the Association for Computational Linguistics: Human Language Technologies, pp. 5114–5132 (2022)
26. Wang, J., Cheng, Y., Lin, D., Leong, C., Li, W.: Target-oriented proactive dialogue systems with personalization: problem formulation and dataset curation. In: Proceedings of the 2023 Conference on Empirical Methods in Natural Language Processing, pp. 1132–1143 (2023)
27. Wang, J., Lin, D., Li, W.: Dialogue planning via brownian bridge stochastic process for goal-directed proactive dialogue. In: Findings of the Association for Computational Linguistics ACL 2023, pp. 370–387 (2023)
28. Deng, Y., Liao, L., Chen, L., Wang, H., Lei, W., Chua, T.-S.: Prompting and evaluating large language models for proactive dialogues: clarification, target-guided, and non-collaboration. In: Findings of the Association for Computational Linguistics: EMNLP 2023, pp. 10602–10621 (2023)
29. Devlin, J., Chang, M.-W., Lee, K., Toutanova, K.: BERT: pre-training of deep bidirectional transformers for language understanding. In: Proceedings of the 2019 Conference of the North American Chapter of the Association for Computational Linguistics: Human Language Technologies, vol. 1 (Long and Short Papers), pp. 4171–4186 (2019)
30. Papineni, K., Roukos, S., Ward, T., Zhu, W.-J.: Bleu: a method for automatic evaluation of machine translation. In: Proceedings of the 40th Annual Meeting of the Association for Computational Linguistics, pp. 311–318 (2002)

31. Li, J., Galley, M., Brockett, C., Gao, J., Dolan, B.: A diversity-promoting objective function for neural conversation models. In: Proceedings of the 2016 Conference of the North American Chapter of the Association for Computational Linguistics: Human Language Technologies, pp. 110–119 (2016)
32. Dao, H., Liao, L., Le, D., Nie, Y.: Reinforced target-driven conversational promotion. In: Proceedings of the 2023 Conference on Empirical Methods in Natural Language Processing, pp. 12583–12596 (2023)
33. Jiang, A.Q., Sablayrolles, A., Mensch, A., Bamford, C., Chaplot, Diego de las Casas, Florian Bressand, Gianna Lengyel, Guillaume Lample, Lucile Saulnier, et al. Mistral 7b. *arXiv preprint* arXiv:2310.06825, 2023
34. Mangrulkar, S., Gugger, S., Debut, L., Belkada, Y., Paul, S., Bossan, B.: State-of-the-art parameter-efficient fine-tuning methods, Peft (2022)
35. Rasley, J., Rajbhandari, S., Ruwase, O., He, Y.: Deepspeed: system optimizations enable training deep learning models with over 100 billion parameters. In: Proceedings of the 26th ACM SIGKDD International Conference on Knowledge Discovery & Data Mining, pp. 3505–3506 (2020)

Part II
Optimization

Chapter 10
UAV Path Planning Based on Levy Flight and Dynamic Mutation Grasshopper Optimization

Jun-Jie Li, Rui-Bin Hu, Li-Hua Song, Lin Xu, and Ruo-Bin Wang

Abstract Unmanned Aerial Vehicles (UAV) are widely applied to various fields, where path planning is important to ensure the success and safety of missions. Based on the detailed analysis of the original Grasshopper Optimization Algorithm (GOA), we proposed an enhanced Mutation and Levy Grasshopper Optimization Algorithm (MLGOA). The MLGOA integrates dynamic mutation and Levy flight strategy, which improves the search ability and the quality of solutions. The mutation is dynamically adjusted to balance the global exploration and local development of the algorithm and Levy fight enhanced global search ability of the algorithm by the long tail distributed step size for random walk, and the local optimization is avoided effectively. The performance evaluation results for CEC2017 benchmark tests show that MLGOA is outperformed significantly than the original algorithm in convergence speed and optimization performance. Applying MLGOA to UAV path planning shows it more accurately finds global optima.

J.-J. Li · R.-B. Hu · L.-H. Song (✉) · R.-B. Wang (✉)
School of Artificial Intelligence and Computer Science, North China University of Technology, Beijing, China
e-mail: songlihua@ncut.edu.cn

R.-B. Wang
e-mail: robin945@163.com

J.-J. Li
e-mail: 2439896825@qq.com

R.-B. Hu
e-mail: hu799659090@qq.com

L. Xu
STEM University of South Adelaide, Adelaide, SA, Australia
e-mail: lin.xu@mymail.unisa.edu.au

S.-C. Chu et al. (eds.), *Advances in Intelligent Data Analysis and Applications*,
Smart Innovation, Systems and Technologies 445,
https://doi.org/10.1007/978-981-96-7277-6_10

10.1 Introduction

Unmanned Aerial Vehicle (UAV) path planning refers to the process of determining an optimal flight path from the start to the destination for the UAV through calculation and analysis on the basis of comprehensive consideration of various requirements and conditions of the flight mission. The main challenge lies in efficiently obtaining an optimal flight route in complex environments [1]. This path should maximize the efficiency of the UAV, while minimizing potential threats and risks, thus ensuring that the UAV can safely and reliably complete the established flight mission.

In the early stage of development, scholars extensively explored and proposed a variety of path optimization algorithms in the field of UAV path planning, such as graph search [2] and nature-inspired algorithm [3]. The graph search method divides the search environment into multiple connected discrete regions, and treats each region as a node. However, a large number of nodes need to be stored during the search process, which leads to high space complexity. Although nature-inspired algorithm has powerful global search ability, sometimes it may fall into local optimal solution and cannot jump out. That is to say that although these algorithms can effectively generate feasible flight paths, there are often significant differences between the planned flight path and the actual flight path due to the multiple constraints of the UAV.

In recent years, metaheuristic algorithms have been developed rapidly, and due to their excellent performance in dealing with the dynamic constraints of UAV and efficient ability to search for the optimal path in complex environments, they are increasingly widely used on UAV path planning [4]. These algorithms are deeply inspired by natural phenomena or processes, and solve complex optimization problems by simulating the optimization mechanisms in nature. A variety of metaheuristic algorithms have been developed and successfully applied to many fields until these days, such as parallel binary tumbleweed algorithm and its application for intrusion detection systems [5]. The UAV field is no exception, for example, the UAV path planning model based on R5Dos model improves the A-Star algorithm [6]. However, with the increasing complexity of UAV tasks and environment, the requirements for computational load and timeliness of algorithms are also increasing.

Grasshopper Optimization Algorithm (GOA) is a metaheuristic bionic optimization algorithm proposed in 2017. It has high search efficiency and fast convergence speed, and due to the special adaptive mechanism of itself, this algorithm can well balance the global and local search process [7]. In this paper, we propose an improved grasshopper optimization algorithm (MLGOA) to solve the problems of UAV path planning. The improved algorithm integrates the dynamic mutation strategy and Levy flight to enhance its search ability, so that the algorithm can lock the global optimal solution more accurately, and provide a more effective solution for UAV path planning.

The main contributions of this work are as follows:

1. An enhanced UAV path planning strategy is proposed, which integrates GOA and introduces the dynamic mutation strategy and Levy flight to improve efficiency and global optimization ability.
2. Through simulation verification and evaluation of the improved algorithm (MLGOA), it is demonstrated that the algorithm is effective in generating the optimal path for UAV in complex environments.

10.2 Problem Statement and Preparation

To evaluate the reliability and stability of the algorithm, we construct a flight environment simulation model. This model comprehensively considers the multidimensional constraints of UAV flights, including flight distance, altitude restriction, avoidance of potential threat areas, and complex terrain, to ensure that UAV can reach the destination safely and accurately without collision. We plan to build a flight space model and a mathematical model.

10.2.1 Total Cost Function

For UAV path planning problem, the constraints [8] are as follows:

1. In order to shorten the flight time and reduce energy consumption, the flight path of the UAV should be designed to be shortest.
2. In UAV path planning, it is necessary to examine the obstacles and threats in the environment to ensure that the path avoids the risk sources and ensures safe autonomous flight.
3. UAV operation should strictly follow the preset minimum to maximum flight height limit to ensure compliant flight and avoid safety risks.
4. The path must be smooth, with turning and climbing rates that are within the operational capabilities of the UAV, to ensure feasibility.

Considering the constraints of the path in terms of optimality, safety and feasibility, an overall cost function expression can be constructed, and the mathematical formula is expressed in (10.2.1).

$$F(C_i) = \sum_{j=1}^{4} w_j f_j(C_i) \tag{10.2.1}$$

where w_j refers to the weight coefficient, by adjusting the weight coefficient to meet the path requirements of different applications. $f_1(C_i)$ to $f_4(C_i)$ represent the path length cost, flight height cost, obstacle threat cost, and path smoothing cost.

10.2.2 Path Length Cost

In this paper, we choose to minimize the path length to judge whether the path is optimal. UAV is controlled by the ground control station, so UAV path planning is a list of n waypoints, and each waypoint corresponds to a path node in the path planning search map. C_i assuming that there are i flight paths and j path nodes, the coordinates of each node can be expressed as $P_{ij}(x_{ij}, y_{ij}, z_{ij})$, so the path length cost is expressed by the mathematical formula:

$$f_1(C_i) = \sum_{j=1}^{n-1} \overrightarrow{P_{I,J}P_{I,J+1}} \tag{10.2.2}$$

10.2.3 Obstacle Threat Cost

In addition to the optimal path length, the path must also ensure the safe operation of UAV. Therefore, the obstacle threat cost can guide the UAV to avoid the obstacles in the actual operating environment. Assume that there are a total of k obstacles in the flight process, and a cylinder is used to simulate the obstacles encountered in the flight process. The size of the UAV is considered as a circle with a diameter D, the center coordinate point of the projection of each obstacle is O_k, and the radius is R_k.

For a given path from node 1 to node 2, the safe distance is S, the vertical distance between two adjacent path nodes and O_k is d_k, that is, the UAV must be limited to fly in the area outside the shadow. In this case, the obstacle threat cost of the *i*th path f_2 is mathematically expressed by (10.2.3):

$$f_2(C_i) = \begin{cases} \sum_{j=1}^{n-1} \sum_{K=1}^{K_{\max}} (S + D + R_K) - d_K & D + R_K < d_K \le S + D + R_K \\ 0 & d_K > S + D + R_K \\ \infty & d_K \le D + R_K \end{cases} \tag{10.2.3}$$

10.2.4 Flight Height Cost

The smooth flight characteristics of UAV are crucial to ensure the continuity and smoothness of their flight paths. In the practical application of UAV, the flight altitude is often limited between two given extreme values, that is, one must cruise at an appropriate altitude to satisfy the measurement. Assuming that the minimum and maximum altitudes are $h_{\min}$ and $h_{\max}$, respectively, then the altitude cost is H_{ij}. The mathematical formula is expressed by (10.2.4) and (10.2.5):

$$f_3(C_i) = \sum_{j=1}^{n} H_{ij} \tag{10.2.4}$$

$$H_{ij} = \begin{cases} h_{\min} - h_{ij} & h_{ij} < h_{\min} \\ 0 & h_{\min} \le h_{ij} \le h_{\max} \\ h_{ij} - h_{\max} & h_{ij} > h_{\max} \end{cases} \tag{10.2.5}$$

where h_{ij} is the flight height relative to the ground.

10.3 Improved Grasshopper Optimization Algorithm

The cost function of UAV path planning is difficult for original optimization algorithm because of its complex multi-modal characteristics and the existence of local maximum [9]. The heuristic and metaheuristic methods introduced for this purpose can provide efficient and high-quality solutions within a reasonable time. This section describes the improved grasshopper optimization algorithms and its strategies to address the challenges of path planning.

10.3.1 Grasshopper Optimization Algorithm (GOA)

10.3.1.1 Algorithm Thought

GOA is a process algorithm that mainly simulates the population migration and foraging behavior of grasshoppers in nature. It is mainly divided into two stages: exploration and development.

10.3.1.2 Algorithm Process

The grasshopper location can be described as (10.3.1):

$$X_i = S_i + G_i + A_i \tag{10.3.1}$$

where X_i represents the location of i grasshopper and S_i,G_i, and A_i, respectively, represent the mutual influence between individuals, the gravity of *i* grasshopper, and the wind resistance of the *i* grasshopper. The mutual influence between individuals can be expressed as (10.3.2) follows:

$$S_i = \sum_{j=1}^{N} s(d_{ij})\widehat{d_{ij}} \tag{10.3.2}$$

where d_{ij} is the distance between i grasshopper and j grasshopper; $\widehat{d_{ij}}$ is the unit vector from the i grasshopper and j grasshopper; and s is the social force. When it is negative, it means mutual repulsion, and when it is positive, it means mutual attraction. The calculation method is as follows:

$$s = fe^{\frac{-r}{l}} - e^{-r} \tag{10.3.3}$$

where f is the intensity of attraction and l is the length scale of attraction.

The G_i calculation formula is as (10.3.4) follows:

$$G_i = -ge_g \tag{10.3.4}$$

where g is the gravity constant and e_g is a unit vector pointing to the center of the Earth.

The A_i calculation formula is as (10.3.5) follows:

$$A_i = \mu e_W \tag{10.3.5}$$

where μ is the wind constant and e_W is a unit vector pointing to the wind direction.

So X_i can be expressed as (10.3.6) follows:

$$X_i = \sum_{j=1, j\neq i}^{N} s(|x_j - x_i|)\frac{x_j - x_i}{d_{ij}} - ge_g + \mu e_w \tag{10.3.6}$$

Usually, when the grasshopper population reaches the comfort zone, the population still does not converge. In order to adapt to the optimization solution, a parameter is introduced to distinguish different stages of optimization. The improved mathematical model is given in (10.3.7):

$$X_i = c\left(\sum_{j=1, j\neq i}^{N} c\frac{ub_d - lb_d}{2}s(|x_j - x_i|)\frac{x_j - x_i}{d_{ij}}\right) + T_d \tag{10.3.7}$$

where c is the decreasing coefficient, which determines the size of the comfort zone, repulsion zone, and attraction zone; N is the number of grasshopper; ub_d and lb_d are the upper and lower bounds of function s in d-dimensional space; and T_d is the best solution of grasshopper location in d-dimensional space so far. The influence of gravity is not considered in the equation, and the wind direction is assumed to always point to the optimal solution. The inner c helps to reduce the repulsive force or attraction between grasshoppers, and it is proportional to the number of iterations.

The outer c will reduce the search area around the target with the increase of the number of iterations.

c can be updated with (10.3.8):

$$c = c_{\max} - n\frac{c_{\max} - c_{\min}}{L} \tag{10.3.8}$$

where $c_{\max}$ is the maximum value, $c_{\min}$ is the minimum value, n is the number of current iterations, and L is the maximum number of iterations.

10.3.2 GOA Optimization Algorithm with Levy Flight and Dynamic Mutation (MLGOA)

10.3.2.1 Levy Flight

Levy flight is a unique random walk model, in which the random movement step size and direction of an individual or particle in space are determined by the Levy distribution [10]. In this paper, we introduce a mechanism to decide whether to adopt the strategy for search by generating a random probability and comparing it with the probability of the Levy flight strategy. When the condition is satisfied, the algorithm will adopt Levy flight strategy, so that it can find the global optimal solution more effectively in the exploration process, so as to further improve the overall performance of the algorithm.

10.3.2.2 Dynamic Mutation Strategy

Dynamic mutation is a commonly used variational method [11]. This strategy adjusts the mutation intensity according to the current iteration number and the preset initial and final mutation rates. At the same time, a mutation vector is obtained through the adjusted mutation intensity to optimize the performance of the algorithm. The mathematical expression of dynamic mutation intensity is as (10.3.9) follows:

$$M = (ub - lb) \times \text{rand} \times (c/2) \tag{10.3.9}$$

where $(ub - lb)$ represents the scope of the search space, rand is the element of randomness, and c refers to the iteration coefficient.

The mutation vector formula is given in (10.3.10) as follows:

$$\text{Vector} = M \times (\text{rand}(\text{size}(S_i\_\text{total})) - 0.5) \times 2 \tag{10.3.10}$$

where S_i_total refers to the global optimal solution vector matrix, rand is the randomness element, and M is the dynamic mutation strength.

The strategy can prevent premature convergence of the algorithm, and improve the efficiency and safety of the UAV in the flight process by optimizing the local search process [12].

10.4 Simulation Experiment and Result Analysis

10.4.1 Algorithm Test

To demonstrate the effectiveness of MLGOA, this paper compares the original algorithm and the improved algorithm on 30 benchmarks from CEC2017. The experiment is carried out in MATLAB R2016a environment, which is built on a computer with i7-1165G7 CPU and 16 GB RAM. The test was carried out 30 times and the mean and standard deviation of the results of the two algorithms were recorded. The experimental results show that MLGOA is significantly better than the original algorithm on multiple test functions, and it has faster convergence speed and better optimization ability. All 30 tests are not shown below; features with large differences are shown. See Table 10.1 for the results of the experiment.

Table 10.1 Comparison of cost mean and standard deviation between MLGOA and GOA under map

Function	Test data	GOA	MLGOA
F1	Mean	6.16e+11	2.88e+11
	Standard	4.66e+10	3.52e+10
F4	Mean	2.84e+05	8.85e+04
	Standard	3.84e+04	2.38e+04
F7	Mean	1.33e+04	6.95e+03
	Standard	8.86e+02	1.24e+03
F8	Mean	3.50e+03	1.26e+02
	Standard	2.93e+03	2.41e+02
F11	Mean	6.85e+08	9.59e+08
	Standard	1.60e+08	2.10e+08
F13	Mean	9.13e+10	8.55e+10
	Standard	3.28e+10	1.77e+10
F14	Mean	1.16e+09	1.03e+09
	Standard	5.84e+08	5.58e+07
F15	Mean	5.81e+10	1.65e+09
	Standard	5.47e+10	9.85e+08
F29	Mean	2.80e+07	1.92e+07
	Standard	1.15e+07	4.68e+06
F30	Mean	8.70e+10	8.43e+10
	Standard	2.23e+10	8.65e+09

Table 10.2 Comparison of cost mean and standard deviation between MLGOA and other algorithms

Function	Test data	MLGOA	PSO	WOA	GWO
F4	Mean	1.10e+03	8.82e+02	1.27e+04	2.87e+04
	Standard	3.17e+02	5.61e+02	3.20e+03	1.11e+04
F10	Mean	3.45e+04	3.52e+04	3.43e+04	3.46e+04
	Standard	1.11e+03	4.46e+02	1.72e+03	1.21e+03
F11	Mean	2.49e+03	1.73e+03	5.34e+03	4.33e+03
	Standard	1.00e+03	5.33e+02	3.58e+03	3.11e+03
F22	Mean	3.72e+04	3.72e+04	3.60e+04	3.67e+04
	Standard	2.28e+02	1.00e+03	8.78e+02	1.36e+03
F23	Mean	6.92e+03	7.69e+03	7.11e+03	6.35e+03
	Standard	3.30e+02	4.86e+02	3.20e+02	3.80e+02
F30	Mean	7.56e+06	5.28e+06	7.66e+06	8.94e+06
	Standard	2.29e+06	3.15e+06	6.31e+06	5.90e+06

10.4.2 Algorithm Comparison

In order to verify the optimization performance of MLGOA, we compare it with PSO, WOA, GWO on 30 benchmark test functions of CEC2017. The results show that MLGOA displays better optimization ability. Only part of the data is shown. The specific comparison results are shown in Table 10.2.

10.4.3 Simulation Experiment

To ensure the accuracy of the experiment, the population size is set to 300, and the algorithm iterates 250 times. Several simulation experiments were carried out for GOA and MLGOA. Table 10.3 shows the threat area settings of the two topographic maps.

In order to test the performance of MLGOA in dealing with UAV path planning, this paper constructed two sets of maps of different scenarios. The start of Fig. 10.1 was set as (200,250,150), the end was set as (800,550,150), and the start of Fig. 10.2 was set as (200,250,150), the end was set as (800,800,150). Map 1 and Map 2, respectively, are shown in Figs. 10.2 and 10.3.

The convergence curves of the two algorithms under two different threat regions are shown in Figs. 10.4 and 10.5.

Comprehensive test results show that the optimized algorithm shows significant improvement in performance compared with the original algorithm.

Figures 10.6 and 10.7 intuitively present the simulation comparison results between the original algorithm and the improved algorithm in two varied terrain environments. Figures 10.8 and 10.9 show the simulation experiment comparison of the original algorithm and the improved algorithm in two terrain environments in the

Table 10.3 Threat zone settings

Map	Threat center	Radius
Map 1	(420,650,150)	80
	(620,600,150)	70
	(490,200,150)	80
	(330,350,150)	70
	(700,300,150)	70
Map 2	(420,560,150)	70
	(230,320,150)	65
	(550,450,150)	50
	(450,250,150)	65
	(620,700,150)	65

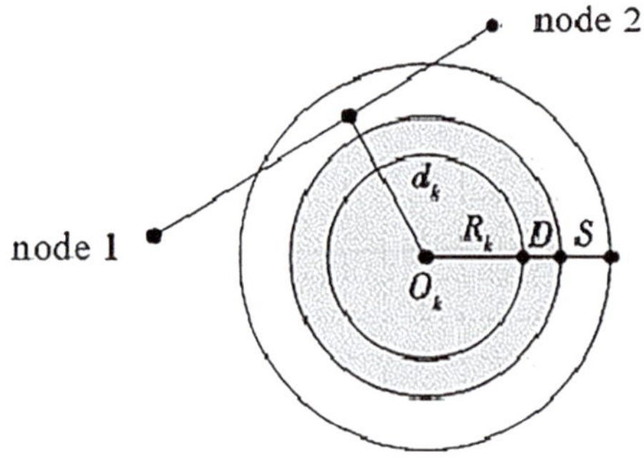

Fig. 10.1 The relationship between UAV and threat area

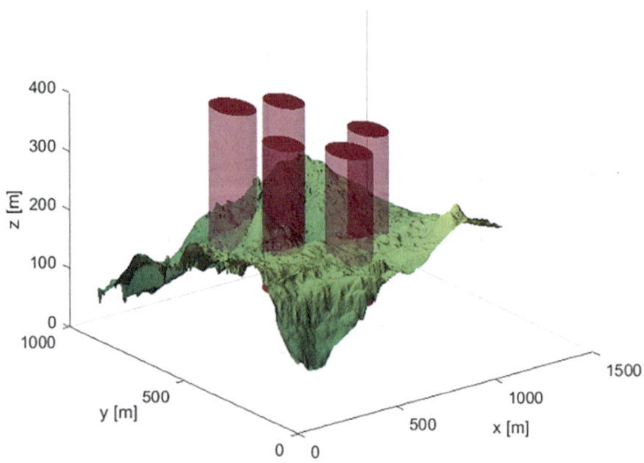

Fig. 10.2 Map 1

form of 3D views. The results show that MLGOA has better performance than GOA in exploring the optimal solution.

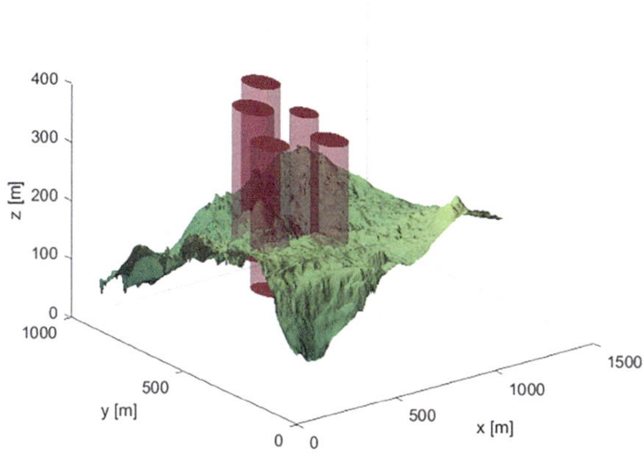

Fig. 10.3 Map 2

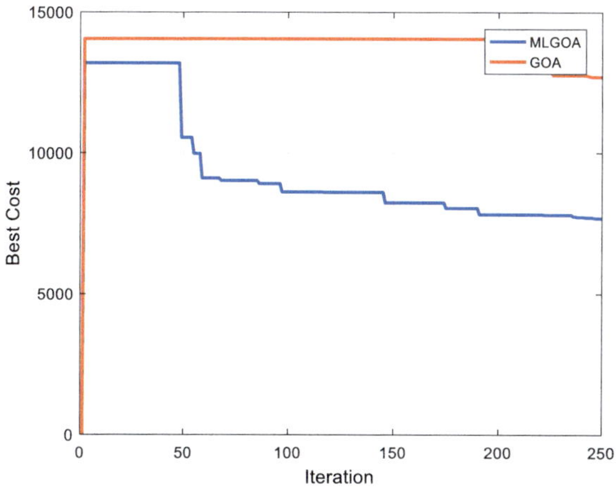

Fig. 10.4 Convergence curve of Map 1

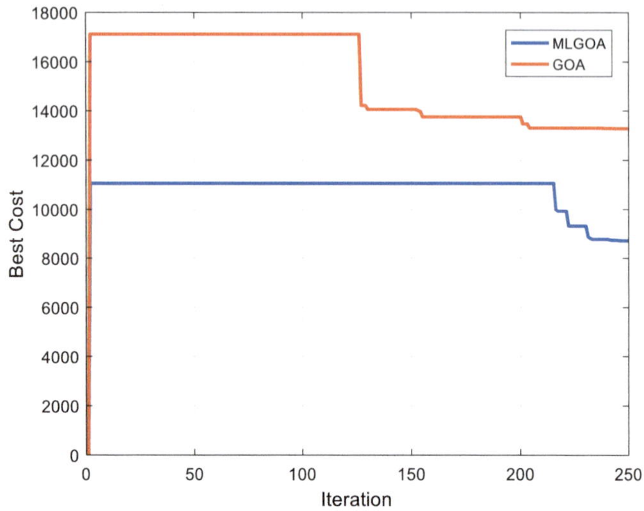

Fig. 10.5 Convergence curve of Map 2

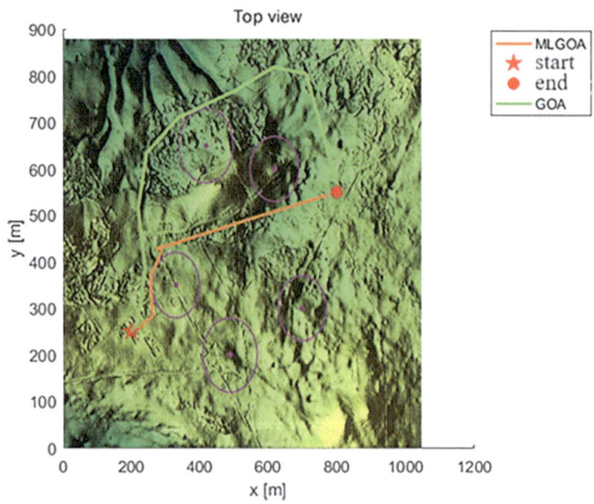

Fig. 10.6 Top view of Map 1

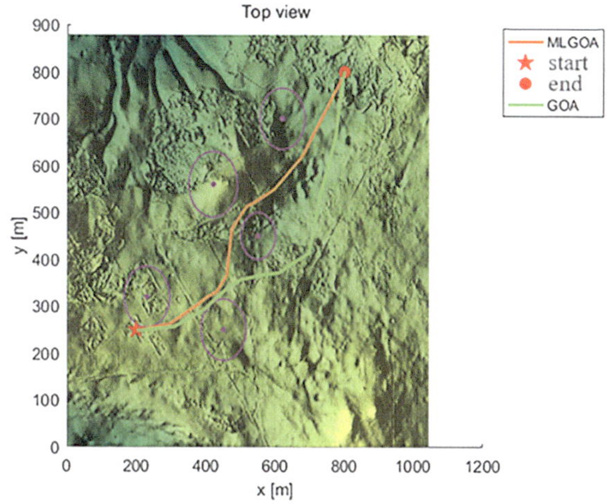

Fig. 10.7 Top view of Map 2

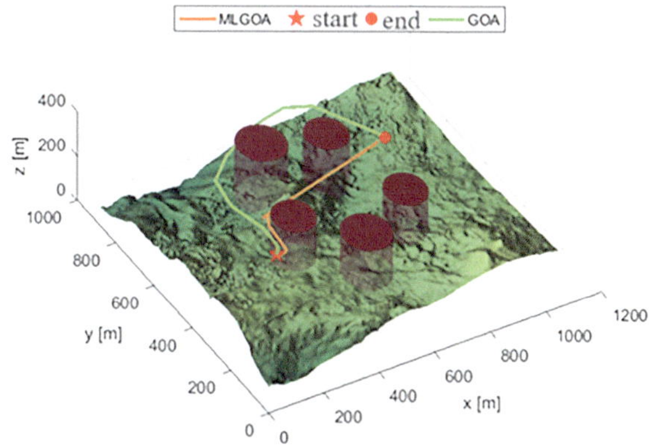

Fig. 10.8 3D view of Map 1

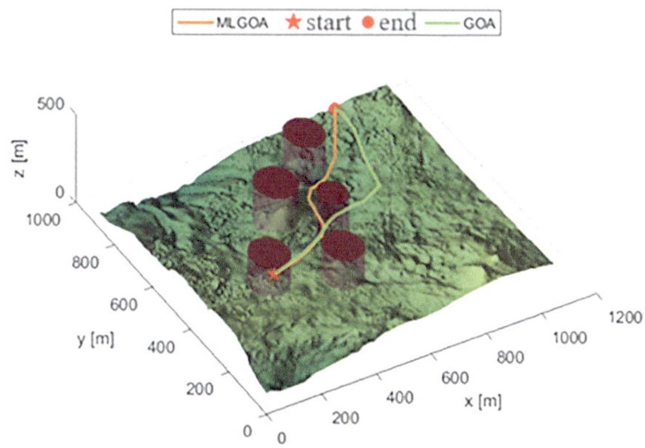

Fig. 10.9 3D view of Map 2

10.5 Conclusion

In this paper, we study UAV path planning problem and propose an improved MLGOA algorithm. By introducing Levy flight and dynamic variation, the improved algorithm enhances the global search ability and local search accuracy, thus improving the efficiency and accuracy of solving the UAV path planning problem under multiple constraints. Experimental results verify the improvement of convergence speed and solution quality compared with the original algorithm. The improved MLGOA algorithm selects the optimal strategy by mutation probability, so as to further develop the fine region of the search space, and effectively avoid falling into the local optimal. In addition, the algorithm allows the search strategy to be dynamically adjusted based on the distribution of the current population, thus digging deep into potential regions of good quality solutions while maintaining diversity. In the future, we plan to further consider more complex and realistic factors to continuously improve the performance of the algorithm, and explore its application potential in a wider range of conditions, aiming to provide a more robust and efficient solution for UAV path planning.

Acknowledgements This work is supported by Beijing College Students Innovation and Entrepreneurship Training Project 2024.

References

1. Wang, R.B., Wang, W.F., Geng, F.D., et al.: UAV path planning in mountain areas based on a hybrid parallel compact arithmetic optimization algorithm. Neural Comput. Appl.Comput. Appl. (2023). https://doi.org/10.1007/s00521-023-08983-2
2. Shenoy, A.K., Pai S.N.: Search graph magnification in rapid mixing of markov chains associated with the local search-based metaheuristics. Mathematics **10**(1), 47–47 (2021)
3. Dhal, G.K., Ray, S., Das, A., et al.: A survey on nature-inspired optimization algorithms and their application in image enhancement domain. Arch. Comput. Methods Eng. **26**(5), 1607–1638 (2019)
4. Genetic Algorithms; Researchers from University of Sao Paulo Report new studies and findings in the area of genetic algorithms (Heuristic and Genetic Algorithm Approaches for UAV Path Planning under Critical Situation). J. Technol. Sci. 157 (2017)
5. Wang, R.B., Geng, F.D., Xu, L.: Parallel binary tumbleweed algorithm and its application for intrusion detection systems (2024)
6. Jian, L., Changyi, L., Weijian, Z., et al.: UAV path planning model based on R5DOS model improved A-star algorithm. Appl. Sci. **12**(22), 11338–11338 (2022)
7. Saremi, S., Mirjalili, S., Lewis, A.: Grasshopper optimisation algorithm: theory and application. Adv. Eng. Softw. 10530–10547 (2017)
8. Neve, G.A.: Application of grasshopper optimization algorithm for constrained and unconstrained test functions. Int. J. Swarm Intell. Evol. Comput. **6**(3), 1–7 (2017)
9. Thoma, A., Thomessen, K., Gardi, A., Fisher, A., Braun, C., et al.: Prioritising paths: an improved cost function for local path planning for UAV in medical applications. Aeronaut. J. **127**(1318), 2125–2142 (2023)
10. Juan, L., Qing, A., Hong, L., et al.: Survey of Lévy flight-based metaheuristics for optimization. Mathematics **10**(15), 2785–2785 (2022)

11. Qingyong, Y., Shu-Chuan, C., Jeng-Shyang, P., et al.: Dynamic multi-strategy integrated differential evolution algorithm based on reinforcement learning for optimization problems. Complex Intell. Syst. **10**(5), 1845–1877 (2023)
12. Shaik, A.L.H.P., Manoharan, M.K., Pani, A.K., Avala, R.R., Chen, C.-M.: Gaussian Mutation-Spider Monkey Optimization (GM-SMO) model for remote sensing scene classification. Remote Sens. **14**(24), 6279 (2022)

Chapter 11
Robotic Arm Trajectory Optimization Using Data-Driven Evolutionary Algorithm

Xiaopeng Wang, Aleš Vysocký, Tomáš Kot, Lingping Kong, Ponnuthurai Nagaratnam Suganthan, Zdenko Bobovský, and Václav Snášel

Abstract Increasing the energy efficiency of industrial robots in component manufacture can decrease running expenses and enhance the sustainability of robot operations. In this work, we use the boosting data-driven evolutionary algorithm with localized data generation (BDDEA-LDG) and a Bezier curve interpolator to optimize the trajectory of the robot arm manipulation. The experiment verified the UR3 robotic arm in 4 real-world cases. The test results indicated that the best result saved 7.63% of the energy, while the worst saved 2.07%, as compared to the movej command (system default). As a result, the proposed algorithm provides competitive results in optimizing the robot's energy consumption.

11.1 Introduction

Energy consumption and environmental deterioration are now pressing challenges due to the rapid growth of industrial automation. Fortunately, countries such as USA, China, India, and the European Union are aware of green manufacturing and have invested heavily in improving energy efficiency and reducing carbon emissions in industrial manufacturing [1, 2]. Currently, robotics has been applied to many industries in life, bringing unprecedented convenience to people's lives. For example, industrial robots are used in assembly line operations such as welding, painting, handling, and assembly. These robots can dramatically increase productivity and reduce human error. Surgical robots help surgeons perform precise and minimally invasive

X. Wang · L. Kong · V. Snášel (✉)
Faculty of Electrical Engineering and Computer Science, VŠB-Technical University of Ostrava, Ostrava, Czech Republic
e-mail: vaclav.snasel@vsb.cz

A. Vysocký · T. Kot · Z. Bobovský
Faculty of Mechanical Engineering, VŠB-Technical University of Ostrava, Ostrava, Czech Republic

P. N. Suganthan
KINDI Center for Computing Research, Qatar University, Doha, Qatar

S.-C. Chu et al. (eds.), *Advances in Intelligent Data Analysis and Applications*,
Smart Innovation, Systems and Technologies 445,
https://doi.org/10.1007/978-981-96-7277-6_11

surgeries in the medical field. Combined with big data and artificial intelligence technologies, assisted diagnostic robots can aid doctors in analyzing and diagnosing conditions. In agricultural applications, robots are used for sowing, fertilizing, weeding, and harvesting. Precision agriculture makes the use of resources more efficient and reduces reliance on manual labor. At disaster sites such as earthquakes and fires, rescue robots can enter hazardous areas for search and rescue, providing vital life support and supplies. There are over 2.7 million industrial robots in the globe. For such a large number, it is necessary to use robots rationally. Improving the energy efficiency of industrial robots is an essential task, which will not only lower manufacturing costs, but also reduce the carbon footprint [3, 4].

In robotic systems, task planning, task scheduling, and motion planning are three key components for achieving robot autonomy and intelligent behavior [5]. Task planning refers to the robot reasoning out a series of specific actions or tasks to be performed from a high-level goal. For example, handling an object is broken down into specific operations grasping the object, moving to a specified location, and dropping the object. Task scheduling is the process of allocating and arranging between multiple tasks and resources, with the aim of enabling tasks to be completed on time and in sequence, given time and resource constraints. For example, coordinating tasks among multiple robots ensures that each robot can perform the task at the appropriate time without conflict. Motion planning is the process by which a robot plans a path for physical movement in an environment that allows it to move safely and efficiently from a starting point to a goal location while avoiding obstacles and conflicts. For example, an industrial robot performs a handling task by precisely planning the trajectory of a robotic arm to avoid collisions and complete the task successfully.

Energy consumption is an important aspect in the design and operation of robotic systems. To improve the range and efficiency of a robot, optimization strategies can usually be divided into hardware, software, and mixed solutions [6, 7]. Energy optimization at the hardware level mainly involves improvements in the physical design of the robot itself, component selection, and energy storage and utilization. These solutions directly improve the physical structure and the efficiency of the energy conversion. Examples include the selection of more energy-efficient motors, the reduction of thermal losses and mechanical friction, and the use of energy-efficient servo motors or brushless motors. Reducing the weight of the robot by using lightweight but robust materials (e.g., carbon fiber, aluminum alloys) reduces the energy required for movement. When the robot performs certain tasks (e.g., decelerating or stopping), regenerative braking can be used to convert kinetic energy into electrical energy and store it back in the battery, extending battery life. Energy optimization at the software level relies on improvements to the robot's control algorithms, task scheduling, and path planning. Intelligent control strategies are used to optimize the energy efficiency of the robot when it is working. For example, reducing unnecessary idle time or inefficient paths while the robot is completing tasks, maximizing the use of each unit of energy. Mixed solutions combine hardware and software optimization strategies to achieve optimal energy consumption optimization. Improvements at the hardware level and intelligent control at the software level can complement each other to fully optimize the energy efficiency of the robot. Software approaches have been widely studied due to their low cost and versatility [8].

In recent years, metaheuristic algorithms have been applied to vehicle routing problems [9], remote sensing scene classification [10], intrusion detection systems [11], prediction of students' academic performance [12], parameter identification of photovoltaic models [13], feature selection [14, 15], and multiple knapsack problems [16]. Some scholars have also used metaheuristic algorithms to optimize the robot's operation trajectory, such as particle swarm optimization, artificial bee colony, teaching-learning-based optimization, and cuckoo search algorithm [17]. Different from the mentioned methods, this study uses the BDDEA-LDG [18] (offline data-driven evolutionary algorithm) combined with the Bezier interpolator to search the energy-saving robotic arm trajectory. The proposed algorithm was verified on the UR3 robotic arm. The major contributions are as follows:

- A new robotic arm trajectory optimization method is proposed, which is based on the BDDEA-LDG combined with the Bezier interpolation.
- In four cases of UR3 robotic arm, the proposed algorithm saves up to 7.63% compared with the system's movej.

This study's remaining sections are structured as follows: A review of the BDDEA-LDG algorithm is given in Sect. 11.2. Section 11.3 is about robotic arm trajectory optimization, including the problem description and experimental verification. Section 11.4 concludes this study and suggests further research directions.

11.2 BDDEA-LDG

In reality, many optimization problems are expensive and computationally intensive, such as blast furnace optimization, trauma system optimization, magnesium furnace optimization, airfoil shape optimization, and air intake ventilation system optimization [19]. These problems are intractable for traditional evolutionary algorithms. Therefore, data-driven algorithms have gradually become a popular and effective optimization strategy, using surrogate models to reduce fitness evaluations (FEs) to drive evolution. A DDEA algorithm's framework includes surrogate model management (SMM) and evolutionary optimization procedure (EOP). SMM manages surrogate models to approximate the fitness function better, while EOP uses the surrogate model to drive the evolutionary process to select the next population generation. DDEA can be classified into two versions: online-DDEA and offline-DDEA, depending on whether the EOP process can call the real fitness function evaluation. Online DDEA is suitable for situations where there is still a small amount of FEs available from expensive fitness evaluations during evolution. In contrast, offline DDEA cannot perform fitness evaluation since accessing or executing the real FEs is too costly. Offline DDEA can be regarded as a particular instance of online DDEA, and it can also be used for problems solved by online DDEA. Currently, researchers have proposed some excellent offline-DDEA algorithms, such as DDEA using selective ensemble (DDEA-SE) [20], DDEA based on model selection (MS-DDEO) [21], boosting DDEA with localized data generation (BDDEA-LDG) [18], DDEA with perturbation-based ensemble surrogates (DDEA-PES) [22], stochastic ranking DDEA with multiple kernels (SRK-DDEA) [23], and active learning with a reliability sampling-based DDEA (ALeRSa-DDEA) [24]. Next, we will provide a detailed introduction to BDDEA-LDG, considering its competitiveness.

The novelty of BDDEA-LDG is that it proposes the localized data generation (LDG) method to overcome data shortage by generating synthetic data to expand the dataset, while combining the boosting strategy (BS) strategy to iteratively integrate multiple surrogates to approximate the real FEs.

11.2.1 Localized Data Generation

LDG synthesizes data near the real evaluation data and expands it into the training dataset for building the surrogate model to improve the model prediction accuracy. The training data set $TD = \{(x_i, f(x_i)) \mid i = 1, 2, \ldots, N\}$ is the real evaluation data pair. x_i is the ith real evaluation data, $f(x_i)$ is its fitness function value, and N is the number of real evaluation data. The newly synthesized data are calculated based on the LDG method as follows:

$$\begin{cases} x_{new} = x_s + \Delta x, \quad |\Delta x| \le \sqrt{\sum_{j=1}^{dim} \frac{(u_j - l_j)^2}{dim}} \cdot 10^{-6} \\ f(x_{new}) = f(x_s) \end{cases} \tag{11.1}$$

where x_s is an element in the subset (S) of TD. S selects the top 50% of real evaluation data with large prediction errors. Δx is a random vector, and dim is the dimension of the problem. The jth dimension's upper and lower bounds are denoted by u_j and l_j, respectively. The fitness value of the new synthetic data is set to $f(x_s)$.

11.2.2 Boosting Strategy

The boosting strategy (BS) iteratively creates a new surrogate model and adds it to the surrogate model set (SMS) until a predefined number of T surrogate models are available. The BS strategy process is as follows: First, the surrogate model M_1 is built based on TD, which is also used by LDG to calculate the prediction error and select $0.5|TD|$ synthetic data. Next, M_2 is built based on TD and $0.5|TD|$ synthetic data. Moreover, M_1 and M_2 are used together by LDG to select new $0.5|TD|$ synthetic data. M_3 is generated using TD and $|TD|$ synthetic data. M_1, M_2, and M_3 are used again in LDG operation. The ith surrogate model M_i is constructed based on $(1 + 0.5(i - 1)) \cdot |TD|$ data. When SMS includes T surrogate models, the BS process is completed. Finally, the predictions of the T surrogate models are averaged to evaluate the fitness values of the solutions.

11.2.3 Framework of BDDEA-LDG

In Fig. 11.1, the BDDEA-LDG framework is displayed. The surrogate model management (SMM) and evolutionary optimization procedure (EOP) are the two components of the algorithm. Establishing the surrogate model is the main goal of SMM. The authors referred to the literature [20] to construct the surrogate models. Each base model is based on radial basis function neural networks (RBFNNs), the activation function is the Gaussian radial basis function, and the number of neurons of the hidden layer is set to the problem dimension. Offline (original) data refers to real evaluation data. After iterative processing of BS combined with LDG strategy, SMS will eventually obtain T surrogate models to predict the fitness values of solutions and then assist EOP in selecting the next generation population. The EOP part is the underlying optimization algorithm, including initialization, variation, evaluation, and selection operations. When the stopping condition is met, the optimal result is output. BDDEA-LDG uses a variant of genetic algorithm, including a simulated binary crossover (SBX), polynomial mutation, and tournament selection [25]. The parameter settings are: the crossover probability is 1, and the mutation probability is $1/dim$.

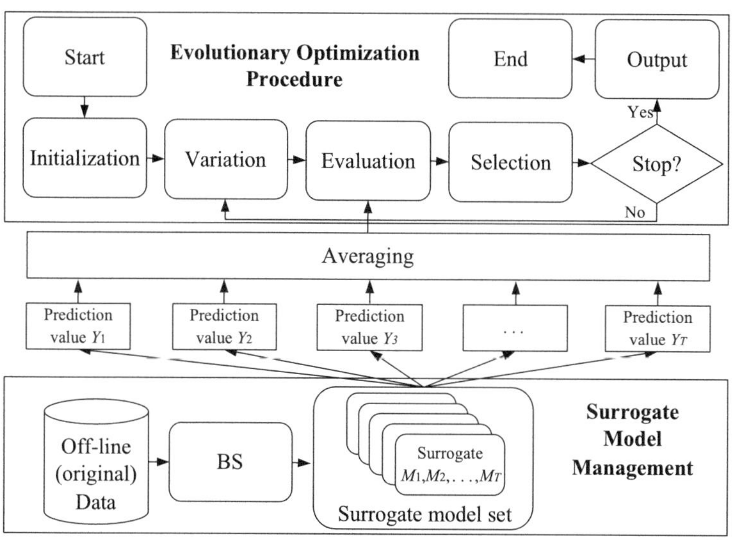

Fig. 11.1 The framework of BDDEA-LDG

11.3 Robotic Arm Trajectory Optimization

In this section, we will introduce the robotic arm trajectory problem and related experimental research.

11.3.1 Problem Description

The robot arm trajectory problem is to find an energy-efficient path from one point to another. In this work, we applied the unloaded UR3 robot arm. An intuitive path operated by movej (provided by the manufacturer) is shown in Fig. 11.2. The operation of UR3 is completed by the cooperation of 6 joints, each of which is controlled by a motor. Figure 11.3 presents a joint search space. The angle of the ith joint is fixed for a given start and end point. Therefore, the key to the problem is to properly control the changes in the joint during movement. In this study, the baseline in this study is movej (black dotted line), and the angle curve of the joint is represented by choosing 5 points between the start and end positions. The blue area is the search space for the ith joint angle. This joint search method can be extended to the robotic arm's 6 joints.

5 angle points can be generated for each joint movement using the optimization process, however the resulting angle trajectory is relatively curved, which frequently causes the robot arm to move abruptly and triggers the protective stop mechanism. Therefore, Bezier curves are introduced to smooth the joint trajectory. A Bezier curve is a smooth, continuous curve defined by a formula using a set of discrete control points. The Bezier curve for $n + 1$ control points P_0, P_1, ..., P_n can be expressed mathematically as [26]:

$$B(t) = \sum_{i=0}^{n} b_{i,n}(t) \cdot P_i \tag{11.2}$$

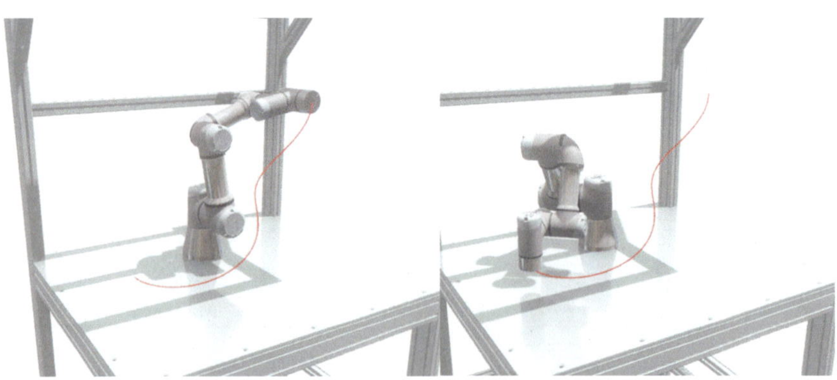

Fig. 11.2 UR3 operates from one point to another

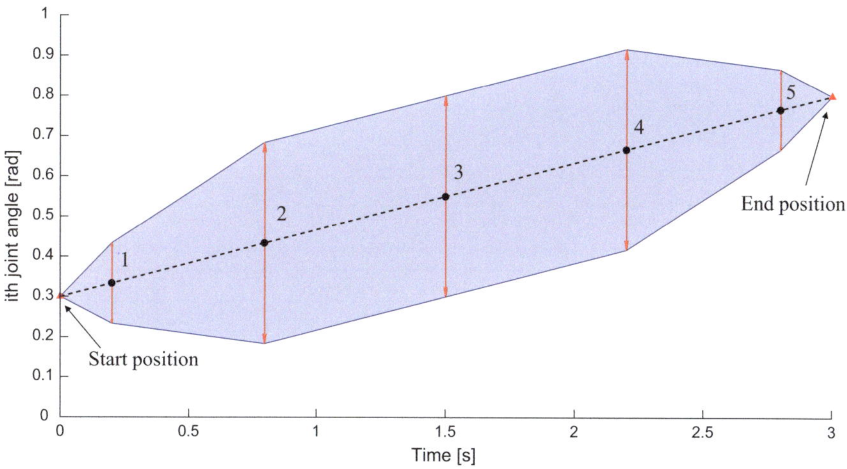

Fig. 11.3 The *i*th joint angle search space

$$b_{i,n}(t) = \frac{n!}{i!(n-i)!} \cdot t^i \cdot (1-t)^{n-i} \tag{11.3}$$

where $b_{i,n}(t)$ is the Bernstein polynomial, and t is a parameter that varies in the interval [0, 1].

Using the voltage and current status information of the robot arm, the energy consumption objective function is calculated as follows:

$$E = \sum_{joints} \int I(t)V(t)dt \tag{11.4}$$

where $I(t)$ and $V(t)$ are the current and the voltage as a function of time, respectively.

11.3.2 Experiment

In the experimental subsection, we used the BDDEA-LDG algorithm combined with the Bezier curve to solve the angle trajectory of the six joints. First, for the BDDEA-LDG algorithm, we randomly sampled 100 data as the offline data set to build surrogate models. Each joint uses movej as the baseline, and the upper bound of the fluctuation range is [0.1, 1, 1, 1, 0.1] and the lower bound is [−0.1, −1, −1, −1, −0.1]. A smaller fluctuation range is set for the first and fifth points of each joint because these two points are close to the start and end points, and a too large setting will trigger the protection mechanism. For the BDDEA-LDG parameter settings, the number of models T established was 50, the population size was 100, and the algorithm was run 31 times. For the UR3 robotic arm, the sampling frequency

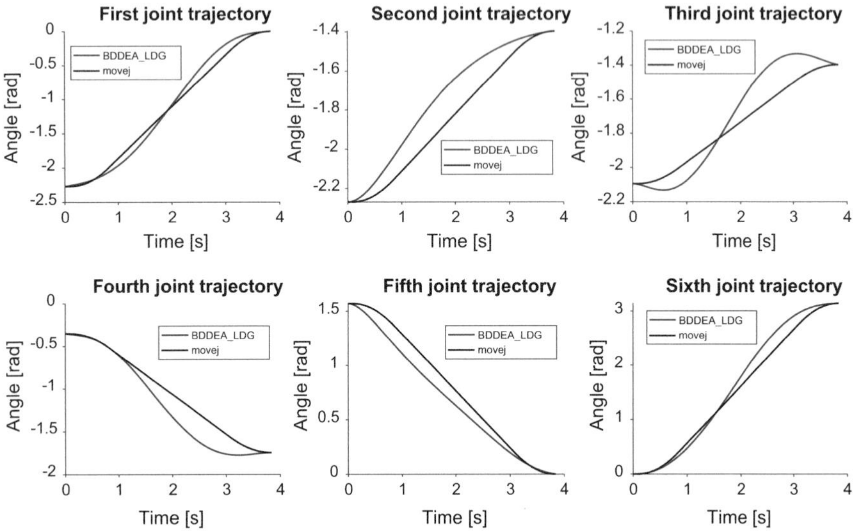

Fig. 11.4 The angle trajectories obtained by BDDEA-LDG and movej in path 1

Table 11.1 Energy consumption for 4 paths

Path	Movej [J]	BDDEA-LDG [J]	Savings (%)
1	247.26	228.40	7.63
2	192.62	186.48	3.19
3	223.05	218.44	2.07
4	235.41	227.37	3.42

was 250 Hz and the voltage was 48 V. We tested four paths. For path 1, the angle trajectories of the 6 joints obtained by BDDEA-LDG and movej are shown in Fig. 11.4. The best results of BDDEA-LDG and the comparison movej are recorded in Table 11.1.

The movej operation is the trajectory of the robot arm joints supplied by the manufacturer. BDDEA-LDG is the trajectory result obtained by the proposed algorithm. The data-driven algorithm can search for more effective joint control trajectories from the test results. Compared with movej in the 4 path tests, BDDEA-LDG saves up to 7.63% of energy and the lowest energy saving is 2.07%. It is proved that the proposed algorithm can solve the robot trajectory optimization problem.

11.4 Conclusion

We contribute a new strategy to optimize the trajectory of the robotic arm in this work. First, we use BDDEA-LDG to search the angles for the joints and then apply the Bezier curve to smooth the joint angle trajectories. The test results on the UR3 robotic arm confirmed that the proposed method is feasible and can save up to 7.63% energy compared with movej. However, the following points deserve further study: The proposed algorithm only compares movej, and the comparison with other offline DDEA algorithms is also worth studying. Furthermore, it can be extended to more robot trajectory optimization, such as testing different types of robots, carrying loads, and avoiding obstacles.

Acknowledgements The authors acknowledge financial support CZ.02.01.01/00/22_008/0004590 by the Czech Republic Ministry of Education, Youth, and Sports in the project "Robotics and Advanced Industrial Production" (ROBOPROX). The research was cofunded by the financial support of the European Union under the REFRESH—Research Excellence For Region Sustainability and High-tech Industries project number CZ.10.03.01/00/22_003/0000048 via the Operational Programme Just Transition.

References

1. Wang, Q., Jiang, X.t., Yang, X., Ge, S.: Comparative analysis of drivers of energy consumption in China, the USA and India–a perspective from stratified heterogeneity. Sci. Total Envir. **698**, 134117 (2020)
2. Thomas, S., Rosenow, J.: Drivers of increasing energy consumption in Europe and policy implications. Energy Policy **137**, 111108 (2020)
3. Vysockỳ, A., Papřok, R., Šafařík, J., Kot, T., Bobovskỳ, Z., Novák, P., Snášel, V.: Reduction in robotic arm energy consumption by particle swarm optimization. Appl. Sci. **10**(22), 8241 (2020)
4. Soori, M., Arezoo, B., Dastres, R.: Optimization of energy consumption in industrial robots, a review. Cognit. Robot. **3**, 142–157 (2023)
5. Szczepanski, R., Erwinski, K., Tejer, M., Bereit, A., Tarczewski, T.: Optimal scheduling for palletizing task using robotic arm and artificial bee colony algorithm. Eng. Appl. Artif. Intell. **113**, 104976 (2022)
6. Carabin, G., Wehrle, E., Vidoni, R.: A review on energy-saving optimization methods for robotic and automatic systems. Robotics **6**(4), 39 (2017)
7. Zhang, M., Yan, J.: A data-driven method for optimizing the energy consumption of industrial robots. J. Clean. Prod. **285**, 124862 (2021)
8. Singh, G., Banga, V.K.: Combinations of novel hybrid optimization algorithms-based trajectory planning analysis for an industrial robotic manipulators. J. Field Robot. **39**(5), 650–674 (2022)
9. Chen, C.M., Lv, S., Ning, J., Wu, J.M.T.: A genetic algorithm for the waitable time-varying multi-depot green vehicle routing problem. Symmetry **15**(1), 124 (2023)
10. Shaik, A.L.H.P., Manoharan, M.K., Pani, A.K., Avala, R.R., Chen, C.M.: Gaussian mutation-spider monkey optimization (GM-SMO) model for remote sensing scene classification. Remote Sensing **14**(24), 6279 (2022)
11. Wang, R.B., Geng, F.D., Xu, L.: Parallel binary tumbleweed algorithm and its application for intrusion detection systems. J. Netw. Intell. (2024)
12. Yue, L., Hu, P., Zhu, J.: Binary particle swarm optimization predicting students' academic performance in college English. J. Netw. Intell. (2024)
13. Wang, X., Chu, S.C., Snášel, V., Kong, L., Pan, J.S., Shehadeh, H.A.: A two-phase quasi-affine transformation evolution with feedback for parameter identification of photovoltaic models. Appl. Soft Comput. **113**, 107978 (2021)
14. Khodadadi, N., Khodadadi, E., Al-Tashi, Q., El-Kenawy, E.S.M., Abualigah, L., Abdulkadir, S.J., Alqushaibi, A., Mirjalili, S.: Baoa: binary arithmetic optimization algorithm with k-nearest neighbor classifier for feature selection. IEEE Access **11**, 94094–94115 (2023)

15. Pan, J.S., Liu, N., Chu, S.C.: A competitive mechanism based multi-objective differential evolution algorithm and its application in feature selection. Knowl.-Based Syst. **245**, 108582 (2022)
16. Wang, L., He, Y., Wang, X., Zhou, Z., Ouyang, H., Mirjalili, S.: A novel discrete differential evolution algorithm combining transfer function with modulo operation for solving the multiple knapsack problem. Inf. Sci. **680**, 121170 (2024)
17. Savsani, P., Jhala, R.L., Savsani, V.J.: Comparative study of different metaheuristics for the trajectory planning of a robotic arm. IEEE Syst. J. **10**(2), 697–708 (2014)
18. Li, J.Y., Zhan, Z.H., Wang, C., Jin, H., Zhang, J.: Boosting data-driven evolutionary algorithm with localized data generation. IEEE Trans. Evol. Comput. **24**(5), 923–937 (2020)
19. Jin, Y., Wang, H., Chugh, T., Guo, D., Miettinen, K.: Data-driven evolutionary optimization: an overview and case studies. IEEE Trans. Evol. Comput. **23**(3), 442–458 (2018)
20. Wang, H., Jin, Y., Sun, C., Doherty, J.: Offline data-driven evolutionary optimization using selective surrogate ensembles. IEEE Trans. Evol. Comput. **23**(2), 203–216 (2018)
21. Zhen, H., Gong, W., Wang, L.: Offline data-driven evolutionary optimization based on model selection. Swarm Evol. Comput. **71**, 101080 (2022)
22. Li, J.Y., Zhan, Z.H., Wang, H., Zhang, J.: Data-driven evolutionary algorithm with perturbation-based ensemble surrogates. IEEE Trans. Cybern. **51**(8), 3925–3937 (2020)
23. Huang, P., Wang, H., Ma, W.: Stochastic ranking for offline data-driven evolutionary optimization using radial basis function networks with multiple kernels. In: 2019 IEEE Symposium Series on Computational Intelligence (SSCI), pp. 2050–2057. IEEE (2019)
24. Pramanik, S., Goswami, A., Roy, A.: Alersa-ddea: active learning with reliability sampling-based evolutionary algorithm framework for solving offline data-driven expensive engineering optimization. Struct. Multidiscip. Optim. **65**(11), 320 (2022)
25. Deb, K., Beyer, H.G.: Self-adaptive genetic algorithms with simulated binary crossover. Evol. Comput. **9**(2), 197–221 (2001)
26. Wang, M., Luo, J., Walter, U.: Trajectory planning of free-floating space robot using particle swarm optimization (pso). Acta Astronaut. **112**, 77–88 (2015)

Chapter 12
Stock Price Forecasting Based on Gated Recurrent Unit and Nutcracker Optimization Algorithm with Non-linear Conversion Factor

Dong-Xing Jing, Xi-Xi Zhou, Shi-Jian Liu, and Zheng Zou

Abstract Accurate stock price prediction is important for investors to reduce investment risks and make informed decisions. Gated Recurrent Unit (GRU) is a deep learning model widely used in stock price prediction due to its superior performance. However, its hyperparameters are usually set based on experience, which is subjective and may lead to poor prediction accuracy and weak generalization ability. In this paper, we propose to optimize the important hyperparameters based on the Nutcracker Optimization Algorithm with Non-linear Conversion Factor (NOA-NCF), to reduce the influence of human factors and enhance prediction accuracy. To substantiate the efficacy of the proposed approach, it has been juxtaposed against other methodologies across six distinct stock index datasets. The experimental outcomes demonstrate that the proposed model achieves superior performance across all datasets, signifying its exceptional capabilities.

12.1 Introduction

The stock market serves as a venue where individuals and organizations can acquire shares of publicly traded companies. Investors and traders leverage the stock market to engage in the buying and selling of stocks, aiming to maximize their profits. Consequently, forecasting stock price movements is of paramount importance. Stock price data is marked by its nonlinearity, non-stationarity, and volatility, which makes

D.-X. Jing
School of Information and Intelligence, Xiangxi Vocational and Technical College for Nationalities, Xiangxi, China

X.-X. Zhou · S.-J. Liu (✉)
Fujian Provincial Key Laboratory of Big Data Mining and Applications, Fujian University of Technology, Fuzhou, China
e-mail: liusj2003@fjut.edu.cn

Z. Zou
College of Computer and Cyber Security, Fujian Normal University, Fuzhou, China

S.-C. Chu et al. (eds.), *Advances in Intelligent Data Analysis and Applications*,
Smart Innovation, Systems and Technologies 445,
https://doi.org/10.1007/978-981-96-7277-6_12

the process of stock price changes susceptible to multiple factors [1]. Research [2] has shown that complex models such as deep learning models can achieve higher accuracy for complex time series prediction problems.

Among various deep neural network models, Recurrent Neural Networks (RNNs) and Gated Recurrent Units (GRUs) are widely used in various time series analysis tasks [3]. The limitation of RNN is that it suffers from the problem of gradient vanishing or explosion. GRU network overcomes this drawback and can handle long-term dependent data effectively. Study [4] demonstrated that GRU can accurately predict stock returns and help investors to get high returns with lower risk, Moreover, in financial market forecasting, the GRU-based model is better and beats the other models by a significant margin [5].

While GRU networks exhibit significant potential for high performance, they are not without their limitations. First, neural network models, including GRU networks, lack interpretability [6]. Second, the values of the model hyperparameters often depend on experience. The issue of interpretability remains an unresolved challenge within the field, whereas this paper primarily addresses the second challenge.

To address the second problem mentioned above, researchers have been exploring ways to solve it. Liu et al. [7] employed grid search to fine-tune the hyperparameters of Long Short-Term Memory (LSTM) networks. The results showed that the prediction accuracy of the proposed model was significantly improved compared to the manually tuned LSTM. Wang et al. [8] proposed EMD-PSO-GRU for time series forecasting, which leverages Particle Swarm Optimization (PSO) to calibrate the GRU network's hyperparameters. The experimental results highlighted that the integrated PSO-GRU model surpasses the benchmark model in performance. Park et al. [9] used a Genetic Algorithm to optimize key parameters, including the number of hidden nodes and learning rate. However, despite these efforts, the optimization algorithms mentioned above are plagued by slow convergence rates and a propensity to converge to local optima, which still leaves the performance of neural networks less than ideal.

This paper proposes a model named NOA-NCF-GRU, where NOA and NCF stand for the Nutcracker Optimization Algorithm and Non-linear Conversion Factor respectively. The NOA-NCF algorithm is specifically engineered to identify the most optimal hyperparameters for the GRU network. Concurrently, an attention mechanism is incorporated to automate the allocation of weights to information features, thereby enhancing the model's predictive accuracy.

12.2 Related Work

12.2.1 GRU Network Based Stock Prediction

The GRU model, as introduced by Cho et al. [10], is an evolution of the LSTM architecture. The GRU streamlines the input and forget gates found in LSTM into a single update gate, which manages the flow of inputs and previous outputs to the subsequent unit. Moreover, it incorporates a reset gate that modulates the extent of historical data to be discarded. This design effectively mitigates the challenges associated with long-range dependencies and limited memory capacity that plague conventional recurrent neural networks. In comparison to LSTM, the GRU boasts a more concise architecture and expedites the training and assembly processes.

Over the recent years, the GRU has gained considerable traction in the realm of financial time series forecasting, delivering commendable accuracy in predictions. Winardi et al. [11] employed GRU networks to forecast fluctuations in nickel stock prices, pitting their model against an array of machine learning and deep learning models. The findings revealed that the GRU model not only achieved the highest prediction accuracy but also markedly surpassed the performance of its counterparts. In another study, Friday et al. [12] introduced the RNN-GRU model for the prediction of stock index prices, opting for a time window of length 10 and incorporating four independent features. Upon comparing their model with the CNN-LSTM model, the results indicated that the RNN-GRU model exhibited robust predictive capabilities.

12.2.2 Hyperparameter Optimization Based on Metaheuristic Algorithms

Metaheuristic algorithms optimize optimization problems by simulating natural phenomena, social behaviors, physical processes, etc., to build mathematical models. These algorithms are particularly useful for fine-tuning the hyperparameters of the GRU model. The primary advantage of metaheuristic algorithms is their ability to circumvent the pitfalls of traditional optimization methods, which often become trapped in local minima, thus failing to identify the global optimum. In contrast, metaheuristic algorithms are adept at escaping local minima, enabling them to scour the entire search space for the optimal solution. This approach allows for a more efficient search for the best hyperparameters, as highlighted in reference [13].

The Whale Optimization Algorithm (WOA), a prevalent metaheuristic approach introduced by Mirjalili et al. [14] in 2016, draws its inspiration from the collective behavior and feeding strategy of whales, and its main idea is to simulate the collaborative feeding behavior of whale groups and find the optimal solution by iteratively searching for solutions in space. The algorithm simulates the whale-specific search method and seining mechanism, which are primarily divided into three pivotal phases: (1) encircling prey, (2) bubble-net attacking, and (3) search for prey.

Although WOA is often applied to optimize the hyperparameters of neural networks, it also suffers from drawbacks such as slow convergence speed and low convergence accuracy [15].

12.2.3 Nutcracker Optimization Algorithm

The Nutcracker Optimization Algorithm (NOA), a metaheuristic methodology put forth by Abdel-Basset et al. [16] in 2023, which is inspired by Nutcrackers. The

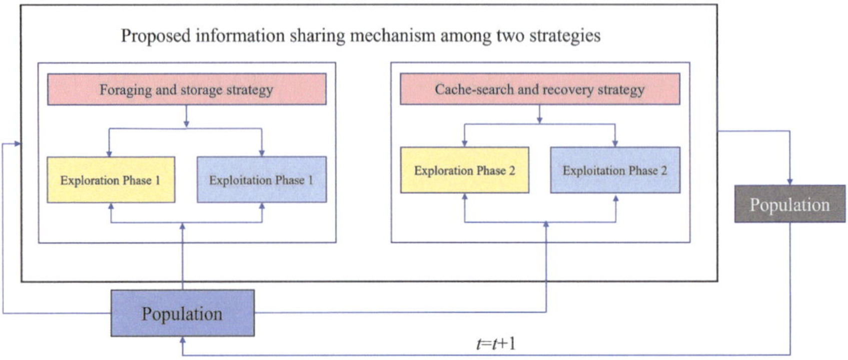

Fig. 12.1 The framework of NOA [16]

authors compared NOA with several highly cited algorithms, such as the Sparrow Optimization Algorithm and the Whale Optimization Algorithm, on multiple CEC datasets. The experimental outcomes revealed that NOA outperformed all other contenders, securing the top rank and demonstrating superior overall performance.

The comprehensive workflow of the NOA is depicted in Fig. 12.1. The algorithm is structured around two principal strategies: (1) the foraging and storage strategy, and (2) the cache-search and recovery strategy. These strategies represent two different behaviors of the star jay occurring at different times. Among them, each strategy contains two behaviors: exploration and exploitation.

12.3 The Proposed Method

12.3.1 Improved Nutcracker Optimization Algorithm

The two strategies of exploration and exploitation of NOA together determine the ability of the algorithm to obtain an optimal solution as well as the speed at which it obtains the optimal solution, and the better the balance between the two strategies, the better the algorithm's performance will be. However, the NOA's transition between these strategies is not optimally balanced, potentially resulting in diminished algorithmic performance. To address this, we enhance the algorithm's efficacy in this paper by integrating a non-linear conversion factor, as detailed in the subsequent steps:

1. Nutcrackers include two behavioral strategies in the foraging and storage phases: Exploration phase 1 and Exploitation phase 1. Meanwhile, a linearly decreasing conversion factor is used to control the conversion between the two behaviors.
2. Nutcrackers contain two behavioral strategies in the cache-search and recovery phase: Exploration phase 2 and Exploitation phase 2. Meanwhile, a constant value is used as a conversion factor to control the transition between the two behaviors.

In this paper, we propose to improve the conversion factors P_{a_1} and P_{a_2} for the exploration and exploitation strategies in the foraging-storage process and the cache search-recovery process. In the original paper, the value of P_{a_1} is taken to be linearly decreasing from 1, and the value of P_{a_2} is set to be a constant of 0.4 according to the authors' suggestion. The use of a nonlinear conversion factor to adjust the above-mentioned conversion factor not only can make it more closely match the behavior of nutcrackers in reality but also can enhance the local search capability of the algorithm.

$$P_{a_1} = P_{a_2} = \frac{1}{2} \cdot \left(1 + \cos\left(\pi \cdot \frac{t}{T_{\max}}\right)\right) \tag{12.1}$$

12.3.2 Stock Prediction

The core of constructing the NOA-NCF-GRU prediction model is the optimization work, and Fig. 12.2 shows the specific algorithmic flow of the prediction model. The optimal values of the critical parameters are determined through the training of the sample data. The specific steps are outlined as follows:

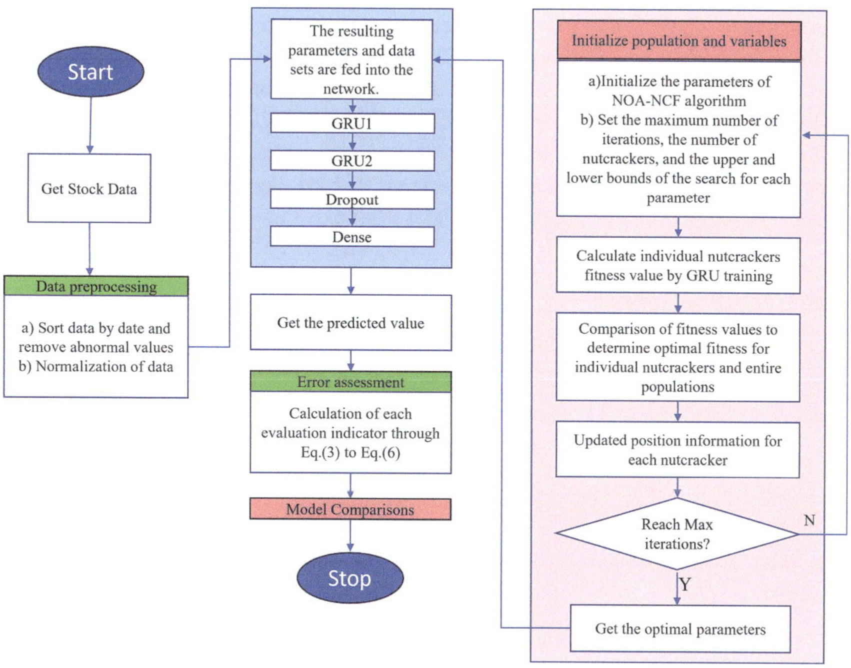

Fig. 12.2 NOA-NCF-GRU prediction model

1. Dataset preprocessing. The input stock data is first sorted by date and outliers are discarded; subsequently, the data is normalized.
2. This entails setting the maximum number of iterations the algorithm will undergo, establishing the quantity of nutcracker agents (n), and defining the upper and lower limits for the search space of each parameter.
3. Initialization of nutcrackers position. Randomly generate nutcrackers $X(\alpha, h_1, h_2, bs, e)$, where α denotes the learning rate, h_1 and h_2 denote the number of nodes in the first and second hidden layers, respectively, bs signifies the batch size, and e represents the number of epochs for training.
4. NOA-NCF continuously optimizes the network parameters through foraging and storage phases as well as cache-search and recovery phases, ceasing only after the maximum iteration count is reached. This process is aimed at identifying the most effective learning rate, the optimal number of neurons in the hidden layers, the ideal batch size, and the perfect number of training sessions.
5. The optimal GRU network parameters obtained from NOA-NCF are then fed into the model. The model is subsequently tested with a test dataset to generate predicted values. Furthermore, the performance of the model is rigorously assessed using a suite of evaluation metrics to gauge its effectiveness.
6. Compare the NOA-NCF-GRU model results with the GRU and NOA-NCF-GRU model results.

The parameters for this study are configured as detailed below: the structure of GRU is given in Table 12.1. The 1st and 2nd layers are GRU layers containing 100 neurons each. The 3rd layer is a Dropout layer with a dropout rate set at 0.35. The 4th layer is a dense layer comprising a single neuron. In the GRU model with the attention mechanism, the 4th layer is the attention layer, and the 5th layer is the fully connected layer. Table 12.2 gives the value ranges of the WOA as well as NOA-NCF optimization GRU hyperparameters as follows: learning rate size range [0.001, 0.01], batch size range [64, 128], Epoch size range [50, 250], and the neuron count in the 1st and 2nd hidden layers varies between [50, 300].

Table 12.1 The structure of GRU

Layer	Parameter
GRU	100
GRU	100
Dropout	0.35
Dense	1

Table 12.2 Hyperparameter alternatives

Hyperparameter	Alternatives
Learning rate	[0.001, 0.01]
Batch size	[64, 128]
Epochs	[50, 250]
Number of neurons	[50, 250]

12.4 Results and Discussion

12.4.1 Dataset and Experimental Environment

The experimental dataset encompasses a selection of prominent stock indices: the Shanghai Composite Index (SSE), Shenzhen Composite Index (SZSE), Hang Seng Index (HSI), Nikkei Index (N225), Bombay Stock Exchange Sensitivity Index (SENSEX), and Dow Jones Index (DJI). All of the above data are obtained through the Tushare repository. The dataset contains a total of 5109 days of stock data between 4/1/2010 and 12/31/2023.

To enhance the precision of our predictions, the acquired dataset undergoes a preprocessing phase. Initially, any rows with missing values are removed from the dataset. Subsequently, the data are normalized to standardize the scale, employing the following normalization technique as detailed in this paper:

$$X_i' = \frac{x_i - \bar{x}}{\max(x) - \min(x)} \tag{12.2}$$

In this equation, x_i denotes the ith variable, $\bar{x}$ is the mean of x_i, $\max(x)$ and $\min(x)$ correspond to the maximum and minimum values of x_i respectively. The dataset is partitioned into a training set and a test set in a 4:1 ratio.

12.4.2 Evaluation Criteria

The model's predictive efficacy in this study is quantitatively evaluated using four key metrics: Root Mean Square Error (RMSE), Mean Absolute Error (MAE), Mean Absolute Percentage Error (MAPE), and the Coefficient of Determination R^2. Let n represent the total number of samples in the test dataset, y_i and $\widehat{y_l}$ denote the true and predicted value of the ith sample point respectively. The calculations for RMSE, MAE, MAPE, and R^2 are calculated as follows.

$$RMSE = \sqrt{\frac{1}{n}\sum_{i=1}^{n}(\widehat{y_l} - y_i)^2} \tag{12.3}$$

$$MAE = \frac{1}{n}\sum_{i=1}^{n}(\widehat{y_l} - y_i) \tag{12.4}$$

$$MAPE = \frac{100\%}{n}\sum_{i=1}^{n}\left(\frac{\widehat{y_i} - y_i}{y_i}\right) \tag{12.5}$$

$$R^2 = 1 - \frac{\sum (y_i - \widehat{y_l})^2}{\sum (y_i - \hat{y})^2} \tag{12.6}$$

In these equations, RMSE, MAE, and MAPE values are inversely proportional to the predictive performance of the model, i.e., lower values represent better performance. The value of R^2 is directly proportional to the predictive performance of the model, i.e., higher values represent better performance.

12.4.3 Results

The results of the comparison experiments of the methods in this paper are shown in Table 12.3, with the best-performing values highlighted in bold. Upon examining the outcomes of the GRU and WOA-GRU, it is evident that the prediction accuracy of the GRU model has been enhanced in the majority of the datasets following the optimization by the WOA algorithm. This improvement demonstrates the optimization algorithm's capability to circumvent the instability in performance that can arise from manual tuning that is either inadequate or excessive. Moreover, it illustrates the algorithm's utility in identifying the optimal parameter combination, thereby enhancing the model's overall performance.

Table 12.3 Comparison of evaluation metrics across models

Stock	Model	RMSE	MAE	MAPE	R^2
SSE	GRU	30.6209	22.8680	0.0070	0.9765
	WOA-GRU	33.2007	25.4800	0.0077	0.9724
	Proposed model	30.3064	22.7861	0.0069	0.9770
SZSE	GRU	177.5774	134.7688	0.0109	0.9892
	WOA-GRU	160.5987	124.1274	0.0102	0.9911
	Proposed model	160.0809	120.5619	0.0098	0.9912
HSI	GRU	355.6344	279.8320	0.0134	0.9902
	WOA-GRU	336.8616	258.2384	0.0124	0.9912
	Proposed model	336.3009	256.8525	0.0123	0.9913
N225	GRU	362.4921	286.8149	0.0099	0.9730
	WOA-GRU	362.4836	285.0683	0.0098	0.9730
	Proposed model	346.7231	269.3286	0.0093	0.9753
SENSEX	GRU	1150.1849	1041.1235	0.0174	0.9430
	WOA-GRU	735.9097	609.1249	0.0103	0.9767
	Proposed model	685.8040	557.6019	0.0094	0.9797
DJI	GRU	520.5093	432.1685	0.0129	0.9005
	WOA-GRU	483.5304	394.9559	0.0118	0.9142
	Proposed model	356.1470	266.3013	0.0080	0.9534

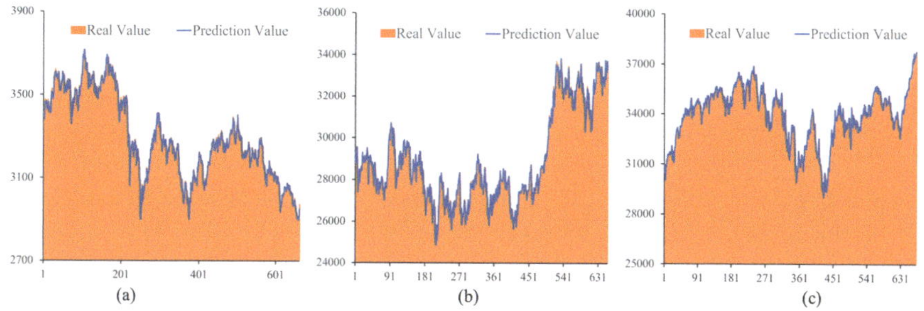

Fig. 12.3 The close price and prediction of SSE (**a**), N225 (**b**), and DJI (**c**) using the proposed algorithm

Figure 12.3 presents the comparative charts of the predictive outcomes for the SSE, HSI, and DJI using the method proposed in this paper. The horizontal axis in Fig. 12.3 represents time, while the vertical axis denotes stock price. It is observable from the figures that the discrepancy between the predicted values and the actual values for the NOA-NCF-GRU on each dataset is minimal. This finding suggests that our model is capable of accurately forecasting the target variables, with a high degree of concordance between the projected outcomes and the actual data.

Meanwhile, the dataset used in this paper covers the stock markets of mainland China, Hong Kong, Japan, India, and the United States, which further proves that the proposed model is universally applicable to predicting stock indices of various countries.

The overarching predictive results demonstrate that the NOA-NCF-GRU model introduced in this paper is particularly effective in handling time-series data. The model not only achieves high prediction accuracy but also exhibits a level of generalization that is commendable.

12.5 Conclusion

In this study, we have presented the NOA-NCF-GRU model, an innovative approach to stock price forecasting that leverages the synergistic power of the Gated Recurrent Unit (GRU) and the Nutcracker Optimization Algorithm with Non-linear Conversion Factor (NOA-NCF). The primary objective was to address the limitations associated with manual hyperparameter tuning in GRU networks, which often leads to suboptimal model performance. Our proposed model employs NOA-NCF to automate and optimize the hyperparameter selection process, thereby enhancing the accuracy and reliability of stock price predictions.

A key contribution of this research is the introduction of the non-linear conversion factor within the NOA algorithm, which has significantly improved the balance between exploration and exploitation, leading to faster convergence and more robust solutions. The attention mechanism integrated into our model further refines the predictions by dynamically assigning weights to relevant features, aligning with the underlying patterns in the data.

The comprehensive experimental evaluation, conducted across six diverse stock indices, has yielded promising results. The NOA-NCF-GRU model has consistently outperformed the traditional GRU and other metaheuristic-optimized models, such as the Whale Optimization Algorithm-GRU (WOA-GRU), across various metrics. These findings underscore the effectiveness of our model in capturing the complex dynamics of financial time series data and producing generalized predictions.

Looking ahead to future research endeavors, incorporating the classification outcomes of relevant online public opinion could serve to enrich the dataset, thereby potentially enhancing the model's predictive accuracy further.

Acknowledgements This work was supported in part by the Natural Science Foundation of Hunan Province (No. 2024JJ7549), the Scientific Research Project of Hunan Provincial Department of Education (No. 24C1236), the Scientific Research Team Project of XiangXi Vocational and Technical College for Nationalities (No. 2024KTZ201), and Fujian Provincial Natural Science Foundation of China (No. 2025J01670).

References

1. Chen, X., Hu, W., Xue, L.: Stock price prediction using candlestick patterns and sparrow search algorithm. Electronics **13**(4), 771 (2024)
2. Kolambe, M., Arora, S.: Forecasting the future: a comprehensive review of time series prediction techniques. J. Electr. Syst. **20**(2s), 575–586 (2024)
3. Kim, H.Y., Won, C.H.: Forecasting the volatility of stock price index: a hybrid model integrating LSTM with multiple GARCH-type models. Expert Syst. Appl. **103**, 25–37 (2018)
4. Fischer, T., Krauss, C.: Deep learning with long short-term memory networks for financial market predictions. Europ. J. Oper. Res. **270**(2), 654–669 (2018)
5. Bhandari, H.N., Rimal, B., Pokhrel, N.R., et al.: Predicting stock market index using LSTM. Mach. Learn. Appl. **9**, 100320 (2022)
6. Wang, L., Ye, Z.: An interpretable framework for stock trend forecasting. J. Phys.: Conf. Ser. **1634**(1), 012026 (2020). IOP Publishing
7. Liu, L., Liang, J., Ma, L., et al.: Gas pipeline flow prediction model based on LSTM with grid search parameter optimization. Processes **11**(1), 63 (2022)
8. Zhang, W., Wang, T.: Short-term power load forecasting model design based on EMD-PSO-GRU. Sci. Progr. **2022**(1), 4755519 (2022)
9. Noh, J., Park, H.J., Kim, J.S., et al.: Gated recurrent unit with genetic algorithm for product demand forecasting in supply chain management. Mathematics **8**(4), 565 (2020)
10. Chung, J., Gulcehre, C., Cho, K., et al.: Empirical evaluation of gated recurrent neural networks on sequence modeling (2014). arXiv:1412.3555
11. Hamzah, H., Winardi, S., Chrismawan, P.E.E., et al.: Robust stock price prediction using Gated Recurrent Unit (GRU). Int. J. Inf. Comput. **5**(1), 29–38 (2023)
12. Friday, I.K., Godslove, J.F., Nayak, D.S.K., et al.: IRGM: an integrated RNN-GRU model for stock market price prediction. In: 2022 International Conference on Machine Learning, Computer Systems and Security (MLCSS), pp. 129–132. IEEE (2022)
13. Gülmez, B.: Stock price prediction with optimized deep LSTM network with artificial rabbits optimization algorithm. Expert Syst. Appl. **227**, 120346 (2023)

14. Mirjalili, S., Lewis, A.: The whale optimization algorithm. Adv. Eng. Softw. **95**, 51–67 (2016)
15. Gharehchopogh, F.S., Gholizadeh, H.: A comprehensive survey: whale optimization algorithm and its applications. Swarm Evolut. Comput. **48**, 1–24 (2019)
16. Abdel-Basset, M., Mohamed, R., Jameel, M., et al.: Nutcracker optimizer: a novel nature-inspired metaheuristic algorithm for global optimization and engineering design problems. Knowl.-Based Syst. **262**, 110248 (2023)

Chapter 13
An Advanced Willow Catkin Optimization Algorithm for Application of Vehicle Routing Problem with Time Window

Jeng-Shyang Pan, Jiale Niu, Qingyong Yang, Xiao-Min Liu, and Shu-Chuan Chu

Abstract The Willow Catkin Optimization (WCO) algorithm combines the characteristics of the dispersal process of willow catkin seeds with the optimization process of the algorithm. Due to its unique optimization principles, it has been proved to outperform some traditional optimization algorithms. However, it still has drawbacks such as low solution accuracy and poor local search capability. To address these deficiencies of the WCO algorithm, this article proposes a Multigroup and Multistrategy Willow Catkin Optimization algorithm (MMWCO). The MMWCO algorithm integrates the Genetic Algorithm (GA) in parallel into the Willow Catkin Optimization (WCO) framework. The algorithm's optimization performance was tested using 30 different types of test functions from CEC2017. Numerical experiment results demonstrate that the MMWCO algorithm outperforms the original WCO algorithm and also shows advantages compared to other intelligent algorithms. Finally, this method was applied to the vehicle routing problem with time windows in transportation. Using Solomon's standard test data, the algorithm achieved favorable results, proving its practicality and feasibility.

J.-S. Pan · J. Niu
School of Computer Science, Nanjing University of Information Science and Technology, Nanjing, China

J.-S. Pan
Department of Information Management, Chaoyang University of Technology, Taichung, Taiwan

Q. Yang
College of Computer Science and Engineering, Shandong University of Science and Technology, Qingdao, China

X.-M. Liu
School of Information and Electronic Technology, Jiamusi University, Jiamusi, China

S.-C. Chu (✉)
School of Artificial Intelligence, Nanjing University of Information Science and Technology, Nanjing, China
e-mail: scchu0803@gmail.com

S.-C. Chu et al. (eds.), *Advances in Intelligent Data Analysis and Applications*, Smart Innovation, Systems and Technologies 445,
https://doi.org/10.1007/978-981-96-7277-6_13

13.1 Introduction

Optimization challenges are common in industrial applications and scientific research. An optimization problem involves finding an appropriate solution under limited conditions to maximize benefits. Currently, with technological advancements, these optimization problems have become increasingly complex. To balance solution quality and computation time, researchers have introduced more intelligent optimization algorithms, such as Particle Swarm Optimization (PSO) [1], Advanced Equilibrium Optimizer (AEO) [2], Particle Swarm Optimization-based Agent Negotiation Framework (PSOAN) [3], and Sine Cosine Algorithm (SCA) [4]. However, no single optimization algorithm can solve all optimization problems, which is the essence of the “No Free Lunch” (NFL) theorem [5].

Inspired by this, Chinese scholar Si-Qi Zhang proposed the Willow Catkin Optimization algorithm (WCO) in 2023 [6]. WCO is a more holistic alternative to conventional swarm intelligence algorithms. However, it converges more slowly, leading to strong global search capabilities but weaker local search abilities. Therefore, this paper introduces a parallel optimization mechanism to enhance the WCO algorithm, known as MMWCO. During problem-solving, multiple populations run in parallel, each using a unique update strategy. Through test function comparisons, MMWCO demonstrates superior solution accuracy and convergence capability compared to WCO, and shows certain advantages over other intelligent optimization algorithms.

This paper ultimately chooses the Vehicle Routing Problem with Time Windows (VRPTW) as the application focus for the algorithm. The MMWCO algorithm employs a real-coded method [7] to convert the discrete VRPTW into a continuous optimization problem. Additionally, the MMWCO algorithm incorporates a relocation operator [8] to enhance solution quality. The algorithm’s effectiveness in addressing VRPTW is validated using the Solomon benchmark dataset.

The remaining parts of this article are organized as follows: Sect. 13.2 reviews the original WCO algorithm and its related improvements. Section 13.3 details the design process of MMWCO, while Sect. 13.4 presents and analyzes the results of numerical experiments. Section 13.5 introduces how MMWCO is applied to VRPTW. Finally, Sect. 13.6 provides the conclusion of the article.

13.2 Related Work

13.2.1 Original Willow Catkin Optimization Algorithm

The iterative update process of individuals in the WCO algorithm [6] is as follows:

$$x_{i+1} = x_i + a \times (u + v) + (2 - a)(P_g - x_i) \tag{13.1}$$

Here, x_i represents the current position of the individual, a is a variable that controls the individual's transition from exploration to exploitation during the iteration and P_g denotes the global optimal solution in the current iteration. The variable a adjusts throughout the iteration process to balance exploration and exploitation:

$$a = c \times e^{-(\frac{t}{T})^2} \tag{13.2}$$

where T represents the maximum number of iterations, c is a constant with a value of 2, and t represents the current iteration count.

In the WCO algorithm, the decision to enter the exploration or exploitation phase is based on the relationship between the distance d_i between the current individual and the global optimal value, and the easy adhesion radius R. If $d_i > R$, the algorithm enters the exploration phase, and the wind component vectors u and v are obtained as follows:

$$\begin{aligned} v &= -ws \times \cos(wd) \\ u &= -ws \times \sin(wd) \end{aligned} \tag{13.3}$$

where ws and wd represent the wind speed and wind direction, respectively. The above equations are the decomposition of wind in a two-dimensional plane. The expressions for calculating ws and wd are as follows:

$$\begin{cases} ws = r \times R \\ wd = r \times 2\pi \end{cases} \tag{13.4}$$

where r is a random number between 0 and 1. If $d_i \leq R$, it indicates that the willow catkins are too close to each other, causing them to cluster together. The expressions for ws and wd in this phase are as follows:

$$DW = 1 - \frac{|p_g - x_i|}{\|x_i - p_g\|} \tag{13.5}$$

$$K = \frac{DW}{\sum_{i=1}^{D} DW_i} \tag{13.6}$$

$$\begin{cases} ws = \mu \times (\sum_{i=1}^{D} K_i |p_g - x_i|) + (1 - \mu) \times r_2 \times R \\ wd = \arccos(\frac{x_i \cdot p_g}{\|x_i\| \times \|p_g\|}) + r_3 \times \frac{\pi}{8} \end{cases} \tag{13.7}$$

where $\|\cdot\|$ represents the Euclidean distance between x_i and P_g, and DW denotes the weight of the total distance between P_g and x_i, μ is a random number ranging from 0.4 to 0.6, and r_2 and r_3 are random numbers in the interval [0, 1].

13.2.2 Genetic Algorithm

The Genetic Algorithm (GA) is a population-based stochastic search technique that mimics the principles of natural selection and genetics as proposed by Charles Darwin. The concept of GA was introduced by Holland [9], and it was first used by De-Jong [10] to solve optimization problems. Real-coded genetic algorithms (RCGA) use real numbers for encoding, which allows for faster convergence to the optimal value compared to binary and Gray-coded genetic algorithms [11]. The specific process of the real-coded genetic algorithm (GA) is as follows:

Two parent candidate solutions, and, are selected using the roulette wheel selection method. D represents the dimensionality of the individuals, and a random vector $\alpha = (a_1, a_1, \ldots, a_D)$ is generated, where each element a_i is a uniformly distributed random number in the interval $[-\gamma, 1 + \gamma]$ (in this article, γ is set to 0.9). Then, the uniform crossover operation of the genetic algorithm is used to generate two new individuals, Y_1 and Y_2.

$$Y_1 = \alpha X_1 + 1 - \alpha X_2 \tag{13.8}$$

$$Y_2 = \alpha X_2 + 1 - \alpha X_1 \tag{13.9}$$

Then, the mutation phase of the genetic algorithm is carried out. For the individual selected for mutation, a binary vector f is first generated to indicate which positions are subject to mutation. Here, r_i is a uniformly distributed random number in the interval [0, 1], and μ represents the mutation probability.

$$f_i = \begin{cases} 1\, r_i < \mu \\ 0\, r_i \geq \mu \end{cases} \tag{13.10}$$

Generate a standard normal distribution random vector $Z = (z_1, z_2, \ldots z_D),$, where each element z_i follows $N(0,1)$. The mutated individual is Y_k.

$$Y_k = X_k + \sigma f Z \tag{13.11}$$

13.3 The Advanced Willow Catkin Algorithm

This section provides a detailed description of how the multigroup and multistrategy mechanism is applied to improve the WCO algorithm.

13.3.1 Algorithm Idea

The original WCO algorithm has good global search capabilities and benefits from fewer parameters and lower complexity. However, it has a slower convergence speed, which results in strong global search abilities but weak local search capabilities. The development capability of the basic WCO algorithm needs further improvement. To address these shortcomings, the concept of multigroup and multistrategy is introduced, and real-coded GA is incorporated to enhance the local search ability of the WCO algorithm. In the Multigroup and Multistrategy WCO algorithm (MMWCO), each population has an equal number of individuals and employs distinct update strategies. Each population optimizes independently with no cross-population interference. On even-numbered iterations, populations communicate and perform individual migrations between them. The worst 20 individuals in each population are replaced by the best 20 individuals from other populations. This process repeats until the algorithm reaches the maximum iteration count or finds the optimal solution.

This article combines the WCO algorithm with the GA algorithm, integrating two population strategies into the MMWCO algorithm. By mixing these two algorithms, the solution diversity is effectively improved, increasing the likelihood of discovering potential solutions. Additionally, different algorithms excel at solving specific types of problems, and combining them allows for leveraging their complementary strengths to tackle more complex issues. The WCO algorithm has superior global search capabilities, while the GA algorithm focuses more on local exploitation. The genetic algorithm can guide the search toward more promising directions.

13.3.2 Algorithm Steps

Based on the above description, the pseudocode for of the MMWCO algorithm is as follows:

Algorithm 1 Pseudocode of MMWCO

```
Input: Population size N, Max iteration T, Fitness function, Dimension D, Upper and Lower bounds UB, LB
Output: Optimal solution individual position and fitness value
Initialize the Populations, Gbest and GlobalBestPos
while (t < MaxIteration or met the mimimum threshold value) do
    for i = 1 : N do
        if i is odd
            //WCO//
            if d_i > R then
                Generate ws and wd using Equation (4)
            else
                Generate ws and wd using Equations (5)–(7)
            End if
        Update Populations using Equations (1) and (2)
        else
            //GA//
            Update Populations using Equations (8) and (11)
        end if
        Get the new optimal individual position with the optimal fitness value
        if t is odd
            Replace the worst 20 individuals in each group by the best 20 individuals from
            another group
        end if
    Get the new optimal individual position with the optimal fitness value
    t = t + 1
end while
```

13.4 Experiments and Analysis

In this section, we used the CEC2017 benchmark to assess the performance of the MMWCO algorithm. For accuracy and fairness, each algorithm was executed 30 times, with the average result taken as the final outcome. All algorithms used a population size of 100 and 1000 iterations to maintain consistency. To better understand the MMWCO algorithm's effectiveness, we tested several algorithms on 50-dimensional cases. The final results were compared with the original WCO and other widely used swarm intelligence algorithms such as PSO [1], SCA [4], WOA [12], SSA [13], and DBO [14]. The mean best values (mean) were calculated with a search space range from -100 to 100. All simulations were performed on the same computer using MATLAB2019b.

The test results of each algorithm for each function were compared with those of the MMWCO algorithm. The "win/ = /lose" notation shows how many times MMWCO outperformed, tied with, or was outperformed by WCO. Bold values highlight the best performing results for each test function in the respective row. The overall comparison results for all benchmark functions are summarized in the table's last row, where bolded values indicate the best performance in each test case.

Table 13.1 presents the test results of the MMWCO algorithm specifically for 50 dimensions. It shows that for the 50-dimensional problems, MMWCO outperformed all other algorithms on 26 out of the 30 test functions. In the 50-dimensional case, compared to WCO, MMWCO was only slightly weaker than WCO on functions F23, F25, F28, and F36. Based on the comparison results from the previous sections, the improved MMWCO demonstrates faster convergence and enhanced exploitation capabilities compared to the original WCO algorithm. Furthermore, compared to other typical swarm intelligence optimization algorithms, MMWCO exhibits superior optimization performance, effectively balancing exploration and exploitation abilities (Fig. 13.1).

13.5 Applying MMSCA to Solve VRPTW

13.5.1 Description and Mathematical Model of VRPTW

Assuming a central warehouse has M ($v = 1, 2,\ldots, M$) vehicles, each having a maximum load capacity of Q, these vehicles provide distribution services to N customers, where $i = 0$ represents the central warehouse. After completing the distribution task, the vehicles must return to the central warehouse. The demand of customer i is q_i ($i = 1, 2,\ldots, N$), and the service time at delivery point i is s_i. The arrival time at customer i is T_i, and service time of customer i window is $[b_i,\ e_i]$. The transportation cost from customer i to j is c_{ij}, and the time taken for the transport is t_{ij} (where $i, j = 0, 1, 2, \ldots, N$). Two binary decision variables are considered, x_{ijv} and y_{iv}. If vehicle v moves from customer i to j, then $x_{ijv} = 1$; otherwise, $x_{ijv} = 0$. If vehicle v serves customer i, then $y_{iv} = 1$; otherwise, $y_{iv} = 0$. The objective function and constraints for the VRPTW problem are as follows:

$$\min f(x) = \sum_{v=1}^{M}\sum_{i=0}^{N}\sum_{j=0}^{N} c_{ij}x_{ijv} \tag{13.12}$$

$$s.t. \sum_{i=1}^{N} q_i y_{iv} \le Q,\ \forall v \tag{13.13}$$

$$\sum_{v=1}^{M} y_{iv} = 1,\ \forall i \tag{13.14}$$

Table 13.1 The 50-dimensional simulation results of CEC2017 benchmark function

Functions		MMWCO	WCO	SCA	PSO	WOA	DBO	SSA
F1	Mean	**6.43E+05**	1.33E+06	4.76E+10	3.4E+09	6.57E+08	1.03E+11	1.03E+11
F2	Mean	**3.43E+31**	8.19E+35	9.38E+67	4.14E+63	4.82E+69	4.57E+78	1.27E+82
F3	Mean	**1.89E+04**	4.75E+04	1.45E+05	4.81E+04	1.78E+05	2.54E+05	1.04E+10
F4	Mean	**612.73**	646.46	9144.60	1157.11	1000.76	2.73E+04	2.20E+04
F5	Mean	**689.78**	747.68	1090.95	879.66	958.94	1139.81	1293.60
F6	Mean	**633.47**	646.83	673.79	660.02	686.00	698.77	713.44
F7	Mean	**1077.72**	1107.49	1675.87	1541.73	1720.39	1980.45	2094.00
F8	Mean	**983.80**	1079.81	1381.31	1187.85	1232.49	1457.65	1625.61
F9	Mean	**4358.30**	1.12E+04	2.69E+04	1.26E+04	2.86E+04	3.36E+04	5.29E+04
F10	Mean	**7953.45**	8067.14	1.48E+04	9580.10	1.14E+04	1.37E+04	1.72E+04
F11	Mean	**1423.35**	1484.38	7851.96	1638.21	2224.56	1.86E+04	2.93E+04
F12	Mean	**3.38E+07**	8.22E+07	1.35E+10	3.23E+09	6.17E+08	5.32E+10	4.42E+10
F13	Mean	8.61E+04	**7.28E+04**	4.17E+09	1.7E+09	5.59E+06	2.48E+10	1.91E+10
F14	Mean	**1.33E+05**	1.36E+05	4.22E+06	9.77E+05	2.60E+06	2.76E+07	4.60E+07
F15	Mean	4.28E+04	**3.78E+04**	5.59E+08	1.32E+08	5.46E+05	3.9E+09	4.15E+09
F16	Mean	**3524.81**	3686.28	5761.03	4056.46	5457.97	7305.48	7809.18
F17	Mean	**3186.55**	3237.35	4581.82	3707.58	4150.76	8589.31	1.52E+04
F18	Mean	1.54E+06	**1.37E+06**	2.26E+07	6.46E+06	1.56E+07	5.42E+07	3.21E+08
F19	Mean	**2.93E+05**	9.31E+05	3.77E+08	9.46E+06	9.29E+06	3.03E+09	3.69E+09
F20	Mean	**3082.50**	3127.32	4033.07	3432.12	3858.31	3876.47	4986.77
F21	Mean	**2470.47**	2535.80	2898.99	2697.96	2921.65	3072.82	3169.90
F22	Mean	**9649.28**	1.00E+04	1.65E+04	1.10E+04	1.26E+04	1.61E+04	1.82E+04
F23	Mean	**2984.94**	3085.63	3581.72	3481.08	3716.64	4044.81	4501.68
F24	Mean	**3118.62**	3213.59	3740.72	3654.45	3754.84	4080.34	4835.40
F25	Mean	**3079.52**	3123.63	7144.95	3261.83	3407.66	1.32E+04	1.34E+04
F26	Mean	6008.24	**5576.41**	1.24E+04	1.09E+04	1.31E+04	1.62E+04	1.53E+04
F27	Mean	**3675.72**	3987.74	4664.66	4117.84	4336.90	5188.60	7183.61
F28	Mean	**3352.78**	3409.11	7338.32	4080.51	4182.53	1.14E+04	1.07E+04
F29	Mean	**5022.41**	5181.27	8013.08	6381.30	8103.73	2.24E+04	2.15E+04
F30	Mean	**1.13E+07**	8.63E+07	9.06E+08	6.65E+07	1.66E+08	4.57E+09	4.72E+09
win/ = /los			26/0/4	30/0/0	30/0/0	30/0/0	30/0/0	30/0/0

$$\sum_{j=1}^{N} x_{ijv} = y_{iv}, \ \forall i, v \tag{13.15}$$

$$\sum_{i=1}^{N} x_{ijv} = y_{jv}, \ \forall j, v \tag{13.16}$$

$$b_i \leq T_i \leq e_i,\ i = 1, 2, \ldots, N \tag{13.17}$$

Equation 13.12 represents the objective function, aiming to minimize the total distance traveled by the delivery vehicles. Constraint Equation 13.13 ensures that each vehicle's capacity is not exceeded. Constraint Equation 13.14 guarantees that every customer is served by a delivery vehicle. Equations 13.15 and 13.16 ensure that each customer is served by only one delivery vehicle. Constraint Equation 13.17 imposes the time window constraint for the service time.

13.5.2 MMSCA Algorithm for VRPTW

VRPTW is a discrete problem; therefore, for the MMWCO algorithm and the original WCO algorithm, real-coded encoding is more appropriate. Each individual generated by MMWCO is an N-dimensional vector, where each value in the vector is a real number. Each real number is constrained within the range $(1, m + 1)$. The integer part of the real number represents the service vehicle, while the decimal part indicates the service sequence of the vehicle on the path.

13.5.3 Objective Function

The algorithm's objective function cannot directly serve as the model's objective function. To address constraint violations simply, a new penalty function method is as follows:

$$\min f(x) = \sum_{v=1}^{M}\sum_{i=0}^{N}\sum_{j=0}^{N} c_{ij}x_{ijm} + \alpha \sum_{v=1}^{M} \max(\sum_{i=1}^{n} q_i y_{vi} - Q, 0) + \beta \sum_{j=1}^{N} \max(T_j - e_j, 0) \tag{13.18}$$

where α and β represent penalty coefficients. In this study, α is set to 10 and β is set to 100.

To further standardize the solution results and reduce penalty costs, a relocation operator [8] was introduced during the algorithm implementation.

13.5.4 Experimental Results and Analysis

In this paper, nine cases were selected from the Solomon standard test database. Each case has 100 customers and 25 available vehicles. Each algorithm's iteration particle count was set to 100, and population size was 100. The continuous test count was set to 30.

Figure 13.2 presents the convergence curve and delivery route map for the MMWCO algorithm solving sample RC103. Table 13.2 contains the results for the MMWCO algorithm, the original WCO algorithm, and the PSO algorithm. The *best** column shows the best known values, while the *GAP* column indicates the deviation between the optimal results obtained by the tested algorithms (represented by *best_t*) and the known optimal results. The calculation formula is provided in Eq. 13.19.

$$GAP = \frac{best\_t - best^*}{best^*} \times 100\% \tag{13.19}$$

According to the results in Table 13.2, the MMWCO algorithm is only outperformed by the PSO algorithm in three instances (R102, R103, and RC102) regarding

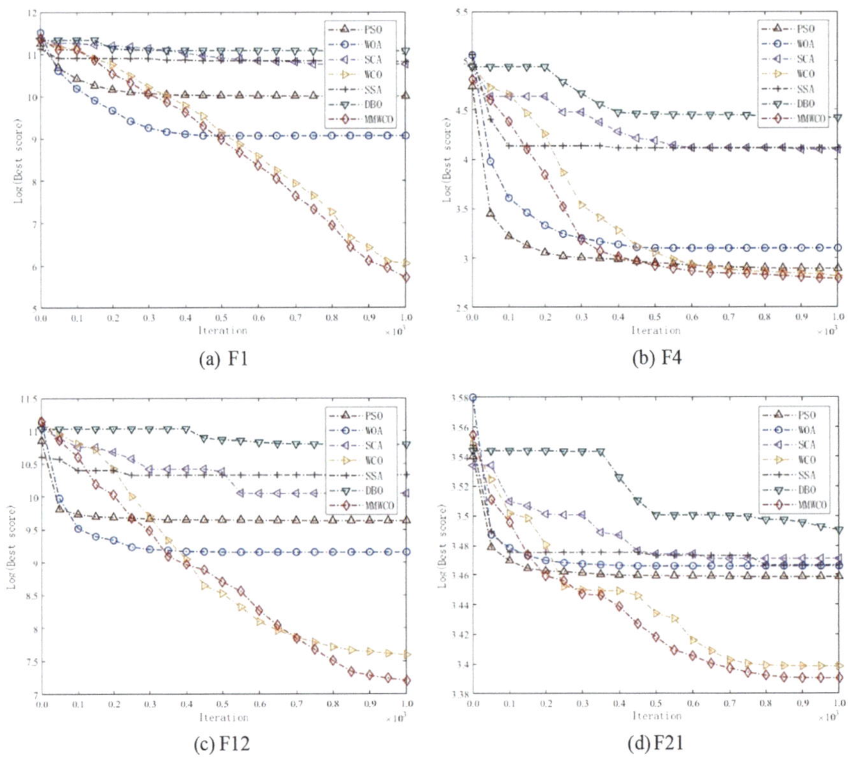

Fig. 13.1 Convergence curves of 50D benchmark functions: F1, F4, F12, F21

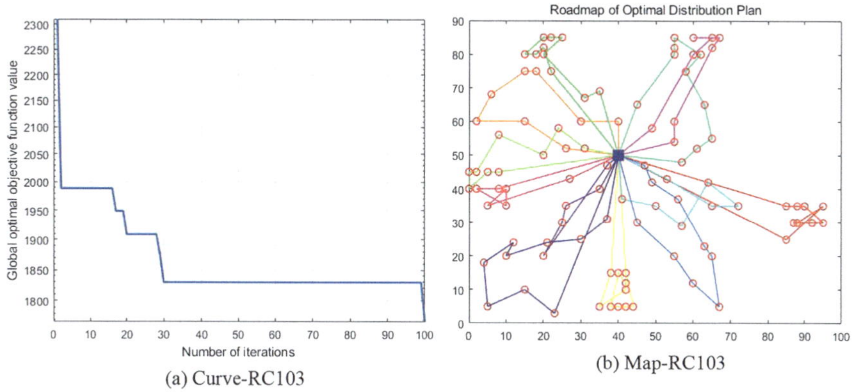

Fig. 13.2 Case RC103 convergence curve and optimal distribution map

the optimal values obtained. For average values, the MMWCO algorithm is only inferior to the other two algorithms in one case (R103). In the remaining cases, the MMWCO algorithm achieves comparable or even better results in terms of both the average solution and the optimal solution compared to the other two population-based algorithms. This demonstrates that the MMWCO algorithm performs better overall than the original WCO and PSO algorithms, offering certain advantages. However, there is still some gap compared to the current best known solutions for R101 and RC103, suggesting room for further improvement in the algorithm. In the future, by incorporating different optimization strategies, the MMWCO algorithm's performance on VRPTW could be further improved.

Table 13.2 Comparison of the test results of MMWCO, WCO, and PSO on the Solomon cases

Tests	Best*	MMWCO			WCO			PSO		
		Mean	Best	GAP(%)	Mean	Best	GAP(%)	Mean	Best	GAP(%)
C101	828.94 [15]	832.143	828.94	0	838.534	828.94	0	834.172	828.94	0
C102	828.94 [15]	842.694	828.94	0	847.341	828.94	0	846.976	828.94	0
C103	828.06 [15]	849.290	828.06	0	883.41	850.626	2.72	903.731	828.06	0
R101	1645.79 [16]	1661.200	1653.176	0.45	1666.645	1653.199	0.45	1670.686	1653.883	0.49
R102	1486.12 [15]	1481.645	1476.536	−0.64	1487.905	1477.625	−0.57	1485.909	1474.684	−0.76
R103	1292.68 [17]	1252.544	1230.217	−4.83	1249.372	1233.554	−4.57	1244.878	1225.344	−5.21
RC101	1696.94 [18]	1706.469	1653.853	−2.54	1709.091	1677.668	−1.14	1715.679	1672.028	−1.47
RC102	1554.75 [18]	1533.784	1504.942	−3.20	1552.088	1511.815	−2.76	1540.015	1492.732	−3.99
RC103	1261.67 [19]	1355.699	1306.952	3.59	1362.068	1312.787	4.05	1367.366	1323.515	4.90

13.6 Conclusion

First, to address the issues in the update process of the WCO algorithm, this paper proposes a MMWCO algorithm. The algorithm consists of multiple populations with the same number of individuals, where each population executes a distinct update strategy. The various strategies include the original WCO strategy and the GA strategy. The WCO strategy enhances the algorithm's global search capability, while still allowing interference in later stages to prevent stagnation. The GA strategy strengthens local optimization and accelerates the algorithm's convergence. Through numerical experiments on test functions, the MMWCO algorithm shows improvements in both accuracy and convergence speed compared to the original WCO algorithm. Finally, the MMWCO algorithm is successfully applied to the VRPTW problem, achieving good results and further validating the algorithm's practicality and effectiveness. Future research will focus on further improving the algorithm and applying it to more practical engineering problems.

Acknowledgments This study was funded by Space-land Collaborative Smart Agriculture Innovation Team Project (grant number 2023-KYYWF-0638).

Disclosure of Interests The authors have no competing interests to declare that are relevant to the content of this article.

References

1. Kennedy, J., Eberhart, R.: Particle swarm optimization. In: Proceedings of the ICNN'95-International Conference on Neural Networks, vol. 4, pp. 1942–1948. Perth, Australia (1995)
2. Pan, J.-S., Zhuang, J., Liao, L., Chu, S.C.: Advanced equilibrium optimizer for electric vehicle routing problem with time windows. J. Netw. Intell. **6**(2), 216–237 (2021)
3. Chen, X., Lu, P., Lin, K.-B., Zeng, J.-Y.: Particle swarm optimization-based agent negotiation framework to support shared decision-making. J. Netw. Intell. **8**(3), 758–775
4. Mirjalili, S.: SCA: a sine cosine algorithm for solving optimization problems. Knowl.-Based Syst. **96**, 120–133 (2016)
5. Wolpert, D.H., Macready, W.G.: No free lunch theorems for optimization. IEEE Trans. Evol. Comput.Evol. Comput. **1**(1), 67–82 (1997)
6. Pan, J.-S., Zhang, S.-Q., Chu, S.-C., Yang, H.-M., Yan, B.: Willow catkin optimization algorithm applied in the TDOA/FDOA joint location problem. Entropy **25**(1), 171 (2023)
7. Wang, W., Wu, B., Zhao, Y., Feng, D.: Particle swarm optimization for open vehicle routing problem. In: Huang, D.S., Li, K., Irwin, G.W. (eds.) Computational Intelligence. ICIC 2006. Lecture Notes in Computer Science, vol. 4114, pp. 1–10. Springer, Berlin, Heidelberg (2006)
8. Prosser, P., Shaw, P.: Study of greedy search with multiple improvement heuristics for vehicle routing problems. Technical report RR96/201. Department of Computer Science, University of Strathclyde, Glasgow, Scotland (1996)
9. Holland, J.H.: Adaptation in Natural and Artificial Systems. University of Michigan Press, Ann Arbor (1975)
10. De-Jong, K.A.: An Analysis of the Behavior of a Class of Genetic Adaptive Systems. Ph.D. thesis, University of Michigan (1975)

11. Deb, K.: Multi-Objective Optimization using Evolutionary Algorithms. John Wiley & Sons, Chichester (2001)
12. Mirjalili, S., Lewis, A.: The whale optimization algorithm. Adv. Eng. Softw.Softw. **95**, 51–67 (2016)
13. Mirjalili, S., Gandomi, A.H., Mirjalili, S.Z., Saremi, S., Faris, H., Mirjalili, S.M.: Salp Swarm Algorithm: a bio-inspired optimizer for engineering design problems. Adv. Eng. Softw.Softw. **114**, 163–191 (2017)
14. Xue, J., Shen, B.: Dung beetle optimizer: a new meta-heuristic algorithm for global optimization. J. Supercomput.Supercomput. **79**, 7305–7336 (2023)
15. Rochat, Y., Taillard, É.D.: Probabilistic diversification and intensification in local search for vehicle routing. J. Heuristics **1**(1), 147–167 (1995)
16. Homberger, J.: Verteilt-parallele Metaheuristiken zur Tourenplanung. Deutscher Universitätsverlag, Wiesbaden (2000)
17. Li, H., Lim, A.: Local search with annealing-like restarts to solve the VRPTW. Eur. J. Oper. Res.Oper. Res. **150**(1), 115–127 (2003)
18. Taillard, É., Badeau, P., Gendreau, M., Guertin, F., Potvin, J.Y.: A Tabu search heuristic for the vehicle routing problem with soft time windows. Transp. Sci. **31**(2), 170–186 (1997)
19. Shaw, P.: Using constraint programming and local search methods to solve vehicle routing problems. In: International Conference on Principles and Practice of Constraint Programming, pp. 417–431. Springer, Berlin, Heidelberg (1998)

Chapter 14
Optimizing Ontology Alignment Through Genetic Programmings

Chengsheng Chi, Xingsi Xue, Huaiyuan Zheng, and Guojun Mao

Abstract Ontology serves as a fundamental technique within the Semantic Web by offering a structured framework for knowledge representation across different fields. However, its practical use is hindered by the challenge of ontology heterogeneity, which results from diverse representations of entities. Ontology Matching (OM) addresses this challenge by detecting alignments between entities with identical meanings in various ontologies, using Similarity Features (SFs). However, the complex dependencies and interactions among SFs make manual construction a challenging combinatorial optimization problem, often resulting in premature convergence and reduced alignment accuracy. In response to this issue, this paper proposes an innovative framework designed to autonomously build and adjust advanced SFs to improve the outcomes of matching. In particular, Multi-Objective Genetic Programming (MOGP) is first used to build diverse high-level SFs, while Single-Objective Genetic Programming (SOGP) is then utilized to refine the accuracy and confidence of the final SF. The experimental findings reveal that our approach reliably yields high-quality alignments, offering notable enhancements in both accuracy and efficiency compared to conventional GP-based OM methods.

C. Chi · G. Mao
School of Computer Science and Mathematics, Fujian University of Technology, Fuzhou, Fujian, China
e-mail: ChengshengChi@163.com

X. Xue (✉)
Fujian Provincial Key Laboratory of Big Data Mining and Applications, Fujian University of Technology, Fuzhou, Fujian, China
e-mail: jack8375@gmail.com

H. Zheng
Library, Fujian University of Technology, Fuzhou, Fujian, China
e-mail: HuaiyuanZheng@fjut.edu.cn

S.-C. Chu et al. (eds.), *Advances in Intelligent Data Analysis and Applications*,
Smart Innovation, Systems and Technologies 445,
https://doi.org/10.1007/978-981-96-7277-6_14

14.1 Introduction

Ontology serves as a pivotal method for the Semantic Web [1], offering a formal knowledge representation for a given domain by outlining specific concepts and their connections. In application scenarios of the latest communication technologies such as cloud computing [2] and inquiry learning [3], ontologies can represent the most appropriate technology for supporting knowledge integration and exchange as they can formally represent information and provide a common meaning for shared resources. Different ontologies frequently characterize entities within the same domain differently, resulting in ontology heterogeneity [4]. Ontology Matching (OM) addresses this challenge by establishing correspondences between different ontologies [5], facilitating knowledge sharing, integration, and interoperability. By resolving differences between ontologies, OM helps mitigate the issue of ontology heterogeneity. Nevertheless, the complex semantic relationships and large-scale entity sets within ontologies [6] present significant challenges to ensuring the efficiency of the OM process. Capturing complex semantic connections requires considering Similarity Features (SFs) across multiple aspects, including structure, semantics, and lexicon [7], which helps improve the accuracy of matching between entities. However, the large number of candidate SFs, coupled with their complex dependencies and interactions [8], makes manual selection and combination of SFs for OM a highly complex combinatorial optimization problem. This leads to a higher likelihood of premature convergence to local optima, posing greater challenges compared to other fields. To tackle these issues, it is imperative to develop flexible, automated techniques to identify the best combination of SFs, subsequently enhancing the efficiency of the matching process.

Genetic Programming (GP) is a type of evolutionary computation method influenced by natural selection [9], making it ideal for automating the SF construction in OM. Its adaptability and robustness enable it to navigate the complex combinatorial space of SF selection, which involves intricate dependencies between features. As a result, GP-based approaches have gained traction in OM by reducing manual effort and improving matching quality. The use of GP for optimization purposes has seen a substantial increase in popularity, as noted by Niazkar et al. [10], who recognize it as a leading algorithm in addressing intricate issues. In OM, GP techniques automatically construct and evolve SFs, capturing structural, semantic, and lexical similarities, which are essential for addressing ontology heterogeneity. However, traditional GP methods often suffer from premature convergence, limiting the diversity of solutions and, thus, the overall quality of ontology alignment [11]. This lack of diversity hampers the discovery of optimal SF combinations, reducing the reliability of the alignment process. Additionally, ensuring the confidence and correctness of the generated SFs remains a challenge, particularly in large-scale OM tasks. Given these challenges, this paper addresses two fundamental questions to improve the quality of ontology alignment:

- How can we increase the diversity of candidate SFs? Achieving greater diversity in SFs is essential to more comprehensively investigate the solution space and bypass local optima.
- How can we ensure the confidence of the constructed SFs? Confidence in the selected and combined SFs is crucial for improving the robustness and correctness of the OM results, especially in highly complex ontologies.

The structure of the paper is as follows: Section 14.2 covers the background information on OM. The designed framework for automatic SF construction is detailed in Sect. 14.3. In Sect. 14.4, the methodology is outlined, while Sect. 14.5 discusses the experimental results along with additional analysis. Lastly, Sect. 14.6 wraps up the study.

14.2 Background

14.2.1 Ontology Matching Problem

Ontology provides a structured framework to depict concepts within a specific field and their interconnections [12], commonly structured as a 3-tuple $(\mathcal{C}, \mathcal{P}, \mathcal{R})$. Here, $\mathcal{C}$ stands for classes, $\mathcal{P}$ for properties, and $\mathcal{R}$ for semantic relationships [13]. The entity set $\mathcal{E} = \mathcal{C} \cup \mathcal{P} \cup \mathcal{I}$ underscores its significance in data integration. Nevertheless, differences in ontologies' detail levels, known as ontology heterogeneity, present obstacles to knowledge exchange [14]. OM helps to address this by aligning semantically similar entities, expressed as a 4-tuple $(e_1, e_2, sim, \equiv)$, with sim indicating the similarity score and $\equiv$ representing equivalence [15]. The quality of matching results is often measured using metrics like recall, precision, and f-measure [16].

The SF construction problem in OM involves generating and selecting features that effectively capture the similarity between entities from different ontologies. These features may reflect various aspects such as structural, semantic, and lexical properties of the entities. Formally, given two ontologies $\mathcal{O}_1$ and $\mathcal{O}_2$, with entity sets $\mathcal{E}_1$ and $\mathcal{E}_2$, respectively, the goal is to construct a set of SFs $\mathcal{SF} = \{sf_1, sf_2, \cdots, sf_n\}$, where each sf_i measures the similarity between entities $e_1 \in \mathcal{E}_1$ and $e_2 \in \mathcal{E}_2$. The objective is to find an optimal combination of these features that maximizes the accuracy of the alignment. This problem is a combinatorial optimization challenge due to the large number of possible SFs and the complex dependencies among them, which complicates the task of selecting and combining features for effective ontology alignment.

14.3 The Framework of Similarity Feature Construction

Figure 14.1 presents the designed two-stage SF construction framework, which combines Multi-Objective GP (MOGP) and Single-Objective GP (SOGP) to enhance both the diversity and confidence of SFs for OM. In the first stage, MOGP evolves the

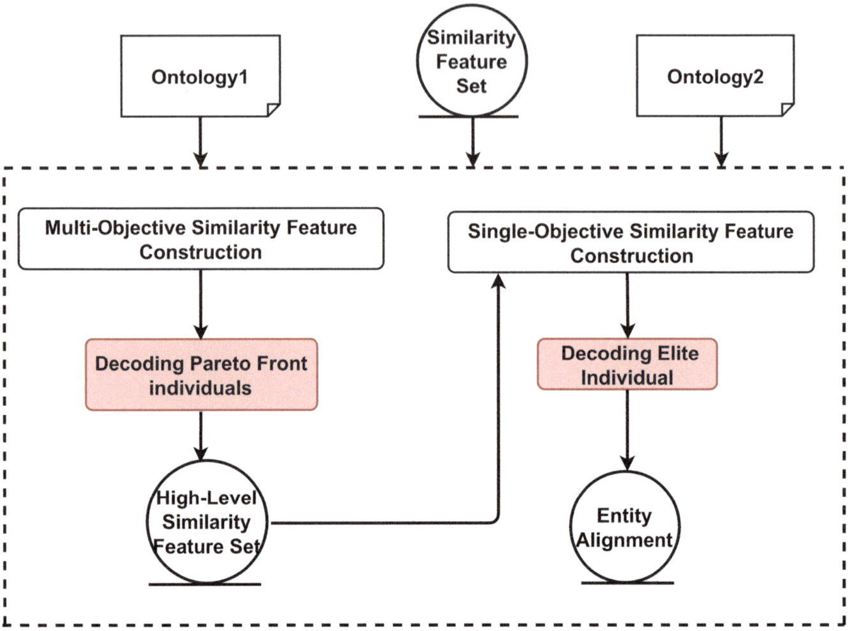

Fig. 14.1 The framework of similarity feature construction

original SF set to automatically construct high-level SFs, ensuring diversity through non-dominated sorting and multi-objective optimization. This stage optimizes both recall and precision simultaneously, which is essential for capturing complex semantic relationships between ontologies. In the second stage, SOGP refines the high-level SFs generated by MOGP, focusing on optimizing the f-measure to improve the overall accuracy and reliability of the ontology alignment results.

14.4 Evolutionary Similarity Feature Construction

Algorithm 1 presents the pseudo-code for the construction of two GPs for SF construction. Initially, a population of multiple solutions is randomly generated, and the individuals are evaluated. In the first stage, MOGP is used to enhance the diversity of solutions. The fitness of individuals is evaluated based on multiple objectives, including tree size, precision, and recall, producing a fitness vector for each individual. Non-dominated sorting is performed, identifying Pareto-dominant individuals through comparisons across these objectives. A tournament selection strategy is used to choose parent individuals from the population, and a crossover operator is applied to exchange sub-trees between selected parents, promoting the combination of good genetic traits. This is followed by a mutation operation, in which random modifications are made to introduce new genetic variations and prevent premature

Algorithm 1: Genetic Programmings for Similarity Feature Construction

Input: Similarity Feature Set $\mathcal{SF}$, SOGP's Maximum Generation $MaximumGen_s$, MOGP's Maximum Generation $MaximumGen_m$, SOGP's Crossover Rate $rate_{c,s}$, MOGP's Crossover Rate $rate_{c,m}$, SOGP's Mutation Rate $rate_{m,s}$, MOGP's Mutation Rate $rate_{m,m}$

Output: Elite Individual $indiv_{elite}$

/* Multi-Objective Genetic Programming */

Initialization Population $\mathcal{P}_m$ with $\mathcal{SF}$;
$evaluation(\mathcal{P}_m)$;
Generation $gen = 0$;
while $gen < MaximumGen_m$ **do**
    $\mathcal{P}'_m = crossover(rate_{c,m})$;
    $\mathcal{P}'_m = mutation(rate_{m,m})$;
    $evaluation(\mathcal{P}'_m)$;
    $non\_dominance\_sorting(\mathcal{P}_m \cup \mathcal{P}'_m)$;
    $\mathcal{P}_m = environment\_selection()$;
    $gen = gen + 1$;
end
Decoding Pareto Front individuals to obtain high-level Similarity Feature set $\mathcal{SF}_h$;

/* Single-Objective Genetic Programming */

Initialization Population $\mathcal{P}_s$ with $\mathcal{SF} \cup \mathcal{SF}_h$;
$evaluation(\mathcal{P}_s)$;
Initialize elite individual $indiv_{elite}$;
$gen = 0$;
while $gen < MaximumGen_s$ **do**
    $\mathcal{P}'_s =$ ***adaptive_crossover***$(rate_{c,s})$;
    $\mathcal{P}'_s = mutation(rate_{m,s})$;
    $evaluation(\mathcal{P}'_s)$;
    $update\&saveElite()$;
    $\mathcal{P}_s = selection()$;
    $gen = gen + 1$;
end
return $indiv_{elite}$;

convergence. Through iterative cycles of selection, crossover, and mutation, the population evolves, and after reaching the termination condition (based on the number of generations or fitness thresholds), Pareto-optimal solutions are output and transformed into high-level SFs.

In the second stage, SOGP refines the high-level SFs produced by MOGP, as well as the original SFs, to improve solution confidence. In SOGP, a novel adaptive crossover operator is introduced to prioritize individuals with smaller tree sizes when fitness values are equal, ensuring that more compact and efficient solutions are favored. As in the first stage, selection, crossover, and mutation operations are performed iteratively until the termination condition is met, resulting in the final output of the optimal SF combination. This two-stage process balances diversity and correctness, ensuring both a broad exploration of potential SFs and a focused refinement of high-quality solutions.

Table 14.1 Comparisons of f-measure (mean and standard deviation) among SOGP, MOGP, and our method across 30 separate trials in each test task

	Test case	SOGP	MOGP	Our approach
Conference	cmt-conf	0.71 (0.03) (+)	0.79 (0.02) (+)	0.83 (0.02) (+)
	cmt-confOf	0.75 (0.02) (+)	0.81 (0.02) (+)	0.85 (0.03) (+)
	cmt-edas	0.71 (0.03) (+)	0.80 (0.03) (+)	0.85 (0.03) (+)
	cmt-iasted	0.72 (0.01) (+)	0.79 (0.02) (+)	0.84 (0.01) (=)
	cmt-sigkdd	0.74 (0.01) (=)	0.82 (0.03) (+)	0.88 (0.03) (+)
	conf-edas	0.62 (0.02) (+)	0.69 (0.01) (+)	0.76 (0.01) (+)
	conf-ekaw	0.72 (0.01) (+)	0.77 (0.03) (+)	0.84 (0.01) (+)
	conf-iasted	0.65 (0.02) (+)	0.75 (0.02) (+)	0.80 (0.03) (+)
	conf-sigkdd	0.71 (0.03) (+)	0.80 (0.03) (+)	0.86 (0.02) (+)
	confOf-ekaw	0.64 (0.01) (+)	0.73 (0.02) (+)	0.79 (0.02) (+)
	confOf-sigkdd	0.77 (0.02) (+)	0.83 (0.02) (+)	0.86 (0.01) (+)
	edas-ekaw	0.72 (0.01) (+)	0.81 (0.01) (+)	0.86 (0.01) (+)
	edas-iasted	0.74 (0.01) (+)	0.83 (0.03) (+)	0.89 (0.02) (+)
	+/–/=	9/0/4	12/0/1	12/0/1

14.5 Experiment

The experiment utilized the Conference datasets of the Ontology Alignment Evaluation Initiative (OAEI)[1] to assess the performance of our method. The chosen fundamental SFs encompass three predominant types within the OM domain [4], i.e., syntax-based, linguistic-based, and structure-based SFs. The GP's setup for SF was determined empirically as follows: a maximum generation count of 1000 for both MOGP and SOGP; a population size of 500 for both MOGP and SOGP; crossover rates of 0.80 for MOGP and 0.85 for SOGP; mutation rates of 0.02 for MOGP and 0.04 for SOGP; and a maximum tree depth of 6 levels for both MOGP and SOGP.

Table 14.1 presents a comprehensive comparison of our method with conventional Evolutionary Algorithms across different test scenarios. In this table, the symbols "(+)/(–)/(=)" denote whether our method surpassed, lagged behind, or matched the performance of other ontology matching techniques, respectively, as determined by the Wilcoxon Signed-Rank test [17]. As shown in this table, our approach consistently demonstrates superior or competitive performance, which is especially prominent in complex matching scenarios, as evidenced by the f-measure obtained over 30 independent runs. Further analysis within the Conference data set highlights our approach's ability to adeptly manage diverse ontological complexities. For example, in the test cases cmt-conf and cmt-confOf, our approach either matches or exceeds the performance of other methodologies, indicating its consistent adaptability and effectiveness across different domains. These outcomes are particularly significant

[1] https://oaei.ontologymatching.org/2024/conference/index.html.

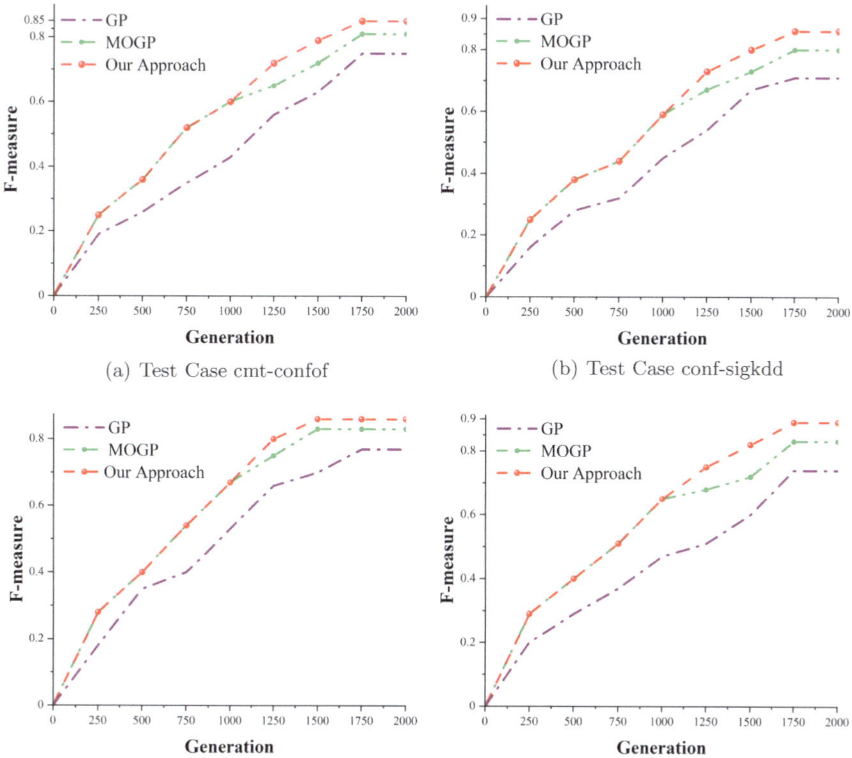

(a) Test Case cmt-confof

(b) Test Case conf-sigkdd

(c) Test Case confof-sigkdd

(d) Test Case edas-iasted

Fig. 14.2 Comparisons among SOGP, MOGP and our approach on the converge graph across four representative test cases

as they validate the practical benefits of our approach's automatic SFs constructing mechanism, which is designed to construct high-level SFs and find the optimal SFs combination. The consistently high performance across a variety of challenging test scenarios not only underscores the accuracy and robustness of our approach's matching results but also confirms the effectiveness of the automatic SFs constructing strategy.

Figure 14.2 compares the performance of our approach and its variants over 30 independent runs across six conference's representative test cases cmt-confof, conf-sigkdd, confOf-sigkdd, edas-iasted. In the cmt-confof test case, our approach quickly converges to a higher f-measure, showing its superior optimization efficiency. In conf-sigkdd, our approach demonstrates quicker convergence and achieves higher accuracy, proving its effectiveness. The test cases confof-sigkdd further emphasize our approach's robustness, as it consistently converges faster and reaches higher f-measure values. Finally, in the test case edas-iasted, our approach achieves the highest f-measure values with rapid convergence, underscoring its optimization capabilities. In general, our approach consistently demonstrates faster convergence to higher f-measure values across all representative test cases, validating its enhanced efficiency and superiority.

14.6 Conclusions

This paper presents a new framework that combines MOGP and SOGP to enhance the efficiency and accuracy of SF construction for OM. In response to the first research question on increasing the diversity of candidate SFs, MOGP is employed to explore the solution space more thoroughly, thus reducing the risk of premature convergence and expanding the variety of potential SF combinations. This increased diversity ensures a broader search capability, effectively capturing complex semantic relationships within large-scale ontologies. Addressing the second question on ensuring the confidence of constructed SFs, SOGP is used to fine-tune the selected SFs, enhancing their accuracy and robustness. This sequential approach not only boosts alignment quality but also ensures the correctness of OM results by optimizing the balance between diversity and confidence in SF construction. The proposed method consistently outperforms traditional GP-based approaches, demonstrating superior performance across various OM tasks by handling complex dependencies and large entity sets more effectively.

Acknowledgements This work was supported by the National Natural Science Foundation of China (No. 62172095).

References

1. Berners-Lee, T., Hendler, J., Lassila, O.: The semantic web: a new form of web content that is meaningful to computers will unleash a revolution of new possibilities. In: Linking the World's Information: Essays on Tim Berners-Lee's Invention of the World Wide Web, pp. 91–103 (2023)
2. Yanamala, A.K.Y.: Emerging challenges in cloud computing security: a comprehensive review. Int. J. Adv. Eng. Technol. Innov. **1**(4), 448–479 (2024)
3. Xue, X., Zhang, J., Yang, C.: An ontology-supported inquiry learning technique. J. Netw. Intell. **6**(1), 16–28 (2021)
4. Xue, X., Wang, Y., Hao, W.: Using moea/d for optimizing ontology alignments. Soft. Comput. **18**(8), 1589–1601 (2014)
5. Trojahn, C., Vieira, R., Schmidt, D., Pease, A., Guizzardi, G.: Foundational ontologies meet ontology matching: a survey. Semantic Web **13**(4), 685–704 (2022)
6. Al-Aswadi, F.N., Chan, H.Y., Gan, K.H.: Extracting semantic concepts and relations from scientific publications by using deep learning. In: International Conference of Reliable Information and Communication Technology, pp. 374–383. Springer (2020)
7. Chen, S., Wang, J., Jiang, F., Lin, C.Y.: Improving entity linking by modeling latent entity type information. In: Proceedings of the AAAI Conference on Artificial Intelligence, vol. 34, pp. 7529–7537 (2020)
8. Xue, X.: Automatic knowledge graph matching via self-adaptive designed genetic programming. Knowl.-Based Syst. **293**, 111,628 (2024)

9. Gen, M., Lin, L.: Genetic algorithms and their applications. In: Springer handbook of engineering statistics, pp. 635–674. Springer (2023)
10. Niazkar, M.: Multigene genetic programming and its various applications. Handbook of Hydroinformatics, pp. 321–332 (2023)
11. Xue, X., Sun, D., Shankar, A., Viriyasitavat, W., Siarry, P.: Efficient large-scale biomedical ontology matching with anchor-based biomedical ontology partitioning and compact geometric semantic genetic programming. J. Ind. Inf. Integr. 100637 (2024)
12. Chen, J., Zhang, J., Xue, X., Huang, Y.: Multi-view ontology alignment visualization (moav): a human-cognition based ontology alignment visualization technique. J. Netw. Intell. **5**(4), 198–210 (2020)
13. Chen, Z., Wang, Y., Zhao, B., Cheng, J., Zhao, X., Duan, Z.: Knowledge graph completion: a review. IEEE Access **8**, 192,435–192,456 (2020)
14. Zhao, X., Jia, Y., Li, A., Jiang, R., Song, Y.: Multi-source knowledge fusion: a survey. World Wide Web **23**, 2567–2592 (2020)
15. Suo, X., Guo, B., Shen, Y., Dai, S., Wang, W., Chen, Y., Zhang, Z.: Krl_match: knowledge graph objects matching for knowledge representation learning. Knowl. Inf. Syst. **65**(2), 641–681 (2023)
16. Shaik, A.L.H.P., Manoharan, M.K., Pani, A.K., Avala, R.R., Chen, C.M.: Gaussian mutation-spider monkey optimization (gm-smo) model for remote sensing scene classification. Remote Sensing **14**(24), 6279 (2022)
17. Harris, T., Hardin, J.W.: Exact wilcoxon signed-rank and wilcoxon mann-whitney ranksum tests. Stand. Genomic Sci. **13**(2), 337–343 (2013)

Chapter 15
EcoSLA-RL: A Reinforcement Learning-Based Approach for Optimizing SLA and Execution Costs in Cloud Data Centers

Chang Wu, Pei-Wei Tsai, Muneeb Ul Hassan, Na Guo, and Yang Lin

Abstract As the demand for economically efficient cloud computing services grows, optimizing VM usage costs and reducing computational expenses are essential for providing better user value. This paper introduces the proposed model, EcoSLA-RL, an algorithm based on reinforcement learning for virtual machine (VM) consolidation in data centers, validated through simulations. Our goal is to provide a robust, user cost-efficient solution by leveraging reinforcement learning for dynamic consolidation, focusing on minimizing computational costs while effectively addressing challenges related to fluctuating workloads and SLA compliance. EcoSLA-RL achieved computation cost reductions of over 10% compared to other algorithms, while maintaining an SLA violation rate as low as 0.053% in large-scale environments.

C. Wu · P.-W. Tsai (✉)
School of Science, Computing and Engineering Technologies, Swinburne University of Technology, Hawthorne, VIC, Australia
e-mail: ptsai@swin.edu.au

M. U. Hassan
School of Information Technology, Deakin University, Geelong, VIC, Australia

N. Guo
School of Mechanical, Electrical and Information Engineering, Shandong University (Weihai), Shandong, China
e-mail: 201069800020@sdu.edu.cn

Y. Lin
University of Pennsylvania, Philadelphia, USA

S.-C. Chu et al. (eds.), *Advances in Intelligent Data Analysis and Applications*,
Smart Innovation, Systems and Technologies 445,
https://doi.org/10.1007/978-981-96-7277-6_15

15.1 Introduction

Cloud computing is now fundamental to contemporary information technology, supporting scalable resource distribution across numerous sectors and applications [1, 2]. Virtual machines (VMs) are integral to cloud infrastructures, offering the flexible and scalable resources needed to support diverse application demands. As cloud computing continues to adapt to an expanding range of complex applications and sectors, it is essential to develop efficient and cost-effective resource management strategies.

With the rising demand for cloud services, managing VM-related computational costs while ensuring Quality of Service (QoS) and adhering to Service-Level Agreements (SLA) presents significant challenges, particularly as workloads fluctuate dynamically [3]. Traditional heuristic approaches often lack the flexibility needed to adapt to these variations efficiently, which can lead to excessive execution costs and resource inefficiencies within data centers.

The core problem this study addresses is the challenge of optimizing VM consolidation in cloud data centers (CDCs) to minimize execution costs while meeting QoS and SLA standards. Current methods face challenges in rapidly adjusting to workload variations, leading to inefficient resource utilization and increased operational expenses.

To address this issue, EcoSLA-RL is introduced as a reinforcement learning approach for consolidating VMs in CDCs. EcoSLA-RL leverages the adaptability of RL to dynamically optimize resource allocation in real time, ensuring QoS and SLA adherence. By employing a structured framework with fixed VM configurations, the algorithm aligns resources with workload demands, effectively reducing cost inefficiencies. In benchmarking against heuristic approaches, EcoSLA-RL demonstrates a scalable, cost-efficient solution for practical cloud environments. In summary, our contribution is as follows:

- This research introduces EcoSLA-RL, a reinforcement learning algorithm designed to improve the efficiency of consolidating virtual machines within cloud data centers, focusing on reducing execution costs while maintaining SLA and QoS compliance.
- Extensive benchmarking reveals that EcoSLA-RL achieves better execution cost savings and resource utilization than traditional heuristic methods, especially in medium- to large-scale workloads.
- Comparative analysis highlights EcoSLA-RL's cost efficiency and low SLA violation rates under fixed physical machine constraints.

15.2 Related Works

Existing VM consolidation algorithms in cloud data centers primarily target improvements in energy efficiency, load distribution, and resource utilization. For instance, the ADVMC algorithm [4] prioritizes VMs for migration using influence coefficients and LSTM models to forecast overloads. Additional algorithms, including RLVMrB [5] and DJ-HA [6], integrate multi-resource optimization and metaheuristic strategies to manage energy use and reduce active hosts, but often overlook VM-related costs. [7] aimed at improving energy efficiency and performance while reducing execution costs. This method migrates long-running virtual machines to more energy-efficient target hosts and employs an optimized resource allocation strategy to enhance resource utilization, thereby reducing execution costs. Zakarya et al. [8] presents TPAPSO-PMWFT, a framework that combines task priority adaptive particle swarm optimization (TPAPSO) with performance monitoring and fault tolerance (PMWFT) to optimize task scheduling in edge data centers. By assessing VM availability and performance, TPAPSO-PMWFT dynamically allocates and migrates tasks to enhance resource efficiency and reduce costs. OSEC [9] employs a real-time Q-learning algorithm to allocate and migrate VMs. Energy consumption and execution costs are optimized through a multi-objective approach, assuming unconstrained host resources. Similarly, the ERLNEIL-MDP [10] incorporates evolutionary strategies and reinforcement learning to boost data processing efficiency, excelling in complex scenarios by leveraging novelty-based fitness selection and imitation-driven experience integration. Balancing energy efficiency and computing costs is essential in modern cloud environments. This research proposes a reinforcement learning-based VM consolidation algorithm that optimizes VM placement under fixed physical machine constraints, enhancing economic efficiency and ensuring SLA and QoS compliance. In future work, security and privacy considerations can also be incorporated [11].

15.3 Framework for Resource Allocation and VM Scheduling

This section presents a comprehensive framework for VM scheduling and resource allocation, designed to optimize data center efficiency and meet SLA requirements. The framework is structured across multiple layers to effectively manage user requests, infrastructure resources, and scheduling processes. By integrating these elements, it enables efficient VM allocation, load balancing, and dynamic resource management, which are crucial for minimizing operational costs and enhancing performance in large-scale cloud environments.

15.3.1 System Model

The proposed framework operates within a three-layer structure for VM scheduling:

- **User layer**: Processes application requests and converts them into VM requirements.
- **Cloud data center layer**: Managed by cloud providers, this layer comprises 3000 PMs to ensure adequate resources for VMs.
- **Scheduling layer**: Responsible for VM allocation, migration, and resource optimization to balance load and reduce costs.

As illustrated in Fig. 15.1 [8], this layered approach supports cost-efficient, SLA-compliant cloud management.

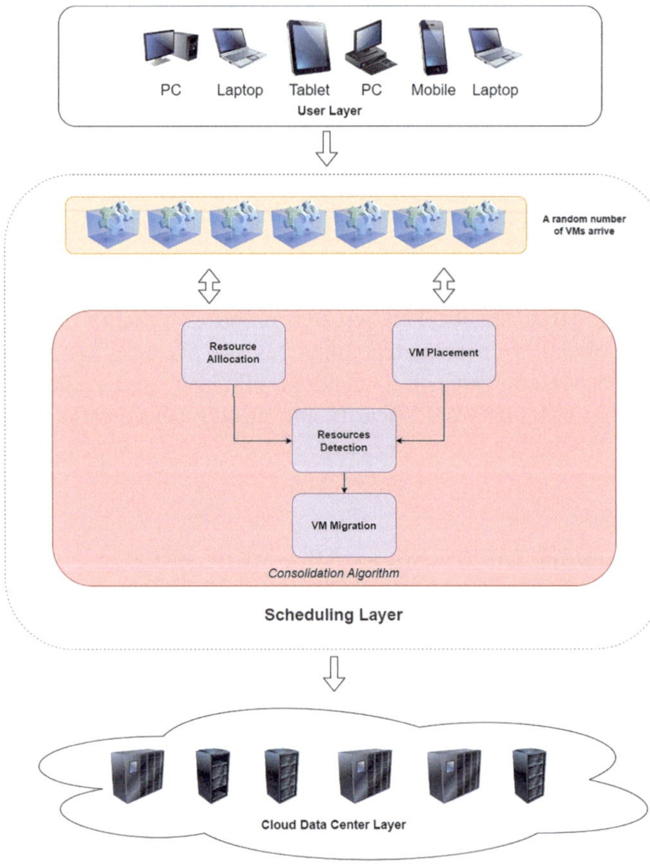

Fig. 15.1 User, scheduling, and data center layer framework

15.3.2 Design of Virtual Machine and Physical Machine Models

Applications requested by various users consist of tasks, each with unique resource requirements. These requests form a list of virtual machines, denoted as

$$VM_t = \left\{ vm_1^1, vm_2^1, \ldots, vm_1^2, \ldots, vm_p^q \right\} \tag{15.1}$$

where VM_t represents the set of VMs at time step t.

$$vm_p^q = \left\{ l_p^q,\ oc_p^q,\ r_{qp}^{\varphi},\ u_{qp}^{cpu}(t) \right\} \tag{15.2}$$

Each vm_q^p corresponds to the p th VM from the q th user request, encompassing information on workload l_p^q, operational cost oc_p^q, resource requirements r_{qp}^{φ}, and CPU utilization $u_{qp}^{cpu}(t)$, which may vary over time due to fluctuating demand.

$$PM = \{pm_1, \ldots, pm_i, \ldots, pm_{3000}\} \tag{15.3}$$

representing a group of 3000 accessible servers within the data center, organized to support various workloads. These PMs are allocated according to resource demands, forming the infrastructure backbone for efficient VM hosting and management.

15.3.3 Execution Cost and Problem Formulation

The execution cost of a VM is determined by its runtime and operational expenses. For each physical machine pm_s, the execution time T_s^{qp} of a VM vm_p^q can be calculated by dividing its workload l_p^q by the MIPS (million instructions per second) rating of the assigned physical machine:

$$T_s^{qp} = \frac{l_p^q}{MIPS_s} \tag{15.4}$$

The aggregate execution cost C_S for pm_s is the total execution time of all assigned VMs, each adjusted by its per-unit operational cost oc_p^q:

$$C_S = \sum_{vm_p^q \in pm_s} T_S^{qp} \cdot oc_p^q \tag{15.5}$$

The objective is to reduce execution expenses across active PMs, within resource constraints:

$$Minimize \sum_{pm_s \in PM_{used}} C_S \tag{15.6}$$

Each physical machine (PM) pm_i has limited capacity, meaning that the total resource demands of the VMs allocated to pm_i must not exceed this capacity.

15.4 Reinforcement Learning Method

The Computational Cost Optimization (CCO) problem involves strategically assigning VMs to PMs in cloud environments, aiming to reduce the operational costs associated with VM execution. This requires allocating resources according to each VM's requirements and optimizing the available resources of the PMs. This study employs reinforcement learning (RL) to address CCO under CPU and memory constraints, ensuring Quality of Service (QoS) for users. The primary goal is to lower execution costs across the data center while balancing resource utilization for cost efficiency and system stability.

15.4.1 State, Action Spaces, and Reward Function Design

The state space in this reinforcement learning model reflects resource availability and utilization across the 3000 servers within the data center. The PMs are divided into three distinct types, indicated by $m \in \{1, 2, 3\}$, with each type having different configurations for CPU cores, memory, and MIPS. The specifications of these three PM types are detailed in Table 15.2.

The definition of the state space is as follows:

$$State = [PM_1, PM_2, \ldots, PM_i, \ldots, PM_{3000}] \tag{15.7}$$

Each PM_i is characterized by its resources remaining for running VMs, represented as

$$PM_i = [m, CPU_i, Memory_i, MIPS_i, Utilization_i], \forall m \in \{1, 2, 3\} \forall i \in \{1, \ldots, 3000\} \tag{15.8}$$

Here, m denotes the type of PM, while CPU_i, $Memory_i$, $MIPS_{i'}$, and $Utilization_i$ represent the remaining CPU, memory, MIPS, and current utilization level of PM_i, respectively. This state space structure enables the model to capture the essential attributes of each PM, facilitating effective resource management and decision-making in the optimization process.

In this reinforcement learning model, the action space represents the selection of one of the 3000 PMs in the data center for VM placement, defined as

$$Action = \{a|a \in \{1,\ 2,\ \ldots,\ 3000\}\} \tag{15.9}$$

where a indicates the index of the chosen PM, enabling efficient VM allocation based on available resources.

The reward function is defined as the negative computational cost with an additional penalty if SLA constraints (CPU and RAM) are not met:

$$r = -C_S - \lambda \cdot PenaltySLA \tag{15.10}$$

Here, C_S is the total execution cost for all VMs on pm_s, and $\text{Penalty}_{\text{SLA}} = 1$ if SLA requirements are not met; otherwise, $\text{Penalty}_{\text{SLA}} = 0$. The parameter λ adjusts the importance of SLA compliance.

This reward structure encourages cost minimization while ensuring adherence to the SLA.

15.4.2 Action Selection and Q-value Update

To achieve a balance between exploration and exploitation during action selection, the $\varepsilon - greedy$ method is applied.

$$\pi(a_t|s_t) = \begin{cases} random\,(A)\ if\ rand\ < \varepsilon \\ a^* \qquad otherwise \end{cases} \tag{15.11}$$

The variable rand denotes a random value in [0, 1], $\in$ is the probability of exploration, and a^* represents the PM selected for optimal placement with the highest Q-value.

The Q-value update for VM placement decisions is defined as

$$Q(s_t, a_t) = (1-\alpha) \cdot Q(s_t, a_t) + \alpha \cdot \left[r_{t+1} + \gamma \cdot \max_a Q(s_{t+1}, a)\right] \tag{15.12}$$

where S_t represents the current state of PM and VM allocations, and a_t is the selected PM for VM placement at step t. Here, α denotes the learning rate, r_{t+1} is the reward obtained after assigning the VM to PM a_t, and γ, the discount factor, manages the trade-off between immediate and future rewards.

15.5 Experiments

User requests with varying resource needs are represented as VMs arriving randomly at the cloud platform, based on five realistic workflows: GENOME, LIGO, Cybershake, MONTAGE, and SIPHT [12].

Virtual Machine: For practical application, seven types of virtual machines (VMs) with varying prices are selected. After preprocessing workflows in the simulator, jobs are allocated to different VMs. A Poisson distribution generates the number and types of VMs arriving at the data center at different times, while a Gaussian distribution produces 10,000 CPU utilization values over 10,000 time steps to simulate workflow fluctuations. Table 15.1 lists the seven VM types used.

Physical Machine: For practical application, three types of physical machines (PMs) with different specifications were selected. A total of 3000 physical machines were used in the simulation, with 1000 PMs of each type. The specifications for each PM configuration are shown in Table 15.2.

We selected three algorithms for comparison in the simulator: AFED-EF [13], EcoCloud [14], and KMI MPCU [15]. Virtual machines (VMs) arrived at each time step following a Poisson distribution with a rate parameter of 4, while their duration depended on task runtime, extending up to thousands of time steps. Throughout execution, VM CPU utilization varied over time in response to task load, modeled by a normal distribution. The experiment was conducted in the Java 17.0.4 environment on an Intel Core i7 processor.

Table 15.1 Specifications for VMs in the simulator

Type	CPU (cores)	Memory (GB)	Price (US dollars)
t2.nano	1	0.5	0.0058
t2.micro	1	1	0.0116
t2.small	1	2	0.0230
t2.medium	2	4	0.0464
t2.large	2	8	0.0928
t2.xlarge	4	16	0.1856
t2.2xlarge	8	32	0.3712

Table 15.2 Specifications for PMs in the simulator

Type	CPU (cores)	Memory (GB)	MIPS	Quantity
FUJITSU RX1330 M4	8	16	2.0	1000
FUJITSU LX1430 M1	64	256	2.6	1000
Inspur NF5280 M5	56	192	3.8	1000

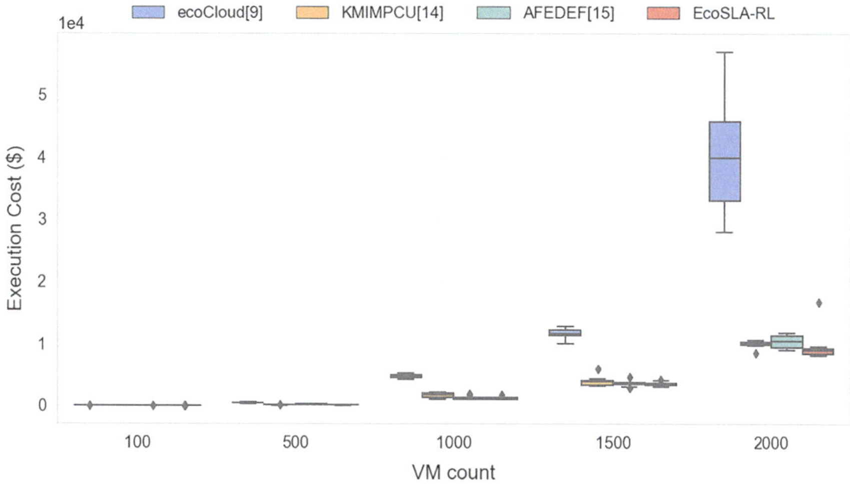

Fig. 15.2 Comparison of execution costs across algorithms under various VM counts

15.5.1 Execution Cost Comparison

Figure 15.2 shows that with an increasing number of VMs, the computation cost for each of the four algorithms progressively rises. Among these algorithms, EcoCloud exhibits the highest execution cost. When the VM count is 100, the computation cost of EcoSLA-RL is reduced by 74.1%, 13.92%, and 17.01% compared to EcoCloud, KMI MPCU, and AFED-EF, respectively, demonstrating significant cost advantages in small-scale computational environments.

With the VM count reaching 2000, EcoSLA-RL continues to show advantages in cost control. In larger scale computational environments, EcoSLA-RL achieves an average execution cost that is about 2.6% less than KMI MPCU and 6.85% below AFED-EF. This indicates that EcoSLA-RL effectively optimizes costs in dynamic workload conditions.

15.5.2 SLA Violation Comparison

As shown in Fig. 15.3, as the number of VMs increases, differences in SLA violation rates become more apparent among the four algorithms. Overall, EcoSLA-RL maintains the lowest SLA violation rate across various computational scales, indicating its effectiveness in QoS management.

At a VM count of 1000, EcoSLA-RL's SLA violation rate is 0.394%, noticeably less than the rates of the other algorithms: EcoCloud has an SLA violation rate of 5.931%, KMI MPCU's rate is 1.239%, and AFEDEF's rate is 2.815%. This lower

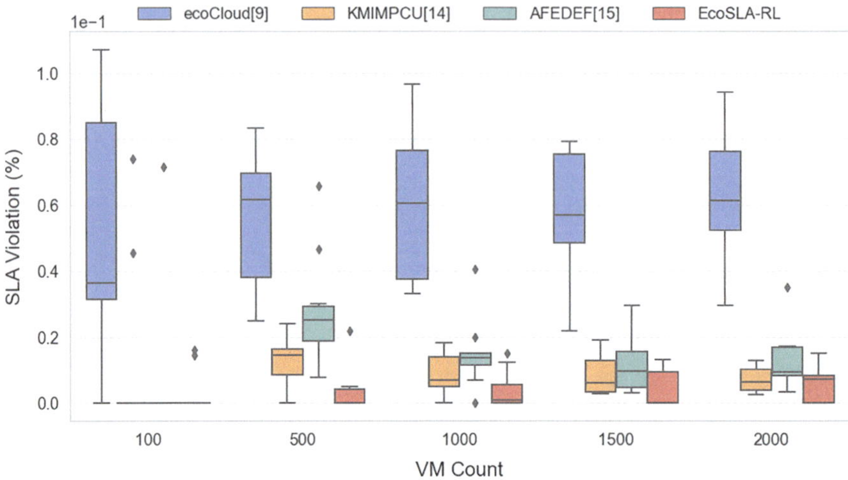

Fig. 15.3 Comparison of SLA violations across algorithms under various VM counts

violation rate suggests that EcoSLA-RL meets SLA requirements under moderate-scale workloads, demonstrating its reliability in dynamic load conditions.

As the VM count rises to 2000, EcoSLA-RL sustains its lower SLA violation rate at 0.053%. In comparison, the SLA violation rates for EcoCloud, KMI MPCU, and AFED-EF are 6.371%, 0.708%, and 1.325%, respectively. These results indicate that EcoSLA-RL is effective in maintaining low SLA violation rates both in smaller scale environments and as computational demands increase.

15.6 Conclusion

EcoSLA-RL has demonstrated its effectiveness as a VM consolidation algorithm, consistently achieving lower SLA violation rates and execution costs across diverse VM scales compared to other approaches like EcoCloud, KMI MPCU, and AFED-EF. By prioritizing efficient resource allocation, EcoSLA-RL upholds QoS standards and SLA compliance, adapting robustly to both moderate and large-scale workloads. These results confirm EcoSLA-RL's potential in balancing cost control with SLA adherence, presenting a promising solution for modern data center optimization.

Future research could focus on enhancing the reinforcement learning training effectiveness of EcoSLA-RL, as well as exploring multi-objective optimization to further improve performance metrics like energy consumption. Such developments would further broaden its applicability and efficiency in increasingly complex and dynamic cloud environments.

References

1. Jennings, B., Stadler, R.: Resource management in clouds: survey and research challenges. J. Netw. Syst. Manag. **23**(3), 567–619 (2015). https://doi.org/10.1007/s10922-014-9307-7
2. Rodriguez, M.A., Buyya, R.: A taxonomy and survey on scheduling algorithms for scientific workflows in IaaS cloud computing environments. Concurr. Comput.: Pract. Exp. **29**(8), e4041 (2017). https://doi.org/10.1002/cpe.4041
3. Khaleel, M.I.: A dynamic weight–assignment load balancing approach for workflow scheduling in edge-cloud computing using ameliorated moth flame and rock hyrax optimization algorithms. Futur. Gener. Comput. Syst. **155**, 465–485 (2024). https://www.sciencedirect.com/science/article/pii/S0167739X24000700
4. Zeng, J., Ding, D., Kang, K., Xie, H., Yin, Q.: Adaptive DRL-based virtual machine consolidation in energy-efficient cloud data center. IEEE Trans. Parallel Distrib. Syst. **33**(11), 2991–3002 (2022). https://ieeexplore.ieee.org/abstract/document/9698981.
5. Ghasemi, A., Toroghi Haghighat, A.: A multi-objective load balancing algorithm for virtual machine placement in cloud data centers based on machine learning. Computing **102**(9), 2049–2072 (2020). https://doi.org/10.1007/s00607-020-00813-w
6. Hybrid meta-heuristic algorithm for optimal virtual machine placement and migration in cloud computing—Infantia Henry—2022—Concurrency and Computation: Practice and Experience—Wiley Online Library. https://doi.org/10.1002/cpe.7353
7. Zakarya, M., Gillam, L.: Managing energy, performance and cost in large scale heterogeneous datacenters using migrations. Futur. Gener. Comput. Syst. **93**, 529–547 (2019). https://www.sciencedirect.com/science/article/pii/S0167739X17326249
8. Kadum, M.M., Deng, X.: Optimizing execution time and cost while scheduling scientific workflow in edge data center with fault tolerance awareness. Nonlinear Eng. **13**(1) (2024). https://www.degruyter.com/document/doi/10.1515/nleng-2024-0015/html. Publisher: De Gruyter
9. Ma, X., Xu, H., Gao, H., Bian, M., Hussain, W.: Real-time virtual machine scheduling in industry IoT network: a reinforcement learning method. IEEE Trans. Ind. Inform. **19**(2), 2129–2139 (2023). https://ieeexplore.ieee.org/document/9910009/
10. Lv, J., Kim, B.G., Slowik, A., Parameshachari, B.D., Kumari, S., Chen, C.M., Li, K.: ERLNEIL-MDP: evolutionary reinforcement learning with novelty-driven exploration for medical data processing. Swarm Evol. Comput. **91**, 101769 (2024). https://www.sciencedirect.com/science/article/pii/S2210650224003079
11. Mei, Q., Xiong, H., Chen, Y.C., Chen, C.M.: Blockchain-enabled privacy-preserving authentication mechanism for transportation CPS with cloud-edge computing. IEEE Transactions on Engineering Management **71**, 12463–12474 (2024), https://ieeexplore.ieee.org/document/9757246/?arnumber=9757246,
12. Juve, G., Chervenak, A., Deelman, E., Bharathi, S., Mehta, G., Vahi, K.: Characterizing and profiling scientific workflows. Future Generation Computer Systems **29**(3), 682–692 (2013), https://www.sciencedirect.com/science/article/pii/S0167739X12001732
13. Zhou, Z., Shojafar, M., Alazab, M., Abawajy, J., Li, F.: AFED-EF: An energy-efficient VM allocation algorithm for IoT applications in a cloud data center. IEEE Transactions on Green Communications and Networking **5**(2), 658–669 (2021). https://ieeexplore.ieee.org/abstract/document/9382007
14. Mastroianni, C., Meo, M., Papuzzo, G.: Probabilistic consolidation of virtual machines in self-organizing cloud data centers. IEEE Transactions on Cloud Computing **1**(2), 215–228 (2013), https://ieeexplore.ieee.org/abstract/document/6681861
15. Zhou, Z., Abawajy, J., Chowdhury, M., Hu, Z., Li, K., Cheng, H., Alelaiwi, A.A., Li, F.: Minimizing SLA violation and power consumption in cloud data centers using adaptive energy-aware algorithms. Future Generation Computer Systems **86**, 836–850 (2018). https://www.sciencedirect.com/science/article/pii/S0167739X17316059

Chapter 16
Evolution of Harris Hawks Optimization: Insights, Innovations, and Applications

Shu-Chuan Chu, Trong-Trieu Hoang, Truong-Giang Ngo, Thi-Kien Dao, and Trong-The Nguyen

Abstract This paper provides a comprehensive overview of the Harris Hawks Optimization (HHO) algorithm, which is inspired by the cooperative hunting behaviors of Harris hawks. We explore the algorithm's foundational principles, detailing its unique exploration and exploitation strategies that mimic natural predatory tactics. The review highlights significant innovations that have enhanced HHO's performance, including hybrid approaches and adaptive parameter tuning. Furthermore, we examine a wide range of real-world applications across diverse industries such as engineering, energy, bioinformatics, and environmental management. By synthesizing insights from various studies, this paper aims to illustrate the versatility and efficacy of HHO in solving complex optimization problems, while also identifying future research directions to further its development.

S.-C. Chu
School of Artificial Intelligence, Nanjing University of Information, Science & Technology, Nanjing, China
e-mail: jan.chu@flinders.edu.au

T.-T. Hoang · T.-G. Ngo
Department of Academic Affairs, Thuyloi University, Hanoi, Vietnam
e-mail: trieuht@tlu.edu.vn

T.-G. Ngo
e-mail: giangnt@tlu.edu.vn

T.-K. Dao · T.-T. Nguyen (✉)
University of Information Technology, Ho Chi Minh City, Vietnam
e-mail: thent@uit.edu.vn

T.-K. Dao
e-mail: kiendt@uit.edu.vn

S.-C. Chu et al. (eds.), *Advances in Intelligent Data Analysis and Applications*, Smart Innovation, Systems and Technologies 445,
https://doi.org/10.1007/978-981-96-7277-6_16

16.1 Introduction

In recent years, the field of optimization has experienced a surge in the development of metaheuristic algorithms [1], which are inspired by natural phenomena and biological behaviors [2]. These algorithms have proven to be effective in solving complex optimization problems across various domains, including engineering, computer science, and environmental management [3]. Among these innovative approaches, the Harris Hawks Optimization (HHO) [4] algorithm has emerged as a prominent method, drawing inspiration from the cooperative hunting strategies of Harris hawks, a species known for their intelligence and teamwork. The HHO algorithm was first introduced by Heidari et al. in 2019, and it quickly gained traction due to its unique mechanism that mimics the dynamic hunting behavior of these birds [4]. Harris hawks exhibit a remarkable ability to adapt their hunting strategies based on environmental conditions and prey behavior, which serves as a powerful metaphor for optimization processes. The algorithm captures this adaptability through a balance of exploration and exploitation, allowing it to efficiently navigate complex search spaces and converge toward optimal solutions [5]. One of the key strengths of HHO lies in its simplicity and ease of implementation, making it accessible to researchers and practitioners alike. The algorithm's performance has been validated across a wide range of benchmark functions and real-world applications, demonstrating its versatility and robustness [6]. As a result, HHO has sparked significant interest in the optimization community, leading to numerous studies that explore its theoretical foundations, variants, and applications [7].

This review paper aims to provide a comprehensive overview of the evolution of Harris Hawks Optimization, focusing on three main aspects: insights gained from its theoretical development, innovations that have enhanced its performance, and real-world applications that showcase its practical utility. By synthesizing existing literature and highlighting key advancements, this paper seeks to contribute to the understanding of HHO and its potential for future research and application. In the following sections, we will delve into the theoretical foundations of HHO, explore the innovations that have emerged since its inception, and examine the diverse applications that illustrate its effectiveness in solving real-world optimization problems. Through this exploration, we aim to provide a holistic view of Harris Hawks Optimization and its significance in the broader context of optimization research.

16.2 Origins of Harris Hawks Optimization

The Harris Hawks Optimization (HHO) algorithm is rooted in the principles of nature-inspired computing, which seeks to emulate the strategies observed in biological systems to solve complex optimization problems [4]. The origins of HHO can be traced back to the study of animal behavior, particularly the cooperative hunting techniques exhibited by Harris hawks (Parabuteo unicinctus). This section explores the biological inspiration behind HHO, its initial formulation, and the foundational concepts that underpin the algorithm.

16.2.1 Biological Inspiration

Harris hawks are known for their unique hunting strategies, which involve teamwork and adaptability. Unlike many other bird species that hunt alone, Harris hawks often hunt in groups, employing sophisticated tactics to capture prey. This cooperative behavior is characterized by several key elements:

- Team Coordination: Harris hawks work together to surround and exhaust their prey, demonstrating effective communication and coordination among group members.
- Adaptive Strategies: The hawks adjust their hunting techniques based on environmental conditions and the behavior of their prey, showcasing a high degree of flexibility and intelligence.
- Exploration and Exploitation: The hunting process involves both exploration of the environment to locate prey and exploitation of opportunities when they arise, reflecting a balance that is crucial for successful hunting.

Figure 16.1 shows the hunting behavior of Harris hawks' inspiration for the HHO algorithm.

These behavioral traits inspired the development of the HHO algorithm, which aims to replicate the hawks' ability to adaptively search for optimal solutions in complex problem spaces.

(a) Low soaring (b) High soaring (c) Swooping

Fig. 16.1 The hunting behavior of Harris hawks

16.2.2 Initial Formulation

The HHO algorithm was first introduced by Heidari et al. in 2019 as a novel meta-heuristic optimization technique. The initial formulation of HHO was designed to mimic the hunting behavior of Harris hawks, incorporating the principles of exploration and exploitation into its search mechanism [4]. The algorithm operates through a series of iterative steps, where potential solutions are represented as "hawks" in the search space. The key components of the initial HHO formulation include the following:

- Initialization: The algorithm begins by randomly generating a population of hawks, each representing a potential solution to the optimization problem.
- Fitness Evaluation: Each hawk's fitness is evaluated based on a predefined objective function, determining how well it performs in relation to the problem at hand.
- Hunting Mechanism: The algorithm simulates the hunting behavior of hawks, where they either explore new areas of the search space or exploit known good solutions based on their fitness levels. This dual approach allows the algorithm to effectively balance exploration and exploitation.

The *initial population* of hawks is generated randomly within the defined search space:

$$X_i^{(0)} = X_{\min} + \mathrm{rand}(0, 1) \times (X_{\max} - X_{\min}), i = 1, 2, \ldots, N \tag{16.1}$$

where $X_i^{(0)}$ is the position of the ith hawk at iteration0; $X_{\max}$, and $X_{\min}$ are the lower and upper bounds of the search space; N is the number of hawks; rand(0, 1) generates a random number between 0 and 1.

Fitness Evaluation: The fitness of each hawk is evaluated using an objective function $f(X)$.

$$f(Xi) = \text{Objective Function}(Xi), i = 1, 2, \ldots, N, \tag{16.2}$$

Hunting Mechanism: The hunting behavior is modeled through a series of equations that dictate how hawks update their positions based on their fitness and the fitness of other hawks. The position update can be expressed as follows:

Exploration Phase: During the exploration phase, hawks search for new positions.

$$X_i^{(t+1)} = X_i^t + r_1 \cdot (X_{best} - \mid X_i^t \mid), \text{ if } p < 0.5, \tag{16.3}$$

where $X_i^{(t+1)}$ is the position of the ith hawk at iteration t; X_{best} is the position of the best hawk found so far; r_1 is a random number in the range [0, 1]; p is a random probability.

Exploitation Phase: During the exploitation phase, hawks refine their search around the best solution:

$$X_i^{(t+1)} = X_{\text{best}} + r_2 \cdot (X_{\text{best}} - \mid X_i^t \mid), \text{ if } p < 0.5, \tag{16.4}$$

where r_2 is another random number in the range [0, 1].

Update Mechanism: The algorithm iteratively updates the positions of the hawks based on the above equations until a stopping criterion is met (e.g., a maximum number of iterations or convergence to a satisfactory solution).

16.2.3 Theoretical Foundations

The theoretical foundations of HHO are built upon established concepts in optimization and swarm intelligence [8]. The algorithm leverages the principles of social behavior, where individuals (hawks) collaborate to achieve a common goal (finding optimal solutions) [4]. The balance between exploration and exploitation is a central theme in many optimization algorithms, and HHO's unique approach to this balance sets it apart from traditional methods [9]. The initial success of HHO in solving benchmark optimization problems demonstrated its potential as a powerful optimization tool. Subsequent research has focused on refining the algorithm, exploring its variants, and expanding its applicability to a wide range of real-world problems.

The origins of Harris Hawks Optimization are deeply rooted in the study of animal behavior, particularly the cooperative hunting strategies of Harris hawks. The algorithm's initial formulation and theoretical foundations reflect a synthesis of biological inspiration and optimization principles, paving the way for its evolution and widespread adoption in the optimization community.

16.3 Innovations and Advances in HHO

Since its introduction, the Harris Hawks Optimization (HHO) algorithm has undergone numerous innovations and advancements that have enhanced its performance, versatility, and applicability in various fields. This section explores significant innovations in HHO, including algorithmic improvements, hybridization with other optimization techniques, and enhancements in parameter tuning strategies.

16.3.1 Enhanced Algorithmic Variants

Enhanced variants of HHO have been developed to improve its performance in solving complex optimization problems. Table 16.1 shows introductory summary of the enhanced algorithmic variants.

Hybrid HHO: One of the most notable innovations in HHO is the development of hybrid models, where HHO is combined with other optimization techniques to

Table 16.1 Enhanced algorithmic variants of Harris Hawks Optimization

Variant	Description	Advantages	References
Hybrid HHO	Combines HHO with other optimization techniques like Genetic Algorithms and Particle Swarm	Improved convergence speed and solution quality	Çetınba [10]
Modified HHO	Implements modifications to the original HHO to enhance exploration and exploitation	Increased stability and robustness in diverse landscapes	Song [11]
Multi-objective HHO	Adapts HHO to solve problems with multiple conflicting objectives using Pareto dominance	Finds a set of optimal trade-off solutions simultaneously	Yüzgeç [12]
Adaptive HHO	Adjusts parameters dynamically based on algorithm performance during the search process	Enhances flexibility and responsiveness to problem landscapes	Hussien et al. [13]
Reactive HHO	Modifies search strategies based on real-time feedback from the environment and solution space	Increases exploration efficiency in dynamic environments	Kumar et al. [14]
Chaotic HHO	Utilizes chaotic maps to enhance the diversity of the search process	Increases exploration capabilities	Gezici et al. [15]

leverage the strengths of both methods. For instance, hybrid approaches that integrate HHO with Genetic Algorithms, Particle Swarm Optimization, and Differential Evolution have been proposed. These hybrids often demonstrate improved convergence speeds and solution quality, particularly in complex and multi-modal optimization problems.

Modified HHO: Researchers have introduced modified versions of HHO to address specific shortcomings observed in the original algorithm. These modifications often focus on enhancing the balance between exploration and exploitation. For example, some variants incorporate adaptive mechanisms that adjust the exploration–exploitation trade-off dynamically based on the performance of the algorithm during the search process. Such modifications have led to improved stability and robustness in diverse problem landscapes.

Multi-objective HHO: The advancements in multi-objective optimization have introduced multi-objective variants of HHO, which are capable of solving problems with multiple conflicting objectives. These models utilize techniques like Pareto dominance and non-dominated sorting to generate a set of optimal trade-off solutions. Multi-objective HHO has been successfully applied in areas such as engineering design, where multiple criteria must be simultaneously optimized.

16.3.2 Innovations in Parameter Tuning

Effective parameter tuning is essential in optimizing the performance of HHO. Innovations in parameter tuning have significantly contributed to the algorithm's efficiency and effectiveness:

Dynamic Parameter Adaptation: Dynamic parameter adaptation involves adjusting the algorithm's parameters during the optimization process based on the current progress of the search. For example, adapting the hawks' exploration rates or the probability thresholds can lead to enhanced performance by allowing the algorithm to respond more flexibly to varying problem landscapes. This adaptability has been shown to improve convergence rates and solution quality.

Meta-optimization Techniques: Meta-optimization techniques, such as using other optimization algorithms for tuning HHO parameters, have also been explored. Techniques like Bayesian optimization and Particle Swarm Optimization have been employed to find optimal parameter settings for HHO, resulting in significant performance improvements. This approach helps mitigate the manual tuning effort often required in conventional optimization processes.

16.3.3 Performance Evaluation Frameworks

The establishment of comprehensive performance evaluation frameworks has been another significant advancement in the study of HHO. Researchers have developed standardized benchmark suites and performance metrics to assess the effectiveness of HHO and its variants systematically. Such frameworks facilitate fair comparisons between HHO and other optimization algorithms, providing insights into their respective strengths and weaknesses.

Table 16.2 indicates a summarizing various performance evaluation frameworks for HHO. The innovations and advances in HHO reflect the algorithm's adaptability and relevance in a rapidly evolving optimization landscape. From hybrid models to dynamic parameter tuning and applications in emerging fields, HHO has proven to be a versatile and powerful tool for solving complex optimization problems. As research continues to evolve, further innovations are anticipated, broadening the scope and enhancing the effectiveness of HHO in real-world applications.

16.4 Real-World Applications of HHO

The HHO algorithm has demonstrated its versatility and effectiveness across a wide range of real-world applications. Its ability to solve complex optimization problems has made it a valuable tool in various fields, including engineering, computer science, finance, and environmental management. This section highlights some of

Table 16.2 Performance evaluation frameworks for Harris Hawks Optimization

Framework	Description	Key features	References
Benchmark suites	Establishes a collection of standard optimization functions used to evaluate the performance of HHO	Includes diverse test functions (e.g., unimodal, multi-modal)	Dutta et al. [16]
Performance metrics	Defines criteria to assess algorithm performance such as convergence rate, accuracy, and robustness	Provides a standardized way to measure algorithm effectiveness	Al-Amodi, et al. [17]
Statistical analysis	Utilizes statistical methods to analyze the performance data collected from HHO experiments	Ensures reliability and validity of results through hypothesis testing	Alabool et al. [5]
Comparative studies	Establishes frameworks for comparing HHO with other optimization algorithms on various benchmarks	Highlights strengths and weaknesses relative to alternative methods	Shehab et al. [8]
Visualization tools	Implements graphical tools to visualize search behavior and convergence patterns of HHO	Helps in understanding the algorithm's effectiveness and dynamics	Kumar et al. [18]

the most significant real-world applications of HHO, showcasing its practical utility and impact. Figure 16.2 shows several real-world applications of HHO.

The versatility of HHO has been further demonstrated through its applications in emerging fields such as data science, bioinformatics, and artificial intelligence. Significant innovations include. Feature Selection in Machine Learning: HHO has been effectively used for feature selection tasks in machine learning, optimizing the selection of relevant features to improve model performance. Scheduling Problems: Innovations have emerged in applying HHO to scheduling and resource allocation problems, where efficient solutions can lead to significant operational improvements in industries such as manufacturing and logistics. Internet of Things (IoT) Optimization: With the rise of IoT, HHO has been adapted to optimize resource management and data processing in IoT environments, highlighting its relevance in contemporary applications.

Engineering Design: HHO has been extensively applied in engineering design optimization, where it is used to enhance the performance and efficiency of various systems. Applications include: *Structural Optimization:* HHO has been employed to optimize the design of structures, such as bridges and buildings, by minimizing weight while ensuring safety and stability [19]. The algorithm effectively navigates the design space to find optimal configurations that meet engineering constraints [20]. *Mechanical Component Design:* In mechanical engineering, HHO is used to optimize the design of components like gears and shafts, focusing on parameters such as material selection, dimensions, and load capacity to improve performance and reduce costs.

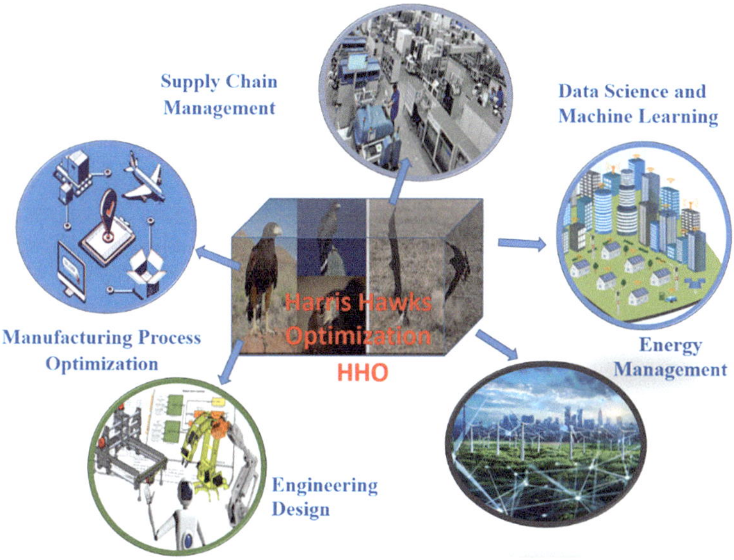

Fig. 16.2 Several real-world applications of HHO

Energy Management: The energy sector has also benefited from the application of HHO in optimizing various processes. Renewable Energy Systems: HHO has been utilized to optimize the design and operation of renewable energy systems, such as wind turbines and solar panels. The algorithm helps in maximizing energy output while minimizing costs and environmental impact [21]. Smart Grid Optimization: In smart grid management, HHO is applied to optimize energy distribution and load balancing, ensuring efficient energy use and reducing operational costs.

Data Science and Machine Learning: In the realm of data science, HHO has found applications in optimizing machine learning models and data processing tasks. Feature Selection: HHO is used for feature selection in machine learning, where it identifies the most relevant features from large datasets to improve model accuracy and reduce overfitting. This application is particularly valuable in fields such as bioinformatics and finance, where datasets can be vast and complex. Hyperparameter Tuning: The algorithm is also employed for hyperparameter tuning of machine learning models, optimizing parameters to enhance model performance and predictive accuracy.

Transportation and Logistics: HHO has been applied to optimize transportation and logistics operations, leading to improved efficiency and cost savings. Vehicle Routing Problems: HHO is used to solve vehicle routing problems, optimizing routes for delivery vehicles to minimize travel time and fuel consumption. This application is crucial for logistics companies seeking to enhance operational efficiency [22]. Supply Chain Optimization: In supply chain management, HHO helps optimize inventory levels, production schedules, and distribution strategies, ensuring that resources are utilized effectively and costs are minimized.

Environmental Management: The environmental sector has also seen the application of HHO in various optimization tasks. Water Resource Management: HHO is employed to optimize the allocation of water resources in agricultural and urban settings, ensuring sustainable use while meeting demand. Pollution Control: The algorithm is used to optimize strategies for pollution control and waste management, helping to minimize environmental impact and comply with regulatory standards.

From engineering design to energy management, data science, transportation, and environmental management, HHO has proven to be a valuable tool for researchers and practitioners alike. As the algorithm continues to evolve, its potential for further applications in emerging fields remains promising, paving the way for innovative solutions to contemporary optimization problems.

16.5 Future Directions and Emerging Trends

As the field of optimization continues to evolve, the Harris Hawks Optimization (HHO) algorithm is poised to adapt and expand its applications in response to emerging challenges and technological advancements. This section discusses potential future directions and trends that may shape the development and application of HHO in the coming years.

Integration with Artificial Intelligence: The integration of HHO with artificial intelligence (AI) techniques is a promising direction for future research. By combining HHO with machine learning and deep learning algorithms, researchers can enhance the algorithm's ability to learn from data and improve its optimization capabilities. This hybrid approach could lead to more efficient solutions in complex environments, particularly in fields such as autonomous systems, robotics, and smart cities.

Development of Adaptive Algorithms: Future advancements in HHO may focus on developing more adaptive algorithms that can dynamically adjust their parameters and strategies based on the characteristics of the optimization problem at hand. Such adaptive mechanisms could improve the algorithm's performance in real-time applications, allowing it to respond effectively to changing conditions and uncertainties in various domains, including finance, healthcare, and environmental management [23].

Multi-modal and Dynamic Optimization: As real-world problems often involve multi-modal and dynamic environments, future research could explore the application of HHO in these contexts. Enhancing HHO's capabilities to handle multi-objective optimization and dynamic changes in the search space will be crucial for its effectiveness in complex scenarios, such as resource management in fluctuating markets or adaptive control in engineering systems.

Enhanced Performance Evaluation Frameworks: The establishment of more comprehensive performance evaluation frameworks will be essential for assessing the effectiveness of HHO and its variants [24]. Future research could focus on developing standardized benchmarks, metrics, and methodologies for evaluating algorithm performance across diverse applications [25, 26]. This will facilitate fair comparisons with other optimization techniques and contribute to the continuous improvement of HHO.

Applications in Emerging Technologies: The rapid advancement of emerging technologies presents new opportunities for the application of HHO. Areas such as the Internet of Things (IoT), blockchain technology, and quantum computing are likely to benefit from HHO's optimization capabilities. For instance, HHO could be used to optimize resource allocation in IoT networks or enhance security protocols in blockchain systems.

Collaborative and Swarm Intelligence Approaches: Future directions may also explore the integration of HHO with other swarm intelligence algorithms, such as Ant Colony Optimization and Firefly Algorithm. Collaborative approaches that leverage the strengths of multiple algorithms could lead to improved optimization performance and robustness. This trend aligns with the growing interest in hybrid optimization techniques that combine various methodologies to tackle complex problems.

As the algorithm continues to evolve, its integration with artificial intelligence, development of adaptive mechanisms, and exploration of multi-modal and dynamic optimization will enhance its relevance in addressing contemporary challenges. By establishing robust performance evaluation frameworks and expanding its applications in emerging technologies, HHO is set to play a significant role in the optimization landscape of the future. Continued innovation and collaboration will be key to unlocking the full potential of HHO in solving complex real-world problems.

16.6 Conclusion

The Harris Hawks Optimization (HHO) algorithm has established itself as a powerful tool in optimization, showcasing significant advancements since its inception. This review has highlighted HHO's evolution, including its innovative variants and diverse real-world applications across engineering, energy management, data science, transportation, and environmental management. HHO's unique approach, inspired by the hunting behavior of Harris hawks, has led to enhancements such as hybrid models and dynamic parameter tuning, improving its adaptability and effectiveness. The establishment of robust performance evaluation frameworks has further clarified its strengths compared to other optimization techniques. Looking forward, HHO's integration with artificial intelligence, development of adaptive algorithms, and exploration of multi-modal optimization present exciting research opportunities. As HHO continues to evolve, it is poised to play a crucial role in addressing complex optimization challenges in various emerging technologies. In summary, HHO's versatility and potential for future advancements make it a valuable asset in the optimization landscape, paving the way for innovative solutions to contemporary problems.

References

1. Pan, J.-S., Wang, G.-L., Chu, S.-C., Yang, D., Snášel, V.: New feature attribution method for explainable aspect-based sentiment classification. Knowl-Based Syst. **304**, 112550 (2024). https://doi.org/10.1016/j.knosys.2024.112550
2. Qiao, Y., Dao, T.K., Pan, J.S., Chu, S.C., Nguyen, T.T.: Diversity teams in soccer league competition algorithm for wireless sensor network deployment problem. Symmetry. **12**, 445 (2020). https://doi.org/10.3390/sym12030445
3. Nguyen, T.-T., Pan, J.-S., Dao, T.-K.: An improved flower pollination algorithm for optimizing layouts of nodes in wireless sensor network. IEEE Access **7** (2019). https://doi.org/10.1109/ACCESS.2019.2921721
4. Heidari, A.A., Mirjalili, S., Faris, H., Aljarah, I., Mafarja, M., Chen, H.: Harris hawks optimization: algorithm and applications. Future Gener. Comput. Syst. **97**, 849–872 (2019). https://doi.org/10.1016/j.future.2019.02.028
5. Alabool, H.M., Alarabiat, D., Abualigah, L., Heidari, A.A.: Harris hawks optimization: a comprehensive review of recent variants and applications. Neural Comput. Appl. **33**, 8939–8980 (2021)
6. Hussien, A.G., Abualigah, L., Abu Zitar, R., Hashim, F.A., Amin, M., Saber, A., Almotairi, K.H., Gandomi, A.H.: Recent advances in harris hawks optimization: a comparative study and applications. Electronics **11**, 1919 (2022)
7. Elgamal, Z.M., Yasin, N.B.M., Tubishat, M., Alswaitti, M., Mirjalili, S.: An improved harris hawks optimization algorithm with simulated annealing for feature selection in the medical field. IEEE Access **8**, 186638–186652 (2020)
8. Shehab, M., Mashal, I., Momani, Z., Shambour, M.K.Y., AL-Badareen, A., Al-Dabet, S., Bataina, N., Alsoud, A.R., Abualigah, L.: Harris hawks optimization algorithm: variants and applications. Arch. Comput. Methods Eng. **29**, 5579–5603 (2022)
9. Qu, C., He, W., Peng, X., Peng, X.: Harris hawks optimization with information exchange. Appl. Math. Model. **84**, 52–75 (2020)
10. ÇetınbaŞ, İ, Tamyürek, B., Demırtaş, M.: The hybrid harris hawks optimizer-arithmetic optimization algorithm: a new hybrid algorithm for sizing optimization and design of microgrids. IEEE Access. **10**, 19254–19283 (2022)
11. Song, M., Jia, H., Abualigah, L., Liu, Q., Lin, Z., Wu, D., Altalhi, M.: Modified harris hawks optimization algorithm with exploration factor and random walk strategy. Comput. Intell. Neurosci. **2022**, 4673665 (2022)
12. Yüzgeç, U., Kusoglu, M.: Multi-objective harris hawks optimizer for multiobjective optimization problems. BSEU J. Eng. Res. Technol. **1**, 31–41 (2020)
13. Hussien, A.G., Amin, M.: A self-adaptive Harris Hawks optimization algorithm with opposition-based learning and chaotic local search strategy for global optimization and feature selection. Int. J. Mach. Learn. Cybern. **13**, 309–336 (2022)
14. Kumar, A., Dhillon, J.S.: Enhanced Harris hawk optimizer for hydrothermal generation scheduling with cascaded reservoirs. Expert Syst. Appl. **226**, 120270 (2023)
15. Gezici, H., Livatyalı, H.: Chaotic Harris hawks optimization algorithm. J. Comput. Des. Eng. **9**, 216–245 (2022)
16. Dutta, A., Jannesari, A.: HHOTuner: efficient performance tuning with harris hawks optimization. Learn. Based Auto-Tuning High Perform. Comput. **137** (2024)
17. Al-Amodi, S., Barik, L., Patra, S.S., Dash, B.B., Mishra, M.R., Pattnayak, P.S.: HHO algorithm for reliable QoS and energy-efficient task scheduling in fog assisted cloud environment. NeuroQuantology **20**, 3235–3248 (2022)
18. Kumar, T.S., Jothilakshmi, S., James, B.C., Prakash, M., Arulkumar, N., Rekha, C.: HHO-based vector quantization technique for biomedical image compression in cloud computing. Int. J. Image Graph. **23**, 2240008 (2023)
19. Dao, T.-K., Nguyen, T.-T.: A review of the bat algorithm and its varieties for industrial applications. J. Intell. Manuf. 1–23 (2024)

20. Nguyen, T.-T., Dao, T.-K., Kao, H.-Y., Horng, M.-F., Shieh, C.-S.: Hybrid particle swarm optimization with artificial bee colony optimization for topology control scheme in wireless sensor networks. J. Internet Technol. **18**, 743–752. https://doi.org/10.6138/JIT.2017.18.4.20150119
21. Pan, J.-S., Zhang, Z., Chu, S.-C., Lee, Z.-J., Li, W.: Application of diversity-maintaining adaptive Rafflesia optimization algorithm to engineering optimisation problems. Symmetry **15**, 2077 (2023)
22. Nguyen, T.T., Pan, J.S., Dao, T.K.: A novel improved bat algorithm based on hybrid parallel and compact for balancing an energy consumption problem. Information (Switzerland) **10** (2019). https://doi.org/10.3390/info10060194.
23. Pan, J.-S., Nguyen, T.-T., Chu, S.-C., Dao, T.-K., Ngo, T.-G.: Network, diversity enhanced ion motion optimization for localization in wireless sensor. J. Inf. Hiding Multimed. Signal Process. **10**, 221–229 (2019)
24. Chu, S.C., Dao, T.K., Pan, J.S., Nguyen, T.T.: Identifying correctness data scheme for aggregating data in cluster heads of wireless sensor network based on naive Bayes classification. Eurasip J. Wirel. Commun. Netw. (2020). https://doi.org/10.1186/s13638-020-01671-y
25. Chen, C.-M., Lv, S., Ning, J., Wu, J.M.-T.: A genetic algorithm for the waitable time-varying multi-depot green vehicle routing problem. Symmetry **15**, 124 (2023)
26. Shaik, A.L., Manoharan, M.K., Pani, A.K., Avala, R.R., Chen, C.-M.: Gaussian Mutation–Spider Monkey Optimization (GM-SMO) Model for Remote Sensing Scene Classification (2022). https://doi.org/10.3390/rs14246279

Chapter 17
A Binary Particle Swarm Optimization Based on Manta Rays Foraging Algorithm for Network Intrusion Detection

Jianhua Liu, Shanglong Li, Yuxiang Chen, and Ge Xu

Abstract Aiming at the problem of network intrusion feature optimization, a binary particle swarm optimization algorithm based on manta rays foraging algorithm is proposed for feature selection of network intrusion detection, which improves the network intrusion detection rate. Firstly, the network intrusion detection rate is used as the objective function of feature selection, and the network state feature is used as the constraint condition to establish the corresponding mathematical model. Then, the binary particle swarm optimization algorithm of manta rays foraging learning strategy is used to find the optimal feature subset. Finally, the machine learning classifier is used to establish the intrusion detection model. The experimental results show that the model can efficiently query the optimal feature subset, and the intrusion detection rate and efficiency are better than the classical intrusion detection model.

17.1 Introduction

Intrusion Detection (ID) is a defense technology that can flexibly respond to network attacks [1]. It can dynamically search and judge intrusions and attacks that occur in the network environment, and timely select response strategies to respond. Intrusion detection is a classification problem of machine learning [2–4], including data acquisition, feature extraction, classification algorithm recognition, and other processes. The feature selection of intrusion detection is to eliminate many unnecessary and

J. Liu (✉) · S. Li · Y. Chen
Department of Computer Science and Mathematics, Fujian University of Technology, Fuzhou, China
e-mail: jhliu@fjnu.edu.cn

J. Liu · G. Xu
Fujian Provincial Key Laboratory of Information Processing and Intelligent Control, Minjiang University, Fuzhou, China

S.-C. Chu et al. (eds.), *Advances in Intelligent Data Analysis and Applications*,
Smart Innovation, Systems and Technologies 445,
https://doi.org/10.1007/978-981-96-7277-6_17

redundant features in the dataset, which is hardly helpful or even may have a negative impact on the detection [5]. So, the appropriate introduction of a feature selection method is a feasible scheme, whose quality directly affects the generalization ability and learning performance of the classifier in machine learning [6].

Feature selection is an NP-hard combinatorial optimization problem, and swarm intelligence optimization algorithms are commonly applied to the feature selection problem in network intrusion detection [7, 8]. Kumar et al. [9] proposed an algorithm based on a spanning tree and pattern matching to enhance the data filtering performance of an intrusion detection system. Vajiheh Hajisalem et al. [10] introduced a hybrid classification approach combining the artificial bee colony algorithm (ABC) with the artificial fish swarm algorithm (AFS), which divides the training dataset and removes irrelevant features. Selvakumar [11] developed a feature selection technique utilizing the firefly algorithm and applied it to network intrusion detection, ultimately achieving better experimental results. Mojtahedi [12] introduced a feature selection method that combines the genetic algorithm with the whale optimization algorithm, effectively extracting features relevant to class labels and detecting misconduct in intrusion detection datasets for wireless networks. Mohammed Abdullahi et al. [13] introduced an enhanced binary manta ray foraging optimization algorithm, combined with feature selection and random forest classification, for network intrusion detection.

In recent years, the binary intelligent algorithm is widely used in feature selection because of its search mechanism suitable for combinatorial optimization problems and high probability of generating high-performance feature subsets [14–16]. To enhance the efficiency and accuracy of intrusion detection systems in specific scenarios, feature selection methods have been improved from various perspectives. However, as research into intrusion behavior deepens and the demands for advanced detection technologies continue to grow, further improvements in the detection efficiency and classification performance of intrusion detection systems remain necessary. Identifying more efficient feature selection methods is still of significant importance. In response to this need, this paper proposes a binary particle swarm optimization (BPSO) algorithm integrated with the learning strategies of the manta ray foraging algorithm (MRFO). The proposed algorithm incorporates three distinct learning strategies of MRFO, known for its strong global search capabilities [17]: chain learning, spiral learning, and tumbling learning. These strategies are embedded into the BPSO through a novel transfer function, resulting in a hybrid algorithm that enables a more thorough exploration of the solution space and exhibits enhanced global search capability. An efficient feature selection method is constructed and applied to network intrusion detection.

17.2 Binary Particle Swarm Optimization

17.2.1 Principle of Binary Particle Swarm Optimization Algorithm

Binary Particle swarm optimization algorithm (BPSO) keeps the velocity updating by extending Particle Swarm Optimization (PSO), where each particle represents a solution of the unknown space, has its own location and moving velocity. The particle updates its velocity by the method of Eq. (17.1).

$$v_i^d(t+1) = \omega v_i^d(t) + c_1 r_1 \left(p\text{best}_i^d - x_i^d(t)\right) + c_2 r_2 \left(g\text{best} - x_i^d(t)\right) \tag{17.1}$$

In Eq. (17.1), $v_i^d(t)$ denotes the velocity of the tth dimension of the ith particle at the tth itherion; ω represents the inertia weight; r_1 and r_2 are random numbers between [0, 1]; c_1 and c_2 both represent learning factors; $p\text{best}_i^d$ represents the individual optimal solution of the current unknown space; $g$best represents the group optimal solution of the current unknown space.

The binary particle swarm optimization algorithm (BPSO) is a discrete algorithm [18], and its position becomes binary code 1 or 0. In the binary particle swarm optimization algorithm, an S-type transfer function is introduced to convert the original velocity values into 0 and 1, which the transfer function is shown in Eq. (17.2).

$$\mathrm{S}\left(v_i^d(t)\right) = \frac{1}{1 + e^{-v_i^d(t)}} \tag{17.2}$$

The transfer function of Eq. (17.2), $\mathrm{S}(v_i^d(\mathrm{t}))$, represents the Sigmoid function, which generates a value in the range [0, 1], and then updates the position of the particle according to the value, as shown in Eq. (17.3).

$$x_i^d(t) = \begin{cases} 1 \text{ if rand } < \mathrm{S}\left(v_i^d(t)\right) \\ 0 \text{ if rand } \geq \mathrm{S}\left(v_i^d(t)\right) \text{ otherwise} \end{cases} \tag{17.3}$$

where rand represents a random number in the range [0, 1], which $x_i^d(t)$ is the position of the current particle.

17.2.2 Manta Ray Foraging Optimization Algorithm

The Manta Ray Foraging Optimization algorithm (MRFO) includes three foraging behaviors, namely chain foraging, spiral foraging, and tumbling foraging. The search phase is described as follows: (1) **Chain foraging stage**. The head to tail of the manta ray is arranged into a foraging chain. At each iteration, every individual, except the first one, is updated based on the best position found so far and the positions of the particles ahead of it. (2) **Spiral foraging stage**. This stage consists of two parts. In the first part, in addition to the first individual, the remaining individuals will move toward the best position found so far, as well as the individuals ahead of them. In the second part, in addition to the first individual, remaining individuals will move toward the individual ahead of them and a randomly generated position within the search space. Throughout the spiral foraging phase, the manta ray will change its probability of entering the first and second parts as the iteration progresses, dynamically adjust the search process, and spirally approach the optimal position. (3) **Rolling foraging stage**. Rolling foraging is a regular, local, random cycle of movement. At this stage, all manta rays will use the best position discovered so far as a reference point, move around this point and roll to a new position.

17.3 Binary Particle Swarm Optimization Algorithm Based Manta Ray Foraging Algorithm

17.3.1 Chain Learning

The particles work like the MRFO algorithm manta rays, which are arranged in a head-to-tail structure, forming a cohesive foraging chain. In addition to the first particle, each particle not only points to the best performing particle in the group, but also points to the particle directly in front of it. According to MRFO algorithm, the velocity update formula of BPSO is modified to Eq. (17.4).

$$v_i^{t+1} = \begin{cases} wv_i^t + c_1r_1(pbest_i - x_i^t) + c_2r_2(gbest - x_i^t) & i = 1 \\ wv_i^t + c_1r_1(pbest_{i-1} - x_i^t) + c_2r_2(gbest - x_i^t) & i = 2, \ldots, N \end{cases} \tag{17.4}$$

where, v_i^t and x_i^t is the velocity and position of the i-th particle in the t-th iteration; the $p$best$_i$ and $g$best are the best position in the individual history and the global best position of the i-th particle, respectively. r_1 and r_2 are two different random numbers uniformly distributed over the interval [0, 1]; c_1 and c_2 is a learning factor, ω is a linear decreasing inertia weight.

17.3.2 Spiral Learning

Spiral learning includes two parts. In the first part, except for the first particle, all particles not only learn from the particles in front of it, but also learn from the best particles. According to MRFO algorithm, the velocity of the new BPSO updates with Eq. (17.5).

$$v_i^{t+1} = \begin{cases} wv_i^t + c_3 r_1(\text{pbest}_i - x_i^t) + c_4 r_2(g\text{best} - x_i^t) & i = 1 \\ wv_i^t + c_3 r_1(\text{pbest}_{i-1} - x_i^t) + c_4 r_2(g\text{best} - x_i^t) & i = 2, \ldots, N \end{cases} \tag{17.5}$$

Because MRFO controls the learning rate of individuals at this stage through the parameter β in the spiral foraging stage. In order to make the particles in the BPSO learn in a similar way, it needs to adjusts the learning factor c_3 and c_4.

From MRFO, a new random position is allocated in the search space as the reference for each individual, which encourages individuals to explore new positions away from the current optimal solution, thereby enhancing the global exploration capability of the algorithm. Therefore, in the second part, the velocity update equation of BPSO is modified as the Eqs. (17.6) and (17.7).

$$x_r = Lw + \text{rand} \cdot (Up - Lw) \tag{17.6}$$

$$v_i^{t+1} = \begin{cases} wv_i^t + c_5 r_1(\text{pbest}_i - x_i^t) + c_6 r_2(x_r - x_i^t) & i = 1 \\ wv_i^t + c_5 r_1(\text{pbest}_{i-1} - x_i^t) + c_6 r_2(x_r - x_i^t) & i = 2, \ldots, N \end{cases} \tag{17.7}$$

Here, x_r denotes a random position generated within the search space, while Lw and Up represent the lower and upper bounds of the search range, respectively; the values of c_5 and c_6 are constants, both set to 2 in this section; additionally, rand refers to a random number uniformly distributed in the interval [0, 1].

17.3.3 Rollover Learning

In the tumbling learning phase, the global best position is used as a pivot, and the particles roll back and forth near the pivot and their individual best positions. Based on the MRFO, the speed update equation of BPSO is designed as Eq. (17.8).

$$x_i^{\text{new}} = \text{pbest}_i^t + 2 \cdot \left(r_1 \cdot g\text{best} - r_2 \cdot p\text{best}_i^t\right) \tag{17.8}$$

Among them, x_i^{new} represents the new position of the i-th particle generated through tumbling learning. Equation (17.8) allows each particle to move to a position between its individual best position and the global best position, facilitating improved exploration and exploitation within this region. As the iteration progresses, the individual optimal positions gradually move closer to the global optimum, and eventually, the particle swarm converges slowly.

Since the particles in this stage only involve changes between positions, a threshold is used to convert the updated position to the binary search space at this stage. The threshold is set to 0.5, as shown in Eq. (17.9).

$$x_i^{t+1} = \begin{cases} 1 \text{ if } x_i^{\text{new}} > 0.5 \\ 0 \text{ if } x_i^{\text{new}} < 0.5 \text{ otherwise} \end{cases} \tag{17.9}$$

Table 17.1 KDD CUP 99 experimental dataset composition

Data type	Training dataset	Testing dataset
Normal	4864	1216
Dos	3914	978
Probe	410	102
R2L	112	28
U2R	52	13
Total	9352	2337

17.4 Network Intrusion Detection Experiment

17.4.1 Experimental Dataset Description

In order to verify the effectiveness of the algorithm that integrates the learning strategy of the manta rays foraging algorithm, the network intrusion detection problem is used to conduct an experiment. The KDD CUP 99 dataset is utilized to test the performance of the algorithm, which is extracted from the real network environment, and the data types include different types of network attack behaviors and the data is uniformly displayed in the list, which is more convenient for result statistics and analysis. The dataset on the official website is given in the form of text, where each line of text represents a data transmission record.

In the dataset, the record of each network connection contains one attack tag and 41 features, and the test set only does not cover the attack tag. There are 23 attack tags, which are divided into two types: legal operation and illegal operation. Only one of the legal operation tags is expressed as normal. The 41 network features are categorized into three types: basic features, content features, and traffic features. In this section, data is randomly sampled from the dataset. The KDD CUP 99 dataset includes 9,352 training samples and 2,337 testing samples. The distribution of the experimental data is presented in Table 17.1.

17.4.2 Data Pre-processing

For the KDD CUP 99 dataset, each attack record contains three string features, so they must be converted into numerical variables. The transformation method can be described by binary values, and the corresponding character features can be represented by integer numbers. For example, in the service type, HTTP can be expressed as 0, FTP can be expressed as 1, SMTP can be expressed as 2, and Telnet can be expressed as 3. In the same way, the label can also be quantified, such as normal can be expressed as 0, dos can be expressed as 1, etc. The above process is a numerical operation.

After numerical processing, the numerical coverage is large, which affects the performance of the algorithm, and the small numerical characteristics are easy to miss. Therefore, numerical variables must be normalized, which are mapped to the range of [0, 1]. The specific mapping rules are based on Eq. (17.10).

$$x\prime = \frac{x - x_{\min}}{x_{\max} - x_{\min}} \tag{17.10}$$

In Eq. (17.10), x represents the original value, $x_{\min}$ and $x_{\max}$ represents the minimum and maximum values in the numerical sample, respectively, which $x\prime$ represents the values after processing.

17.4.3 Parameter Setting

For the parameter settings: the particle population size is set to 20, the maximum number of iterations is set to 100, the algorithm is run independently 20 times, and the average result is taken as the final optimization outcome.

17.4.4 Experimental Result Analysis

To demonstrate the effectiveness of the BPSO-MRFL algorithm proposed in this paper for intrusion detection, this section compares and analyzes five swarm intelligence algorithms across five metrics: fitness value, classification accuracy, number of selected features, precision, and recall. The formulas for calculating precision and recall are provided in Eqs. (17.11) and (17.12).

$$\text{accuracy} = \frac{T_{\text{positive}}}{T_{\text{positive}} + F_{\text{positive}}} \tag{17.11}$$

$$\text{recall} = \frac{T_{\text{positive}}}{T_{\text{positive}} + F_{\text{Negative}}} \tag{17.12}$$

In the Eqs. (17.11) and (17.12), T_{positive} represents the number of correctly detected intrusion samples, F_{positive} denotes the number of normal samples misclassified as intrusion samples, while F_{Negative} represents the number of intrusion samples not detected.

17.4.5 Experimental Results of KUD CUP 99 Dataset

Table 17.2 presents the experimental results of BPSO-MRFL in comparison to GA, BCS, BFA, and BALO across five metrics on the KDD CUP 99 dataset. As shown in Table 17.2, the BPSO-MRFL algorithm achieves a classification accuracy of 99.8% on this dataset, with an average of 7.05 selected features, outperforming the other four algorithms across all five indicators. This demonstrates the superior performance of the BPSO-MRFL algorithm in solving network intrusion detection problems.

To better illustrate the optimization performance of these five algorithms, Fig. 17.1 displays the iterative curves of classification accuracy for each algorithm on the KDD CUP 99 dataset. From Fig. 17.1, it is evident that while the BPSO-MRFL algorithm has a slower convergence rate, it achieves higher accuracy at convergence compared to the other algorithms, suggesting that it has advantages in avoiding local optima. Figure 17.2 shows the box plot of classification accuracy for the five algorithms on the dataset, where it can be observed that BPSO-MRFL exhibits greater classification stability than the other algorithms.

Table 17.2 The experimental results of KDD CUP 99 dataset on different algorithms

Algorithms	Fitness	Nicety of grading	Select the number of features	Accuracy	Recall
GA	0.0430	0.9583	7.24	0.9590	0.9575
BCS	0.0342	0.9682	11.34	0.9692	0.9677
BFA	0.0248	0.9780	12.32	0.9791	0.9776
BALO	0.0053	0.9977	12.41	**0.9990**	**0.9972**
BPSO-MRFL	**0.0037**	**0.9980**	**7.05**	**0.9990**	**0.9972**

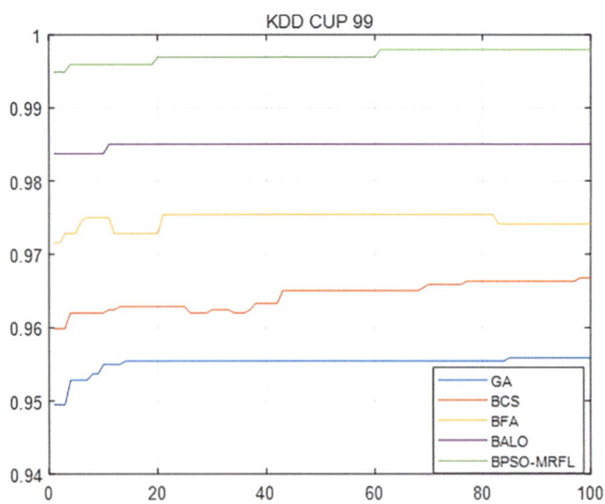

Fig. 17.1 Iteration curves of different algorithms on KDD CUP 99 dataset

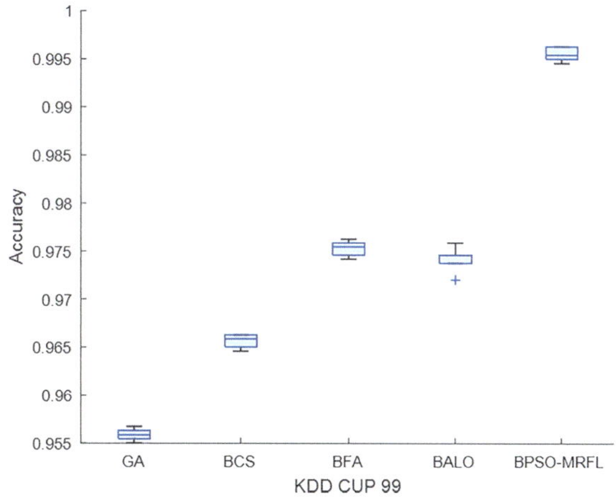

Fig. 17.2 Boxplots of different algorithms on KDD CUP 99 dataset

17.5 Conclusion

In summary, the BPSO-MRFL algorithm proposed in this paper has greater advantages in solving network intrusion detection problems, which significantly improves the classification accuracy, reduces the number of selected features, and has good stability.

Acknowledgements This paper is funded by the Open Fund Project of Fujian Provincial Key Laboratory of Information Processing and Intelligent Control (Minjiang University) (No. MJUKF-IPIC202402).

References

1. Liao, H.J., Lin, C.H.R., Lin, Y.C., et al.: Intrusion detection system: a comprehensive review. J. Netw. Comput. Appl. **36**(1), 16–24 (2013)
2. Mishra, P., Varadharajan, V., Tupakula, U., et al.: A detailed investigation and analysis of using machine learning techniques for intrusion detection. IEEE Commun. Surv. Tut. **21**(1), 686–728 (2018)
3. Wensheng, G., Lili, C., Shicheng, W., et al.: Anomaly rule detection in sequence data. IEEE T Knowl. Data EN **35**(12), 12095–12108 (2023)
4. Tung-Tso, T., Han-Yu, L., Wei-Ning, H., et al.: Anomaly detection through outsourced revocable identity-based signcryption with equality test for sensitive data in consumer IoT environments. IEEE T Consum. Electr. (2024)
5. Turukmane, A.V., Ramkumar, D.: M-MultiSVM: an efficient feature selection assisted network intrusion detection system using machine learning. Comput. Secur. **137**, 103587 (2024)
6. Fei, Y., Xiaodong, W., Longfu, H.: Enhancing clustering performance: a fuzzy subspace clustering method with local correlation and sparse feature selection. J. Netw. Intell. **9**(1), 427–442 (2024)
7. Huan-Huan, C., Ya-Juan, L., Cheng, P., et al.: Employment prediction models based on weighted feature selection and semi-supervised machine learning. J. Netw. Intell. **9**(2), 971–987 (2024)

8. Theng, D., Kishor, K.B.: Feature selection techniques for machine learning: a survey of more than two decades of research. Knowl. Inf. Syst. **66**(3), 1575–1637 (2024)
9. Kumar, G.R., Mangathayaru, N., Narsimha, G.: An Approach for intrusion detection using novel Gaussian based Kernel function. J. Univers. Comput. Sci. **22**, 589–604 (2016)
10. Vajiheh, H., Shahram, B.: A hybrid intrusion detection system based on ABC-AFS algorithm for misuse and anomaly detection. Comput. Netw. **136**, 37–50 (2018)
11. Selvakumar, B., Muneeswaran, K.: Firefly algorithm based feature selection for network intrusion detection. Comput. Secur. **81**, 148–155 (2019)
12. Mojtahedi, A., Sorouri, F., Souha, A.N., et al.: Feature selection-based intrusion detection system using genetic whale optimization algorithm and sample-based classification (2022)
13. Ibrahim, H.H., Mohammed, A., Mansur, M.A., et al.: An improved binary manta ray foraging optimization algorithm based feature selection and random forest classifier for network intrusion detection. Intell. Syst. Appl. **16**, 200114 (2022)
14. Priya, S., Pradeep, M.K.: Binary bat algorithm based feature selection with deep reinforcement learning technique for intrusion detection system. Soft. Comput. **27**, 10777–10788 (2023)
15. Alawad, N.A., Abed-Alguni, B.H., AI-Betar, M.A., et al.: Binary improved white shark algorithm for intrusion detection systems. Neural Comput. Appl. **35**, 19427–19451 (2023)
16. Saheed, Y.K.: A binary firefly algorithm based feature selection method on high dimensional intrusion detection data. In: Illumination of Artificial Intelligence in Cybersecurity and Forensics, pp. 273–288. Springer International Publishing, Cham (2022)
17. Gharehchopogh, F.S., Ghafouri, S., Namazi, M., et al.: Advances in manta ray foraging optimization: a comprehensive survey. J. Bionic Eng. **21**, 953–990 (2024)
18. Nezamabadi-pour, H., Rostami-Shahrbaba, M., Farsangi, M.: Binary particle swarm optimization: challenges and new solutions. CSI J. Comput. Sci. Eng. **6**, 21–32 (2008)

Part III
Artificial Intelligence

Chapter 18
Generative Adversarial Networks Based on Dual-Domain Fusion for Image Steganography Cost Learning

Tao Zhang, Li Liu, Anhong Wang, Yingchun Wu, and Zhifeng Xing

Abstract Generative adversarial networks have demonstrated significant potential in automatically learning image embedding costs. However, the existing U-Net-based generator structure uses the same convolution repeatedly when extracting image features, resulting in a lack of diversity of extracted features and serious information loss with the increase in network depth. To address this issue, this paper introduces a generative adversarial network model based on dual-domain fusion. This model introduces a wavelet convolution branch in the generator to capture various frequency features and spatial information, thereby enriching the amount of information that the model can obtain. Meanwhile, we have also designed a multi-level feature fusion module that efficiently integrates the features obtained from the encoding end, the wavelet convolution branch, and the upsampling decoding end. Through multi-level feature extraction and fusion, the ability of the network to capture image details is greatly improved, and the reconstruction ability of the decoding end is enhanced, optimizing the performance of the generator. We confirmed the effectiveness of the proposed method in resisting steganalyzers through multiple experiments.

18.1 Introduction

Image steganography [1] is an algorithm that embeds secret information into images, with the primary requirement that the visual quality remains unchanged before and after embedding. This technology finds extensive application in fields such as privacy and copyright protection. In our increasingly digital society, the risks of personal information and state secrets being compromised have heightened the importance of researching image steganography. A key direction in image steganography research is the distortion minimization framework. The goal of this research direction is to

T. Zhang · L. Liu (✉) · A. Wang · Y. Wu · Z. Xing
Taiyuan University of Science and Technology, Taiyuan, China
e-mail: skies5315@sina.com

S.-C. Chu et al. (eds.), *Advances in Intelligent Data Analysis and Applications*,
Smart Innovation, Systems and Technologies 445,
https://doi.org/10.1007/978-981-96-7277-6_18

design appropriate distortion functions to guide the image embedding process, in order to minimize the total embedding cost.

With the rapid development of deep learning [2], neural networks [3] have emerged as effective tools for image embedding cost learning. In 2017, the ASDL-GAN [4] algorithm was the first to introduce Generative adversarial networks [5–7] into image steganography cost learning, generating an embedding probability map by automatically learning regions suitable for embedding information in images. The embedding probabilities are subsequently converted into embedding costs, guiding the generation of stego images, which are then fed alongside the cover images into the steganalyzer for adversarial training. While ASDL-GAN exhibits longer training times and inferior performance compared to some traditional algorithms, it lays the foundation for further exploration in deep learning within this field. Then UT-GAN [8] is proposed, which uses a U-Net [9] structure in the generator and improves the discriminator part. At the same time, a differentiable double-Tanh activator is introduced in the embedded simulator section. Compared to ADSL-GAN, UT-GAN has shown significant performance improvements and has reduced training time. Huang et al. [10] proposed Steg-GMAN, which employs dual steganalyzers structure in the discriminator and introduces a new training strategy to update the parameters of both the generator and the discriminator, further enhancing its performance over UT-GAN. We believe that to effectively train Generative Adversarial Networks, the performance of the generator and discriminator should remain balanced. Based on this concept, we believe that it is essential to further enhance the generator's performance.

We propose a generative adversarial network based on dual-domain fusion. We argue that the U-Net structure in existing algorithms uses only the same convolution operation at the encoding end, which limits its ability to extract diverse features. In addition, increasing network depth may lead to information loss, adversely affecting the model's overall performance. In response to this problem, we introduce a wavelet convolution branch within the generator, which more effectively extracts multi-frequency features. To optimize the feature fusion process, we introduce a multi-level feature fusion module that integrates features from the encoding end, the wavelet convolution branch, and the decoding end after upsampling, thus improving the feature reconstruction capability at the decoding end. We conducted various experiments to confirm the efficacy of the proposed approach. The specific contributions of this paper are detailed as follows:

1. By introducing a branch that utilizes wavelet convolution, we aim to extract multiple frequency features from images, thereby enhancing the network's capacity to manage complex image characteristics. Furthermore, wavelet convolution facilitates a larger receptive field, enabling the network to capture contextual information more effectively.
2. At the decoding end, we introduce a multi-level feature fusion module designed to extract a diverse range of features through multi-level convolution and subsequently fuse these features. This method improves the network's capacity to extract detailed and global features, strengthening its reconstruction capability.

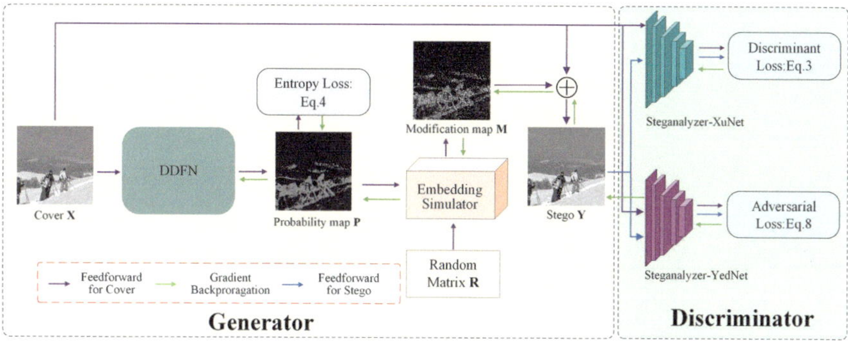

Fig. 18.1 The general framework of the paper

3. We conducted various experiments to confirm the efficacy of the proposed approach.

18.2 Our Method

18.2.1 Overall Structure

We propose a generative adversarial network based on dual-domain fusion. The overall framework of this paper is illustrated in Fig. 18.1 and consists of three main components: the generator, the embedding simulator, and the discriminator. In this paper, we refer to the improved generator as a dual-domain fusion network (DDFN). This network enhances its ability to capture edge and texture information by introducing an additional branch that utilizes wavelet convolution [11] to extract features at various frequencies and spatial scales. In addition, we fuse the features extracted by wavelet convolution with the features from the encoding end and those obtained from upsampling at the decoding end. In order to integrate these three features more effectively, a multi-level feature fusion module is designed in this paper. The introduction of this module improves the accuracy of the reconstruction features of the decoding end, enables the generator to obtain more accurate probability maps, and improves the ability of the stego image generated according to the probability maps to resist steganalyzers.

18.2.2 Wavelet Convolution Branch

The dual-domain fusion network is shown in Fig. 18.2. The proposed algorithm introduces an additional wavelet convolution branch that performs convolution operations from various perspectives, extracting information across different frequencies

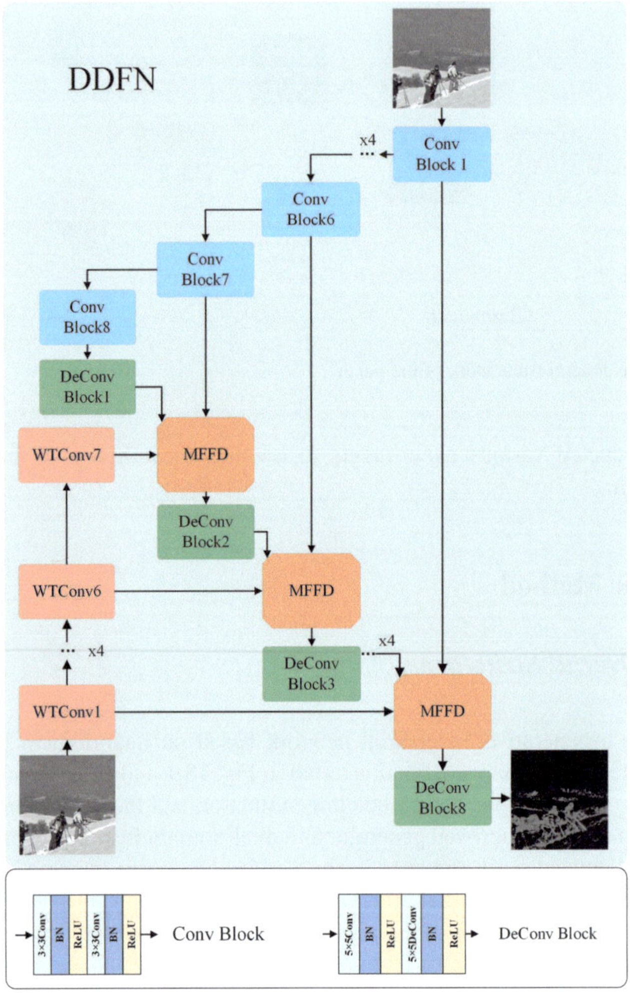

Fig. 18.2 Structure of dual-domain fusion network

and spatial dimensions. This approach enriches the edge and texture information in the network.

The implementation of wavelet convolution is initially grounded in the Haar wavelet transform, which involves applying a low-pass filter along with three distinct high-pass filter convolution kernels. Convolution is performed with a stride of 2 using these kernels, enabling the extraction of four different frequency feature maps. These feature maps are then concatenated along the channel dimension, after which a 1×1 convolution kernel is employed to adjust the number of feature channels. The advantages of wavelet convolution include its ability to capture a broader frequency spectrum, achieve a larger receptive field, and extract more detailed information at various levels. Moreover, wavelet convolution does not introduce a significant number of additional parameters.

18.2.3 Multi-level Feature Fusion Module

To effectively integrate the features obtained from the encoding end, wavelet convolution branch, and upsampling at the decoding end, this paper proposes a multi-level feature fusion module. As illustrated in Fig. 18.3, this module first concatenates these three features along the channels, resulting in the feature representation R. Then, we process R through a 1 × 1 convolution layer followed by LeakyReLU, resulting in the feature representation R1. Next, R1 is divided into two processing paths, which can be described as follows:

$$R_2 = \text{LeakyReLU}(\text{conv}_1(R_1)) \tag{18.1}$$

$$R_{2\_1} = \text{LeakyReLU}(\text{conv}_3(R_1) + R_1) \tag{18.2}$$

conv_1 represents a 1 × 1 convolution, conv_3 represents a 3 × 3 convolution. Subsequently, we obtain R3, R3_1, R4, and R4_1 by repeating the aforementioned operations twice. The feature R4_1 is processed through a 3 × 3 convolution layer followed by a LeakyReLU to obtain R5. Finally, we merge the channels of R2, R3, R4, and R5 to obtain R6, adjust the number of channels of R6 using a 1 × 1 convolution, add it to the input feature R, and then apply LeakyReLU to obtain the final result.

In the multi-level feature fusion module, a series of convolutional layers are employed to extract features at various levels, providing a rich informational foundation for subsequent feature fusion and enhancing the model's capability to represent complex features. Simultaneously, multiple residual connections facilitate the transmission of input features to deeper network layers, effectively mitigating the

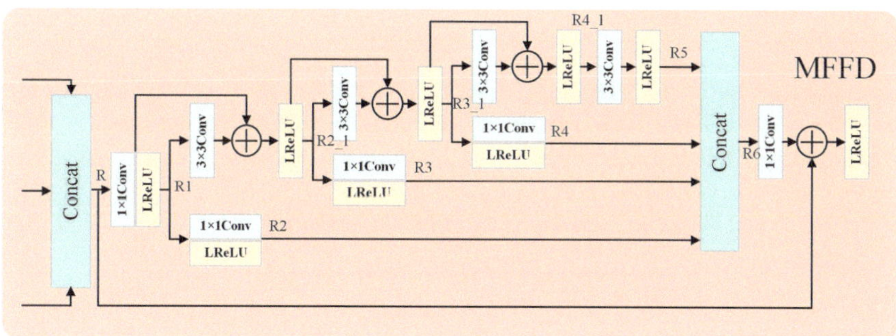

Fig. 18.3 Structure of multi-level feature fusion module

vanishing gradient problem commonly associated with deep networks. The multi-level feature fusion module efficiently extracts and fuses features from the encoding end, the wavelet convolution branch, and the decoding end, thereby enhancing the reconstruction capability of features at the decoding stage. This improvement increases the accuracy of the generated probability maps, subsequently strengthening the resilience of the generated stego images resisting steganalyzers.

18.2.4 Embedding Simulator

The embedding simulator is designed to simulate the information embedding process by utilizing a random matrix of the same size as the cover image to represent the secret message. Due to the non-differentiability of STC [12] encoding, we employ the hyperbolic tangent function to approximate this encoding, resulting in modified probability maps that are subsequently added to the cover image to generate the stego image.

To evaluate the model's performance, we convert the embedding probabilities into embedding costs and then apply STC encoding for secret embedding, which generates the final stego image.

18.2.5 Discriminator

The discriminator part of this paper employs the dual steganalyzers proposed by Huang, namely Xu-Net [13] and Yedroudj-Net [14]. Due to the differing performance of the two steganalyzers, the weaker one is used to update the discriminator's parameters, while the stronger one is used to update the generator's parameters.

18.2.6 Loss Function

This paper primarily includes two types of loss functions: discriminator loss and generator loss, which will be explained in detail in the following sections.

For the discriminator, we utilize a cross-entropy loss function, defined as follows:

$$L_D = \max_{i \in \{1,2\}} \left(-z_0 \log D_i(X) - z_1 \log D_i(Y)\right) \tag{18.3}$$

where $D_i(X)$ and $D_i(Y)$ represent the results obtained from the cover image and stego image after passing through the ith steganalyzer model, while z_0 and z_1 denote the correct labels for the cover image and stego image, respectively.

The generator loss consists of two components, the first component constrains the probability map for generating a specific embedding payload, defined as follows:

$$L_G^1 = (c - H \times W \times q)^2 \tag{18.4}$$

where c represents the actual embedding capacity, H and W are the height and width of the image, and q is the given embedding rate. To ensure the payload, the actual embedding capacity c is calculated based on the embedding probability, as follows:

$$c = \sum_{i=1}^{H}\sum_{j=1}^{W}\left(-p_{i,j}^{+1}\log_2 p_{i,j}^{+1} - p_{i,j}^{-1}\log_2 p_{i,j}^{-1} - p_{i,j}^{0}\log_2 p_{i,j}^{0}\right) \tag{18.5}$$

$$p_{i,j}^{+1} = p_{i,j}^{-1} = \frac{p_{i,j}}{2} \tag{18.6}$$

$$p_{i,j}^{+1} + p_{i,j}^{-1} + p_{i,j}^{0} = 1 \tag{18.7}$$

where $p_{i,j}$ represents the embedding probability of pixel $x_{i,j}$, $p_{i,j}^{+1}$ and $p_{i,j}^{-1}$ are the probabilities of the pixel value being modified by $+1$ and -1, respectively, while $p_{i,j}^{0}$ denotes the probability that the pixel value remains unchanged.

The second part is an adversarial loss, defined as follows:

$$L_G^2 = -\min_{i\in\{1,2\}}(-z_0 \log D_i(X) - z_1 \log D_i(Y)) \tag{18.8}$$

In summary, the total loss of the generator can be expressed as

$$L_G = \alpha \times L_G^1 + \beta \times L_G^2 \tag{18.9}$$

the hyperparameters in the formula are set as $\alpha = 10^{-7}$ and $\beta = -1$.

18.3 Experiment

18.3.1 Experimental Setup

To train and evaluate the model proposed in this study, we utilize the BOSSBase dataset [15], which consists of 10,000 images, each with a resolution of 512 × 512. Using MATLAB for image processing, we cropped these images into 40,000 images with a resolution of 256 × 256, forming the BOSSBase training set. For the testing set, we use the BOW2 dataset [16], which contains 10,000 images with a resolution of 256 × 256. All training and testing procedures were conducted on a single NVIDIA

RTX 3090 GPU, and the experimental setup was implemented using the PyTorch framework.

During training, we set the initial learning rate to 0.0001 and implemented a learning decay strategy, reducing the learning rate to 0.4 of its original value every 20 epochs. The training lasted for 72 epochs with a batch size of 24. The evaluation metric used in this paper is the error rate, where a higher error rate indicates better resistance to steganalysis.

18.3.2 Performance Comparison with Other Algorithms

In this section, we perform a comparative analysis of the proposed steganography algorithm alongside two deep learning-based algorithms, Steg-GMAN [10] and UT-GAN [8] as well as a traditional algorithm, HILL [17], across various embedding capacities. We also assess the resistance of each algorithm against steganalysis using three different steganalyzers: Xu-Net [13], Yedroudj-Net [14], and CovNet [18]. The experimental results are summarized in Table 18.1, with bold text highlighting the best performance in each category.

Analyzing the data presented in Table 18.1, we observe that the proposed method demonstrates optimal performance across all three steganalyzers and four different embedding capacity conditions. Specifically, when faced with the CovNet, our method achieves a significant performance improvement of 1.33% compared to the advanced Steg-GMAN algorithm. This result indicates that our method has a clear advantage in enhancing the concealment of steganographic images.

Table 18.1 Experimental comparison with other advanced algorithm

Steganalyzer	Steganography scheme	Embedding capacities				Average gap
		0.2	0.3	0.4	0.5	
Xu-Net	HILL	42.12	36.68	31.61	26.9	8.13
	UT-GAN	45.1	42.91	40.1	36.61	1.28
	Steg-GMAN	45.96	43.49	40.54	36.70	0.79
	Our	**47.01**	**43.87**	**41.72**	**37.23**	/
Yedroudj-Net	HILL	35.53	27.55	21.5	17.52	12.58
	UT-GAN	39.72	34.55	28.08	24.47	6.4
	Steg-GMAN	44.26	41.09	34.21	28.61	1.07
	Our	**45.69**	**42.19**	**35.51**	**29.04**	/
CovNet	HILL	30.76	24.58	19.34	15.15	12.06
	UT-GAN	35.58	30.54	24.2	20.16	6.9
	Steg-GMAN	39.61	37.79	30.83	24.51	1.33
	Our	**42.69**	**38.76**	**31.63**	**24.99**	/

Table 18.2 The performance of the improved algorithm in this article under different embedding capacities

Steganalyzer	Steganography scheme	0.3	0.4	0.5
CovNet	Base	37.79	30.83	24.51
	Method 1	37.93	31.09	24.94
	Method 2	37.60	31.53	24.65
	Our	38.76	31.63	24.99

18.3.3 Ablation Experiment

To evaluate the effectiveness of the proposed algorithm, we designed ablation experiments. We refer to the unmodified baseline model as "base", the version incorporating the wavelet convolution branch as "method 1", and the version featuring the multi-level feature fusion module as "method 2". In these experiments, we used CovNet to analyze how these modifications affect the algorithm's performance. The experimental results are summarized in Table 18.2.

The data presented in Table 18.2 shows that at three different embedding capacities, the model's performance improves with the addition of the wavelet convolution branch alone. However, introducing only the multi-level feature fusion module results in a slight decrease in performance at an embedding capacity of 0.3 bpp. Notably, when both enhancement methods are applied together, the model's performance surpasses that of either method used individually. This finding suggests that the combination of the wavelet convolution branch and the multi-level feature fusion module significantly enhances the model's resistance to steganalysis.

18.4 Conclusion

In this paper, we propose a generative adversarial network based on dual-domain fusion. Introducing a wavelet convolution branch aims to enhance the information the network can extract. Wavelet convolution captures feature information across various frequencies and spatial scales, providing a more diverse feature representation. To further improve feature fusion, we design a multi-level feature fusion module that integrates features from multiple levels, thereby enhancing the reconstruction capability of the generator at the decoding stage. A series of experiments demonstrate that the stego images generated by the dual-domain fusion network exhibit improved resistance to steganalyzers.

Acknowledgements This work was supported in part by the National Natural Science Foundation of China (No. 62072325), the Fundamental Research Program of Shanxi Province (202103021224272, 202103021224278), and Taiyuan University of Science and Technology Scientific Research Initial Funding (20212039).

References

1. Wang, D., Yang, G., Chen, J., et al.: GAN-based adaptive cost learning for enhanced image steganography security. Expert Syst. Appl. **249**, 123471 (2024)
2. Xu, P., Wang, K., Hassan, M.M., et al.: Adversarial robustness in graph-based neural architecture search for edge AI transportation systems. IEEE Trans. Intell. Transp. Syst. **24**(8), 8465–8474 (2023)
3. Zhao, Y., Ullah, F., Chen, C.M., et al.: Efficient Malware detection using hybrid approach of transfer learning and generative adversarial examples with image represenation. Expert Syst. (2024)
4. Tang, W., Tan, S., Li, B., et al.: Automatic steganographic distortion learning using a generative adversarial network. IEEE Signal Process. Lett. **24**(10), 1547–1551 (2017)
5. Goodfellow, I., Pouget-Abadie, J., Mirza, M., et al.: Generative adversarial nets. In: Advances in Neural Information Processing Systems, vol. 27
6. Luo, L., Cai, J., Fan, Z., et al.: Generative adversarial networks based on cooperative games. J. Netw. Intell. **9**(1), 88–107 (2024)
7. Wang, J., Liu, D.H.: Artistic font style transfer based on deep convolutional generative adversarial networks. J. Netw. Intell. **9**(3), 1693–1705 (2024)
8. Yang, J., Ruan, D., Huang, J., et al.: An embedding cost learning framework using GAN. IEEE Trans. Inf. Forensics Secur. **15**, 839–851 (2019)
9. Ronneberger, O., Fischer, P., Brox, T.: U-net: convolutional networks for biomedical image segmentation. In: Medical image computing and computer-assisted intervention-MICCAI 2015, pp. 234–241. Springer (2015)
10. Huang, D., Luo, W., Liu, M., et al.: Steganography embedding cost learning with generative multi-adversarial network. IEEE Trans. Inf. Forensics Secur. (2023)
11. Finder, S.E., Amoyal, R., Treister, E., et al.: Wavelet convolutions for large receptive fields (2024). arXiv:2407.05848
12. Filler, T., Judas, J., Fridrich, J.: Minimizing additive distortion in steganography using syndrome-trellis codes. IEEE Trans. Inf. Forensics Secur. **6**(3), 920–935 (2011)
13. Xu, G., Wu, H.Z., Shi, Y.Q.: Structural design of convolutional neural networks for steganalysis. IEEE Signal Process. Lett. (2016)
14. Yedroudj, M., Comby, F., Chaumont, M.: Yedroudj-net: an efficient CNN for spatial steganalysisc. In: 2018 IEEE International Conference on Acoustics, Speech and Signal Processing, pp. 2092–2096. IEEE (2018)
15. Bas, P., Filler, T., Pevný, T.: Break our steganographic system: the ins and outs of organizing BOSS. In: International Workshop on Information Hiding, pp. 59–70. Springer (2011)
16. BOWS-2. http://bows2.ec-lille.fr
17. Li, B., Wang, M., Huang, J., et al.: A new cost function for spatial image steganography. In: IEEE International Conference on Image Processing, pp. 4206–4210. IEEE (2014)
18. Deng, X., Chen, B., Luo, W., et al.: Fast and effective global covariance pooling network for image steganalysis. In: Proceedings of the ACM Workshop on Information Hiding and Multimedia Security, pp. 230–234 (2019)

Chapter 19
A Novel Deep Learning Method for Bearing Fault Diagnosis Based on FMD and Transformer

Kai Xu, Gang Xiang, Yang Yu, Yuanyuan Sheng, Mutao Liu, Xizhi Yue, Huanyu Liu, and Junbao Li

Abstract Bearing faults are one of the most common faults in industrial equipment, and their effective detection is crucial for extending the service life of the equipment. Research on bearing faults based on vibration signals has rapidly developed in recent years. Extracting representative features from the original signal completely and effectively is key to solving such problems, and existing methods have not fully addressed this issue. This paper proposes a bearing fault detection model based on Feature Mode Decomposition (FMD) and Transformer, which decomposes the original complex signal into several simple basic modal signals, enabling effective feature extraction. The ability of a transformer to handle long-distance information can effectively process one-dimensional bearing vibration signals. Experimental results show that the classification accuracy of FMD-Transformer is over 95% on the Ottawa bearing vibration dataset. Combining FMD and neural network models provides new ideas and directions for bearing fault research.

19.1 Introduction

Fault diagnosis technology is crucial for equipment maintenance, enhancing service life, safety, and efficiency [1, 2]. Bearings, vital in rotating machinery, significantly impact system performance and reliability [3]. Analyzing bearing signals can identify hidden issues, optimize bearing structures, and boost fault resistance [4]. Fault diagnosis methods fall into two categories: mechanistic model-based and data-driven

K. Xu · Y. Yu · Y. Sheng · X. Yue · H. Liu (✉) · J. Li
Harbin Institute of Technology, Harbin, China
e-mail: liuhuanyu@hit.edu.cn

G. Xiang
Beijing Aerospace Automatic Control Institute, Beijing, China

M. Liu
Changping District, Beijing, China

S.-C. Chu et al. (eds.), *Advances in Intelligent Data Analysis and Applications*, Smart Innovation, Systems and Technologies 445,
https://doi.org/10.1007/978-981-96-7277-6_19

[5]. While high-precision models yield good results, their complex establishment limits practical applications [6]. However, with advancements in deep learning and industrial big data, data-driven methods have gained widespread uses [7, 8].

Fault diagnosis typically involves three steps: feature extraction, feature selection, and fault classification [9]. Feature extraction is crucial, as it aims to derive fault-related features from complex signals. Traditional methods rely on time-frequency domain transformations, which limit the information dimension of the extracted features and complicate subsequent tasks [10]. Modal decomposition has emerged as an effective way for signal multi-component analysis [11]. Common methods include Empirical Mode Decomposition (EMD), Ensemble Empirical Mode Decomposition (EEMD), Variational Mode Decomposition (VMD), and Local Mode Decomposition (LMD). EMD, introduced by Huang et al. [12], adaptively decomposes signals into intrinsic mode functions, effectively handling non-linear and non-stationary signals, though it suffers from mode aliasing and endpoint effects [13, 14]. EEMD, proposed by Wu et al. [15] in 2013, mitigates mode aliasing by average processing after adding a white noise sequence, which suffers from high computational complexity. VMD offers a variational solution to detect center frequency and bandwidth, which avoids mode aliasing. However, it is sensitive to preset parameters [16]. LMD decomposes signals into product functions, and the envelop signals and pure frequency modulation signals are separated through iterative processing [17]. While modal decomposition methods have advanced signal processing research, they still face challenges related to computational complexity and other shortcomings.

Current research on bearing faults focuses on three main areas as follows. First, signal processing, where the complex operating environment and noise hinder effective fault feature extraction; existing methods still have limitations in feature extraction [18, 19]. Second, artificial intelligence-related methods. Machine learning has advanced fault classification through convolutional neural networks (CNN), recurrent neural networks (RNN), and other models leveraging extensive bearing data [20]. Third, the multi-method fusion enhances prediction and classification performance by integrating multiple models, offering better generalization than single models [21]. Recently, combining machine learning with signal processing-based feature extraction has emerged as a novel approach in bearing fault research [22]. Inspired by this, a fusion model based on FMD and Transformer is proposed in this paper, which achieves great accuracy on bearing fault classification.

19.2 Theoretical Basis

19.2.1 Feature Mode Decomposition

Traditional EMD faces challenges such as mode aliasing and an insufficient theoretical basis. The use of white noise in EEMD undermines its performance. In contrast, though theoretically robust, the VMD fails to adequately capture the pulse and shock features of signals [16]. Existing methods struggle to isolate distinct components due to the pulse and periodic nature of the signals. To improve signal processing and adaptability, the FMD is introduced. FMD decomposes the original signal into basic modal components with time-frequency characteristics for analysis. It has three key attributes. (i) FMD effectively captures the pulse and periodic nature of signals, allowing for precise decomposition of fault modes [23]. (ii) It uses Finite Impulse Response (FIR) filters to extract relevant signal modes while removing redundant ones, improving decomposition thoroughness. (iii) FMD decomposes mechanical fault information without prior knowledge. The signal processing can be represented mathematically by Eq. (19.1).

$$\begin{aligned} &\underset{\{f_k(l)\}}{\arg\max}\left\{CK_M(\mathbf{u}_k)=\sum_{n=1}^{N}\left(\prod_{m=0}^{M}u_k(n-mT_s)\right)^2\Big/\left(\sum_{n=1}^{N}u_k(n)^2\right)^{M+1}\right\} \\ &\text{s.t.}\,u_k(n)=\sum_{l=1}^{L}f_k(l)x(n-l+1) \end{aligned} \tag{19.1}$$

where $u_k(n)$ is the decomposition mode number k, f_k is the filter number k, T_s is the sample measurement period, and M is the shift order.

Compared with the common initialization method, the Hanning window can locate the fault period after two iterations, which significantly improves efficiency. The FIR filter is initialized by the Hanning window. After that, the filtered signal can be obtained by the following Eq. (19.2).

$$\begin{aligned} &\boldsymbol{u}_k=\boldsymbol{X}*\boldsymbol{f}_k \\ &\begin{bmatrix} u_k[1] \\ \vdots \\ u_k[N-L+1] \end{bmatrix}=\begin{bmatrix} x(1) & \cdots & x(L) \\ \vdots & \ddots & \vdots \\ x(N-L+1) & \cdots & x(N) \end{bmatrix}\begin{bmatrix} f_k(1) \\ \vdots \\ f_k(L) \end{bmatrix} \end{aligned} \tag{19.2}$$

where k represents the filter number and * represents the convolution operation. Update the filter parameters by the original signals $\mathbf{X}$ and $\mathbf{u}_k$. After finishing the iteration, a $CC_{k\times k}$ matrix is constructed for every two patterns, dropping the pattern with the smaller CK value in the two patterns. After repetitions, the FMD mode decomposition is completed after the preset mode number is obtained.

$$CC_{pq}=\frac{\sum_{n=1}^{N}\left(u_p(n)-\overline{u}_p\right)\left(u_q(n)-\overline{u}_q\right)}{\sqrt{\sum_{n=1}^{N}\left(u_p(n)-\overline{u}_p\right)^2}\sqrt{\sum_{n=1}^{N}\left(u_q(n)-\overline{u}_q\right)^2}} \tag{19.3}$$

VMD addresses mode aliasing in digital mode decomposition but falls short by prioritizing bandwidth minimization over fault characteristics, leading to suboptimal performance on bearing fault signal decomposition. In contrast, FMD effectively extracts mechanical fault features while considering the pulse characteristic and periodicity of the raw signal. Compared to VMD, FMD excels in mechanical signal decomposition and analysis. This paper introduces FMD into bearing fault classification, resulting in improved diagnostic outcomes due to its advanced fault feature extraction capabilities.

19.2.2 Transformer

Early sequential transduction models primarily relied on intricate cyclic or convolutional neural networks. The sequential nature of state sequences inherent in these recursive models hinders the possibility of parallelization [24]. Consequently, serial processing cannot be layered effectively, leading to diminished efficiency and accuracy. In 2017, Google introduced the Transformer model, which leverages attention mechanisms to circumvent the limitations of inefficient and redundant circular structures [25]. This model establishes dependencies between inputs and outputs, as well as between the encoder and decoder, without the need for traditional RNN and CNN architectures.

Transformer achieves its capability for global feature extraction through stacked attention mechanisms. The attention mechanism functions as a mapping of a query and a set of key values to an output, known as scaled dot-product attention, with its computational process represented by Eq. (19.4).

$$\text{Attention}(Q, K, V) = \text{softmax}\left(\frac{QK^T}{\sqrt{d_k}}V\right) \tag{19.4}$$

The implementation of dot-product attention can be achieved through a highly optimized matrix multiplication algorithm, resulting in enhanced speed and reduced memory usage in practical applications. In contrast to a singular attention function, multi-head attention enables the model to concurrently concentrate on information from various locations across different subspaces, thereby improving overall model performance. The multi-head attention (MHA) mechanism can be articulated as Eq. (19.5).

$$\begin{gathered}\text{MultiHead}(Q, K, V) = \text{Concat}(\text{head}_1, ..., \text{head}_h)W^O \\ \text{wherehead}_i = \text{Attention}(QW_i^Q, KW_i^K, VW_i^V)\end{gathered} \tag{19.5}$$

The Transformer utilizes an encoder-decoder architecture and incorporates an attention mechanism to address the sequence dependency issues inherent in traditional RNN networks, demonstrating outstanding performance in long-range feature extraction [25]. Due to their remarkable feature extraction abilities, Transformers have been extensively applied and studied in the domains of natural language processing and computer vision, yielding impressive outcomes.

19.3 Proposed Model

This paper proposes a model for bearing fault diagnosis, termed the FMD-Transformer, which integrates the FMD and Transformer methodologies. The model's architecture is depicted in Fig. 19.1. During the data preprocessing and modal feature extraction, the original signal is sampled with a specified frequency to create data samples, from which the fundamental modal signal is derived through FMD. The time-frequency domain features extracted from each mode are then concatenated to form a one-dimensional feature vector. The neural network architecture comprises a series of stacked Transformer Encoders followed by a fully connected layer. The encoder employs the multi-head attention mechanism to excavate relationships among various segments of the input vector, assigning distinct weights to each position to emphasize critical components. To address issues of gradient vanishing and to expedite model convergence while minimizing the risk of overfitting, residual connections and layer normalization are implemented in our model. Ultimately, a feedforward neural network is utilized to further refine the sequences, enhancing the model's expressions and learning capacities. The output generated from multiple iterations of the Transformer encoder serves as the feature vector for the input sample, which is then processed through the fully connected layer to yield the classification outcome. The detailed parameters of the proposed model's network structure are presented in Table 19.1.

The input signal requires preprocessing to generate the input data sample. FMD is utilized to extract the fundamental mode signals for further analysis. The FMD processing workflow is depicted in Fig. 19.2, and the key parameter configurations

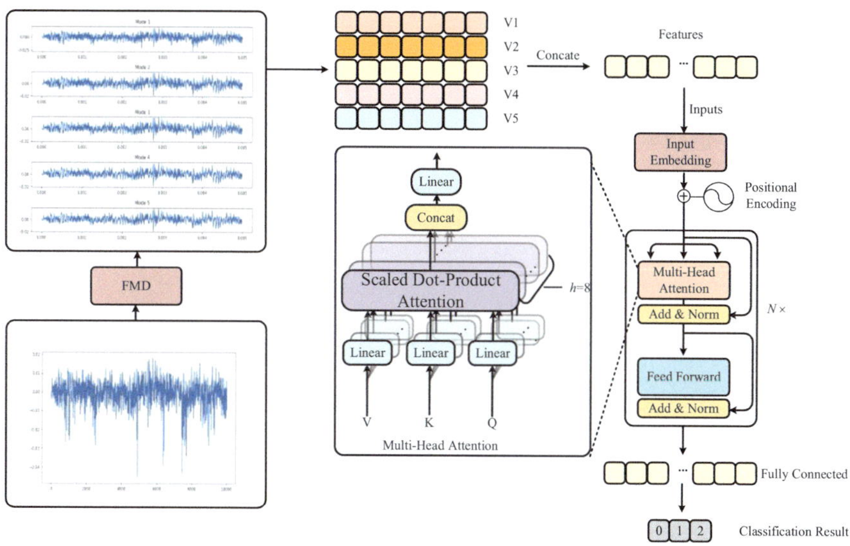

Fig. 19.1 Schematic diagram of the proposed FMD-transformer model

Table 19.1 Parameters of the FMD-Transformer model

Layers	Name	Parameters
0	FMD	$N = 5, L = 100, I = 10, \theta = 0.05$
1	Input layer	Dim = (1, 35)
2–7	Transformer encoder 1–6	$d\_model = 512$, $n\_head = 8$
8	Fully connected	Input = (1, 512), output = (1, 3)
9	Outputs	Label = (1, 1)

are listed in Table 19.1. Each fundamental mode is analyzed in time-frequency domains, with consistent features selected to form the corresponding feature vector. The learning and inferencing process for each sample depends on the feature vector rather than the sample itself. Subsequent steps follow standard transformer-based deep learning models and are not detailed anymore. The forward propagation process can be formulated as follows

$$f(\mathbf{x}) = \text{Concat}(\text{TTF}(\text{FMD}(\text{Split}(\mathbf{x})))) \tag{19.6}$$

$$y = \text{Transformer}(\text{Feature}(\boldsymbol{x})) \tag{19.7}$$

where TTF represents feature extraction in the traditional time-frequency domain, and Concat represents the concatenation of different modal feature vectors.

The extraction of features from modal signals in the time-frequency domain primarily encompasses modal energy, root mean square (RMS) value, peak factor, pulse width, spectral peak frequency, spectral center frequency, and spectral bandwidth. The features were computed based on the equations provided in Table 19.2.

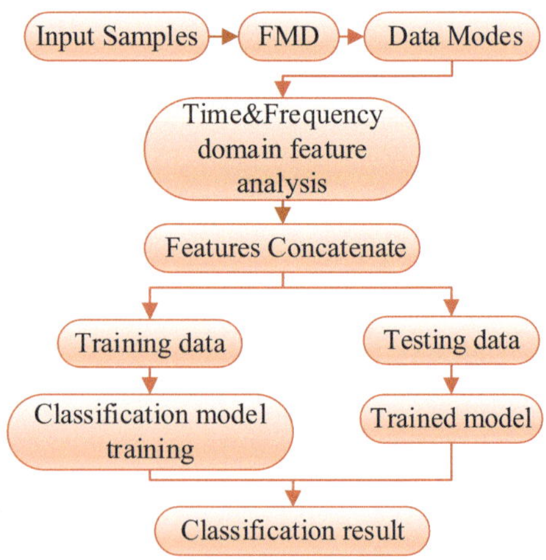

Fig. 19.2 Fault diagnosis process of FMD-transformer model

Table 19.2 Feature extraction methods in the time-frequency domain

Features	Equations
Modal energy	$M_{\text{Energy}} = \sum_i x_i^2$
Rms	$M_{\text{Rms}} = \sqrt{\frac{1}{n}\sum_i x_i^2}$
Peak factor	$M_{\text{PF}} = \frac{\max(\lvert x \rvert)}{M_{\text{Rms}}}$
Pulse width	$M_{\text{PW}} = \max(\{i \lvert \lvert x_i > \theta\}) - \min(\{i \lvert \lvert x_i > \theta\}) + 1$
Spectral peak frequency	$M_{\text{PFre}} = \text{Fre}(\arg\max_i(\text{Mag}_i))$
Spectral center frequency	$M_{C\text{Fre}} = \frac{\sum_i \text{Fre}_i \times \text{Mag}_i}{\sum_i \text{Mag}_i}$
Spectrum bandwidth	$M_{\text{SB}} = \text{Fre}_{U_{ind-1}} - \text{Fre}_{L_{\text{ind}}}$ $L_{\text{ind}} = \arg\max_i\left(\sum_{j=0}^{i} \text{Mag}_j^2 \geq 0.95 \times \frac{\sum_i \text{Mag}_i^2}{2}\right)$ $U_{\text{ind}} = \text{len}(\text{Mag}) - \arg\max_i\left(\sum_{j=0}^{i} \text{Mag}_{\text{len}(\text{Mag})-j}^2 \geq 0.95 \times \frac{\sum_i \text{Mag}_i^2}{2}\right)$

Note i is the sampling point, x_i is the amplitude of the signal at each sampling point, Fre is the spectrum sequence of the modal signal, Mag is the amplitude sequence corresponding to the spectrum sequence, and θ is the preset threshold according to the dataset

19.4 Experiments and Results

19.4.1 Experiment Setup and Dataset

This study focuses on bearing faults, which utilize a vibration signal dataset from Ottawa, Canada. The dataset includes signals under four varying speed conditions: A (speed increase), B (speed decrease), C (increase then decrease), and D (decrease then increase). At each speed, three fault states are included: healthy, inner ring defect, and outer ring defect. The signal collection is repeated three times for each speed and fault state, yielding a total of 36 signals, each sampled at 200,000 Hz for 10 s. Each signal in the dataset is split into multiple segments with 0.05 s intervals, yielding 200 samples per signal and 7200 samples in total across 36 bearing signals. As outlined in Chaps. 3 and 7 time-frequency features are extracted as feature vectors for each sample. The 7200 samples are split into a training set and a testing set in a 7:3 ratio.

The following experiments were implemented in 64-bit PyCharm with a computer with I5-12400F at 2.5 GHz (6 cores) and 16 GB of memory. During the experiment, the learning rate was 0.001 and the batch size was 512.

19.4.2 Results and Analysis

Experiments on FMD-Transformer. FMD, a novel signal mode decomposition technique, has garnered considerable interest. This paper evaluates its effectiveness by integrating FMD with a back propagation network (BP), CNN, and long-short-term memory (LSTM) for bearing fault classification using the dataset mentioned above. Similar to the FMD-Transformer, the data preprocessing and signal feature extraction process, as described in Chap. 3, uses the resulting signal feature vectors as inputs to BP, CNN, and LSTM networks for subsequent classification tasks. For comparison, the control group generates feature vectors from mean, variance, extreme values, and standard deviation in the time domain. TTF is used to represent feature extraction in the conventional time-frequency domain, aligning with FMD.

Table 19.3 presents the classification accuracy and loss achieved through the combination of TTF, FMD, and various classification models. It is evident that when TTF is employed for feature extraction, the classification performance of LSTM significantly surpasses that of other models, even outperforming the Transformer. We hypothesize that, on one hand, the proposed LSTM model is specifically designed to address the issues of gradient vanishing or explosion faced by traditional RNN networks when handling long sequence data, enabling LSTM networks to be more adept at processing sequential signals such as bearing vibration signals compared to other neural network architectures. On the other hand, the strengths of Transformer are manifested in its capacity to extract global information. For the feature vectors derived from bearing vibration signals, additional preprocessing techniques may be necessary to fully leverage the capabilities of the Transformer.

When utilizing FMD for feature extraction, the classification accuracy across all models generally exceeds that of TTF. The classification loss is markedly diminished when FMD is applied, aligning with the classification results presented in Table 19.3. Indicating that FMD is more effective in extracting critical fault features from such data, thereby providing more valuable inputs for the model.

To further evaluate the efficacy of the FMD-Transformer in the context of this dataset classification task, this study conducts additional experiments. Table 19.4 presents the classification accuracy rates for the TTF-Transformer and FMD-Transformer models across various fault types. Both models attain a 99% accuracy rate in identifying inner circle faults. However, the FMD-Transformer model demonstrates significantly superior accuracy compared to the TTF-Transformer model

Table 19.3 Comparison of classification results and total loss

Classification model	Accuracy/loss	Classification model	Accuracy/loss
TTF + BP	78.6%/0.76	FMD + BP	91.4%/0.62
TTF + CNN	82.1%/0.40	FMD + CNN	92.5%/0.31
TTF + LSTM	**88.9%/0.26**	FMD + LSTM	93.1%/0.16
TTF + Transformer	81.67%/0.32	FMD + Transformer	**95.1%/0.05**

Table 19.4 Classification accuracy of each fault mode

Fault mode	Accuracy of TTF-transformer (%)	Accuracy of FMD-transformer (%)
Healthy	75	97
Inner fault	99	99
Outer fault	77	89

under healthy conditions and for outer ring faults. This suggests that the integration of the FMD feature extraction technique with the Transformer architecture can substantially enhance model performance, particularly in specific fault types.

Experiments on the Parameter of FMD. FMD is a multi-parameter modulation decomposition technique. We are particularly interested in the correlation between its parameters and decomposition efficacy. Based on the experimental findings of Miao et al. [23], FMD demonstrates robustness in the selection of k within the interval of [2, 24], in which k is recommended to be selected within the range of [7, 12]. In the model presented in this paper, the quantity of modal numbers directly dictates the size of the vibration signal feature vector. Therefore, to investigate the relationship between the number of modes and the precision of bearing fault classification, we conducted a series of experiments by varying the number of modes within the range of [2, 19].

Figure 19.3 shows the results of various network models as a function of the number of modes. As the number of modes, n increases from 2 to 10, accuracy improves for each model. This indicates that within a certain range, FMD is more thorough in signal decomposition and more effective in feature extraction with the increase of n. However, beyond $n = 10$, network performance stabilizes with only minor fluctuations, which is consistent with the original FMD study. Thus, we can conclude that increasing modes can improve model performance up to a threshold, but this comes with greater computational complexity. Therefore, the choice of mode number n requires a compromise between decomposition performance and computational complexity.

19.5 Conclusions and Analysis

This paper focuses on investigating bearing faults. Though VMD addresses the modal aliasing issue, it lacks precision in fault characteristic decomposition. This paper introduces the FMD-Transformer for bearing fault diagnosis, presenting FMD as an alternative to EMD, EEMD, and VMD. FMD shows excellent analytical and decomposition capabilities for mechanical fault signals, effectively overcoming VMD's limitations, which is essential for feature extraction in fault diagnosis. Experimental results show that FMD significantly improves feature extractions, with the FMD-Transformer achieving over 95% accuracy in classification tasks. This research highlights the importance of efficient modal decomposition methods in feature extraction,

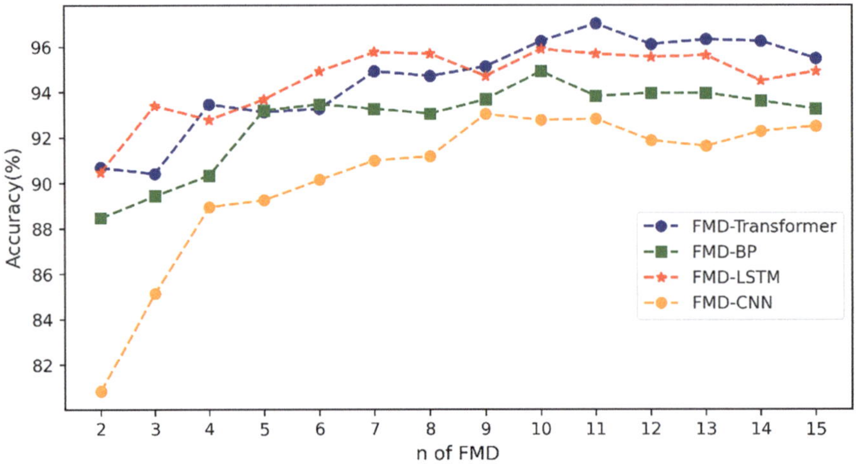

Fig. 19.3 The effect of modal number on accuracy

indicating that model accuracy can be enhanced by examining various target signals and employing different modal decomposition techniques. Additionally, results show that model accuracy improves with an increased number of modes within a certain range. However, our study has some drawbacks. Firstly, we use a simple three-class classification task to validate FMD's efficacy, which requires furthermore validation. Future research could investigate more complex signals and tasks for further validation. Secondly, this research prioritizes methodologies over structural considerations, suggesting that future studies could optimize model architectures to improve performance.

References

1. Xu, G., Liu, M., Jiang, Z.: Bearing fault diagnosis method based on deep convolutional neural network and random forest ensemble learning. Sensors **19**(5), 1088 (2019)
2. Sun, Y., Tao, H., Stojanovic, V.: Autoregressive data generation method based on wavelet packet transform and cascaded stochastic quantization for bearing fault diagnosis under unbalanced samples. Eng. Appl. Artif. Intell. **138**, 109402 (2024)
3. Liang, Z., Cao, J., Ji, X.: Fault severity assessment of rolling bearings method based on improved VMD and LSTM. J. Vibroeng. 21292 (2020)
4. Li, M., Huang, J., Shi, P.: A secondary optimization strategy in stochastic resonance modeling for the detection of unknown bearing faults. Chaos Solitons Fractals **189**, 115576 (2024)
5. Huang, Y., Xu, Z., Jie, Z., Shan, T., Cao, L., Wang, J., Shen, Y.: Bearing fault diagnosis based on time-varying targeted enhancement of stress wave signal. In: IEEE International Conference on Sensing, Diagnostics, Prognostics, and Control (SDPC), pp. 57–62 (2024)
6. Nguyen, V.H., Seshadrinath, J., Wang, D., Nadarajan, S., Vaiyapuri, V.: Model-based diagnosis and rul estimation of induction machines under inter-turn fault. IEEE Trans. Ind. Appl. **53**, 2690–2071 (2017)

7. Gong, W., Chen, H., Zhang, Z.: A novel deep learning method for intelligent fault diagnosis of rotating machinery based on improved CNN-SVM and multichannel data fusion. Sensors (Basel) **7** (2019)
8. Wang, X., Lv, J., Slowik, A., Parameshachari, B.D., Li, K., Chen, C.-M., Kumari, S.: DLLF-2EN: energy-efficient next generation mobile network with deep learning-based load forecasting. IEEE Trans. Netw. Serv. Manag. 1932–4537 (2024)
9. Sheng, Y., Liu, H., Li, J.: Bearing performance degradation assessment and remaining useful life prediction based on data-driven and physical model. Meas. Sci. Technol. **34**(5), 055002 (2023)
10. Li, H., Zhang, Q., Qin, X., Sun, Y.: Fault diagnosis method for rolling bearings based on short-time Fourier transform and convolution neural network. J. Vib. Shock **37**, 124–131 (2018)
11. Desjardins, L.S., Lau, T.D.: Enhanced operational modal analysis and change point detection for vibration-based structural health monitoring of bridges. J. Infrastruct. Intell. Res. **3**(4), 100121 (2024)
12. Huang, N., Shen, Z., Long, S.R., Wu, M.C., Shih, H.H., Zheng, Q., Yen, N., Tung, C., Liu, H.H.: The empirical mode decomposition and the Hilbert spectrum for nonlinear and non-stationary time series analysis. Proc. R. Soc. London. Ser. A: Math. Phys. Eng. Sci. **454**, 903–995 (1998)
13. Yao, X., Wang, J., Cao, L., Yang, L., Shen, Y., Wu, Y.: Fault diagnosis of rolling bearing based on VMD feature energy reconstruction and ADE-ELM. In: IEEE International Conference on Sensing, Diagnostics, Prognostics, and Control (SDPC), pp. 81–86 (2022)
14. Hu, M., Wang, C., Zhuang, C., Wang, Y.: Bearing fault diagnosis method based on data augmentation and MCNN-LSTM. In: IEEE 6th Information Technology, Networking, Electronic and Automation Control Conference (ITNEC), pp. 663–671 (2023)
15. Wu, Z., Huang, N.: Ensemble empirical mode decomposition: a noise-assisted data analysis method. Adv. Data Sci. Adapt. Anal. **1**, 1–41 (2009)
16. Dragomiretskiy, K., Zosso, D.: Variational mode decomposition. IEEE Trans. Signal Process. **62**(3), 531–544 (2014)
17. Smith, J.S.: The local mean decomposition and its application to EEG perception data. J. R. Soc. Interface. **2**(5), 443–454 (2005)
18. Sheng, Y., Liu, H., Li, L.: A hybrid method of frequency-weighted energy operator and power spectrum fusion to detect bearing faults. Rev. Sci. Instrum. **94**(5), 15 (2023)
19. Takao, T., Yu, T., Shizuko, H.: Three-dimensional finite-difference time-domain simulation of moving sound source and receiver with directivity. Appl. Phys. **62**(SJ) (2023)
20. Rani, S., Memoria, M., Almogren, A.: Deep learning to combat knee osteoarthritis and severity assessment by using CNN-based classification. BMC Musculoskelet. Disord. **25**(1), 817 (2024)
21. Han, Y., Hong, B.: Deep learning based on Fourier convolutional neural network incorporating random kernels. Electronics **10**(16) (2021)
22. Ghorbel, A., Eddai, S., Limam, B.: Bearing fault diagnosis based on artificial intelligence methods: machine learning and deep learning. Arab. J. 1–18 (2024)
23. Miao, Y., Zhang, B., Li, C., Lin, J., Zhang, D.: Feature mode decomposition: new decomposition theory for rotating machinery fault diagnosis. IEEE Trans. Ind. Electron. **70**(2), 1949–1960 (2023)
24. Lipton, Z.C., Berkowitz, J., Elkan, C.: A critical review of recurrent neural networks for sequence learning. Comput. Sci. **1506**, 00019 (2015)
25. Vaswani, A., Shazeer, N.M., Parmar, N., Uszkoreit, J., Jones, L., Gomez, A.N., Kaiser, L.: Attention is all you need. Neural Inf. Process. Syst. **30**(1), 6000–6010 (2017)

Chapter 20
Deep Learning Based Virtual Human Posture Detection and Tracking

Tianyou Xue and Yi-Jui Chiu

Abstract It is evident that there is still scope for enhancement in the domain of human body posture and movement detection systems. These systems are founded upon the principles of deep learning algorithms. The identified areas for improvement encompass the quantity of data sets, the precision of detection, and the real-time detection performance. Firstly, we extracted frames from action videos to obtain a datasets. Then, we used roboflow to annotate joint anchor points on the datasets. Next, YOLOv8 model was employed for training and testing on the datasets, and the performance of two different YOLOv8 models was compared. The findings of the experimental study demonstrate that YOLOv8n-pose outperforms the YOLOv8s-pose model in terms of training speed and mean average precision (mAP). YOLOv8n-pose maintains recognition accuracy while minimizing model size, parameter count, and runtime memory usage. Experimental results show that YOLOv8n-pose has an average accuracy of 80.1% and a precision of 98% when the number of parameters is only 3.3 M and the GPU memory occupancy is only 1.47 G during training. This ensures a favourable balance between training precision and the velocity of the training process.

20.1 Introduction

The amount of data available on computer networks has increased significantly in recent years. Furthermore, there has been a notable increase in computing power. These factors have driven the continuous development of deep learning techniques. The ability to detect and track the posture of individuals in video images has become a crucial aspect of human-computer interaction. Regarding posture detection and tracking, Liu et al. [1] used five machine learning algorithms to conduct sitting posture detection experiments. They selected an office environment for a long-term

T. Xue · Y.-J. Chiu (✉)
School of Mechanical and Automotive Engineering, Xiamen University of Technology, Xiamen, Fujian Province, China
e-mail: chiuyijui@xmut.edu.cn

S.-C. Chu et al. (eds.), *Advances in Intelligent Data Analysis and Applications*, Smart Innovation, Systems and Technologies 445,
https://doi.org/10.1007/978-981-96-7277-6_20

sitting experiment and provided a long-term sitting history record. The findings of the experimental study demonstrated that the method in question is capable of accurately identifying five distinct sitting postures, with an overall accuracy rate of 98.07%. Ma et al. [2] proposed a fall detection system that uses homography projection to obtain global multi-view information of the scene. The algorithm was tested on a public datasets of fall detection based on multiple cameras. The results showed that the algorithm has high recognition efficiency and accuracy. Tlili et al. [3] proposed a system for monitoring seated individuals in real time. It comprises an intelligent belt with inertial sensors. These collect and transmit posture data to a cloud server. The cloud server is responsible for the processing of the data collected. Wang et al. [4] proposed an empirical research method for AI-driven sports pose detection in sports training, demonstrating significant adaptability and robustness for weightlifting pose detection. A novel detection technique based on cumulative inter-frame differences is also introduced, which is capable of pinpointing the region where the weightlifter's pose has changed. Li et al. [5] devised two preliminary detection systems based on identity-preserving human pose detectors. A series of experiments demonstrated that these detectors accurately detect human poses in CHP datasets. The results show that these baseline detectors can effectively detect human poses on the CHP datasets. Kulikajevas et al. [6] proposed novel Deep Recurrent Hierarchical Network (DRHN) model based on MobileNetV2 has been put forth, offering enhanced flexibility through the reduction or elimination of the pose detection issue associated with the constrained observation of the human torso in the depth frame. They achieved a sitting posture recognition accuracy of 91.47% at a rate of 10 fps. Huang et al. [7] put forth a system for the detection of student postures that employs the use of intelligent surveillance cameras and advanced computer vision technology, with the incorporation of the open-source model OpenVINO, which is designed to facilitate the accurate identification of student postures. The Message Queuing Telemetry Transport (MQTT) protocol is used to facilitate data transmission, effectively establishing a seamless posture detection work in a classroom environment. Guo et al. [8] proposed a new privacy-preserving task between human detection and human pose estimation by introducing the IPHPDT (Identity-Preserving Human Pose Detection in Thermal Images) benchmark. The results show that three baseline detectors can effectively perform accurate pose detection on the IPHPDT datasets. Cun et al. [9] proposed a method for detecting and recognising the relationship between sitting posture and sitting discomfort in aircraft passengers. Multiple machine learning methods were utilised to identify eight sitting postures with the help of pressure data from seat panels and backrests. The results showed that the highest level of classification accuracy, 89.26%, was achieved through the utilisation of the Support Vector Machine (SVM) approach with a kernel function based on Radial Basis Theory. Zhang et al. [10] proposed a model for representing human poses in unintentional falls, the 'five-point inverse dynamics model'. They employed an enhanced two-branch M-CNN to facilitate the extraction and construction of the inverse dynamics structure of human poses in intricate scenes. The experimental results in real-world scenes show that the method is robust, versatile and has high detection accuracy. Zhou et al. [11] proposed a strategy to improve the effectiveness and transferability of adversarial attacks against UAVs. Their research shows that even the smallest perturbations can significantly reduce model accuracy. This is a significant advance in the understanding of transferability of adversarial attacks.

Many related researches have also been done in our lab, Sun et al. [12] proposed a Thin Plate Sample Interpolation (TPS) method to complete the missing data. The results show that the mean error of the linear interpolation method for three samples with small missing data is 1.16, 2.79 and 1.07 times of the TPS method, respectively. The mean error of the linear interpolation method for the three samples is 1.22 times as high as 1.80 t. Chiu et al. [13] proposed a hysteresis loop based algorithm to select the duty cycle perturbation. Finally, by comparing with the conventional algorithm, the following results were obtained: transient tracking speed of 47 ms, steady state oscillation of 0.23 W, and relative error of 0.12%. These results demonstrate the superiority of the algorithm.

In summary, video virtual human posture detection and tracking based on deep learning is of considerable value and importance in the field of intelligent transport, sports training, and smart home. The objective of this paper is to present a deep learning approach for the detection of human posture by taking the posture of virtual characters instead of real people as the research object. Firstly, the raw data is collected by recording the video and frame extraction is performed on the video to build and image datasets. Secondly, the calibration of the avatar and its key nodes is carried out in Roboflow tool. Finally YOLOv8n-pose model is used to train and test the datasets and two different models are compared and comprehensively evaluated.

20.2 Target Audience and Parties

The target object datasets of this study is selected as the source of virtual characters and virtual scenes in UNCHARTED LEGACY. Since the character modelling and character movements are very similar to those of real people, the pose detection results of avatars have the same value as the pose detection of real people. The avatar datasets collection has the advantages of repeatable collection, high motion completion, and cost savings.

The datasets in this paper uses the recording procedure in the NVIDIA driver to obtain the original 3 min, 60-frame video datasets with appropriate resolution, the video is frame-pumped by python script at 30-frame intervals, and the result of frame-pumping is a datasets of 600 images in jpg format. The data set comprises 80% training images and 20% test images in addition, when testing the video samples, a 50 s homologous video is selected as the test sample. In the collection of the datasets, the brightness of the scene, the angle of the character, the character's movement and other factors are fully considered. The gesture characteristics of the characters in the scene are comprehensively collected to restore the impact of environmental factors on the results of recognition when detecting and tracking the gestures of virtual characters.

The datasets annotation is done by using roboflow tool to annotate the key points of characters and bones in the images. The TXT tag file is obtained, and the tag file corresponds to the image file one by one, and the content of the tag file corresponds to the location of the human body in the image as well as the key anchor point location information.

20.3 Posture Detection and Tracking Algorithms

The YOLOv8 algorithm used in this study has a number of advantages such as high efficiency and accuracy. This algorithm was developed by Ultralytics, Inc. It integrates and improves on the advantages of previous generations of YOLO algorithms. The network architecture of YOLOv8 reflects that this algorithmic design contains the core architecture of CSP's backbone (backbone) and Neck. And based on this, Use C2f components instead of C3 components to further lighten the weight, At the same time, the idea of CSP and SPPF components are retained. Improvements to PAN-FPN still use PAN while removing the convolutional structure in the upsampling stage. Decoupled-Head idea is introduced in the training and inference processes of neural networks. The VFL loss function is employed for the purpose of predicting the classification loss, while the DFL loss function and CIOU loss function are utilised for the prediction of regression loss. Thus the YOLOv8 algorithm has higher accuracy and smaller time cost during the training tests for target detection and pose estimation. The model's performance has been enhanced when confronted with tasks of a complex nature. The function of the Neck module is to extract and fuse the features of the target object that have been detected. Firstly, the multi-scale feature maps P3, P4 and P5 output from the feature extraction network (Backbone) are input into the PAN-FPN structure to achieve effective multi-scale feature fusion. Next, the feature map designated as P5 is upsampled and merged with the P4, resulting in the generation of an intermediate designated as F1. Subsequently, F1 is upsampled once more following processing by the C2f layer and fusion with P3, resulting in the generation of T1. Subsequently, T1 is convolved with F1 in order to achieve fusion, resulting in F2. This is then processed by the C2f layer to generate T2. Following additional processing, a second iteration of T2 is produced. Subsequently, T2 is convolved with P5 and fused to generate F3. F3 is then processed by the C2f layer to finally generate T3 therefore, that the T1, T2 and T3 are the outputs of the Neck module, which provide fused multiscale features to support the subsequent detection tasks. Figure 20.1 illustrates the flowchart of target feature extraction and fusion for the Neck component.

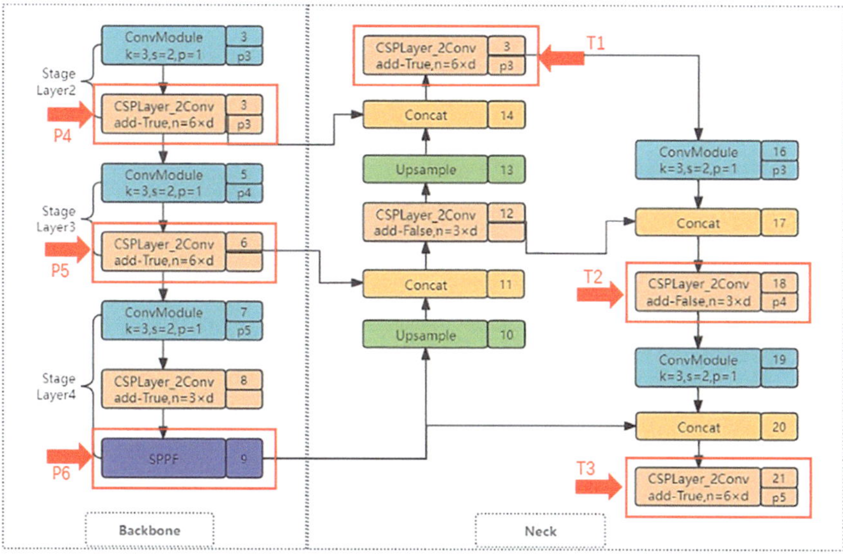

Fig. 20.1 Neck flowchart of target feature extraction and fusion of components

20.4 Model Training and Measurement

20.4.1 Experimental Environment Configuration

This deep learning-based research model for avatar pose detection and tracking was trained and tested using the PyTorch 1.13.1 version, as well as a parallel computing platform and the CUDA 11.6 deep learning framework. The parameters employed for the training and testing phases are presented in Table 20.1.

Table 20.1 Training and testing setup parameters

Training and testing environment parameters	Value of parameter
Image size	640
Batch size	10
Optimizer	AdamW
Momentum	0.937
Epochs	100

20.4.2 Modelling Training

Model training firstly requires the collected datasets and the corresponding label files to be passed into the YOLOv8-pose model at the same time, parameters such as the training set as well as the validation set path and the number of samples to be trained are specified in the yaml file. Second, the training process dynamically adjusts the model parameters through the batch gradient descent algorithm. In the forward propagation process, the training model extracts key features from the input images, compresses and abstracts the data information embedded in the images layer by layer in the convolutional and pooling layers into a feature map, and is used to generate the confidence and prediction frames for the targets detected in the image. Then, YOLOv8 calculates the loss by localisation loss function, confidence loss function, category loss function together and calculates the gradient of different parameters in the loss function by back propagation algorithm. Finally, the model parameters are updated again by the Adam optimiser according to the gradient direction and the learning rate, which finally makes the loss function take the smallest value as possible.

The trends of different loss functions during model training are shown in Fig. 20.2. Figure 20.2a represents the loss of the prediction frame regression, which indicates the error that occurs when anchoring the prediction frame of the target object, and the trend of the image shows that the prediction frame loss decreases steadily with the increase of the number of sample training times, indicating that the model's prediction accuracy of the bounding box increases gradually with the increase of the number of iterations during the model training process, and that the prediction frame loss function decreases drastically when the number of training times is close to 100, indicating that the training of the model has basically completed the optimal learning of the parameters. Figure 20.2b represents the loss of pose prediction. The loss value is relatively high when the number of training sessions is low, indicating that the model has a large error in predicting the pose at this point and is less accurate. As the model training proceeds, The loss value declines at a rapid rate and then tends to decrease steadily. It shows that the initial stage of model training has been able to accommodate tasks of posture estimation with lower precision requirements. Figure 20.2c represents the loss of target detection. The model tends to have errors when detecting targets, and the image shows that the target detection loss has a significant decreasing trend at the initial time, and then the decreasing trend tends to stabilise, indicating that the model's recognition judgement of the detected object gradually converges. Figure 20.2d represents the classification loss, the decreasing trend of this loss curve in the initial stage is also obvious, quickly reaching and keeping at a low level, indicating that the model is more accurate in the classification task of the target. In summary, the trends of the loss function in Fig. 20.2 all show a decreasing trend during model training, which indicates that the prediction model has been continuously optimised during the sample training and the pose estimation has been improving.

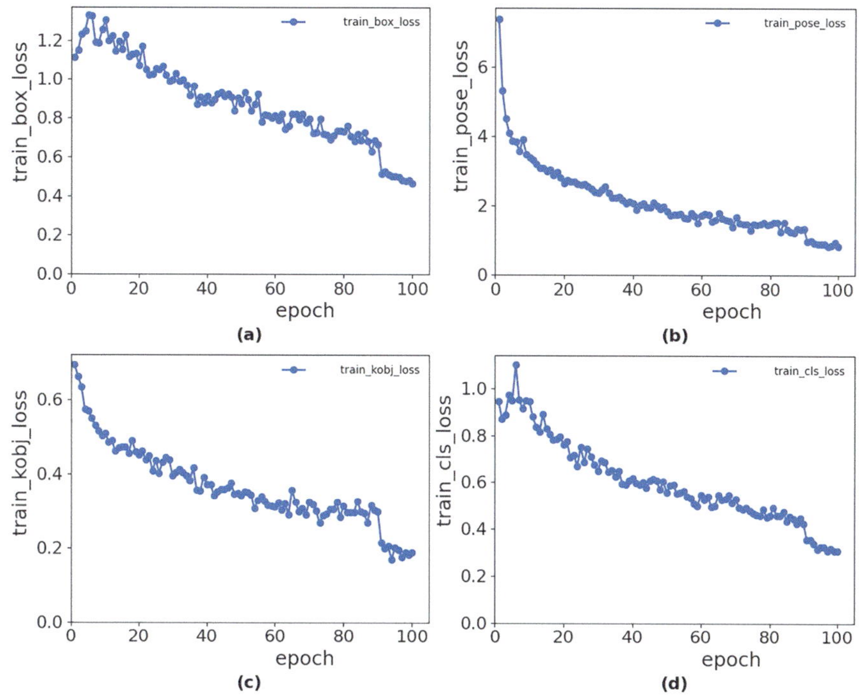

Fig. 20.2 Loss function downward trend

20.4.3 Posture Detection and Tracking Test

The best model weights file is derived after 100 times of model training. The test experiments use the best model weights file for attitude detection and tracking of the target, and the best model weights file has the highest test detection accuracy. At the same time, the best model weight file also saves the structure definition of the training model, as well as the optimiser state, learning rate scheduler state and other information during the training process. In this study, pose detection and tracking is done for both image and video formats. The video example pose detection and tracking procedure is illustrated in Fig. 20.3 and the virtual human images sample pose detection test results are illustrated in Fig. 20.4.

20.4.4 Test Experiments

Experimental Evaluation Criteria. To visually compare the recognition capabilities of the models, the mean average precision (mAP), the single training time of the samples, the memory occupation of the samples for training, and the model inference

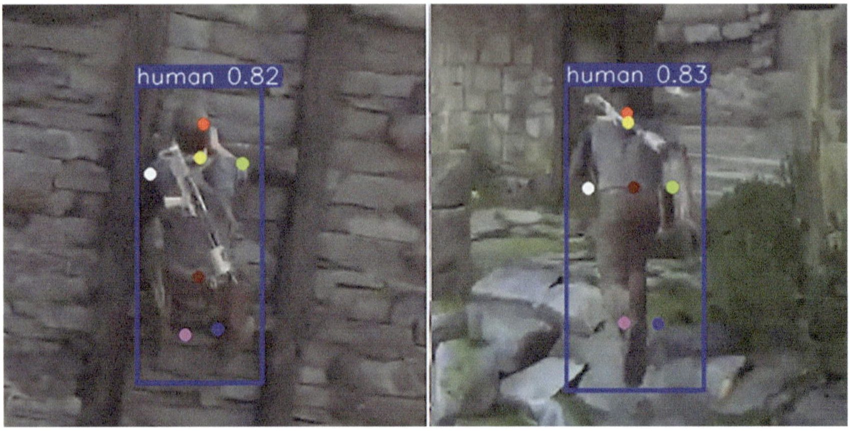

Fig. 20.3 Video sample pose detection process

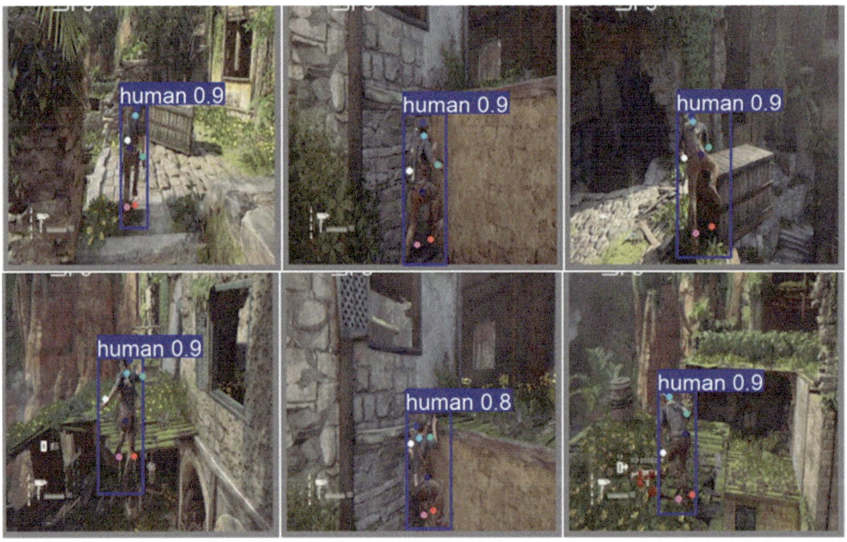

Fig. 20.4 Virtual human posture detection

time are jointly used as the assessment basis for this study. The formula for mean average precision is shown in (1).

$$\mathrm{mAP} = \frac{1}{n}\sum_{i=1}^{n}\int_{0}^{1}\left(\frac{\mathrm{Tp}}{\mathrm{Tp}+\mathrm{Fp}}\times 100\%\right)\mathrm{d}\left(\frac{\mathrm{Tp}}{\mathrm{Tp}+\mathrm{Fn}}\times 100\%\right) \tag{20.1}$$

In the expression Tp and Fp represent the positive and negative samples identified using the model, and Fn represents the model identification errors, that is positive and negative sample identification errors.

Test Experiment Results. Figure 20.5 shows the average accuracy mean value under different IoU thresholds for the sample training process, the greater the mean accuracy, the higher the detection accuracy of the model on the target. Figure 20.5a shows that the mean value of the average accuracy decreases when the model is trained for fewer times, and as the number of times the model is trained increases, the mean value of the average accuracy of the detection of the prediction frames is close to 100% and stable after 40 samples are trained. It shows that this model has a high and stable performance in detecting the samples when threshold of the IoU is set to 0.5. Figure 20.5b demonstrate that the mean value of the average accuracy of the detection of the prediction frames fluctuates a lot when the number of times the model is trained is less than 20, and the value of the average accuracy of the prediction frames stabilises and converges to 80% after the number of times the model is trained is more than 40 times. The average accuracy of the detection decreases relative to the threshold setting of 0.5 when threshold of the IoU is set from 0.5 to 0.95. Figure 20.5c shows the mean average accuracy of detection of joints at an IoU threshold of 0.5, which fluctuates when the number of trainings is less than 20, rises significantly to 100% when the number of trainings is from 20 to 60, and stabilises after the number of trainings is more than 60. Figure 20.5d shows the average accuracy mean for detection of joints at IoU thresholds from 0.5 to 0.95. Average accuracy also increases steadily with the number of training sessions, eventually reaching 80%. It shows that the model still performs well under more stringent criteria.

20.5 Model Comparison

In this study, the two models YOLOv8n-pose and YOLOv8s-pose are compared horizontally under the same hardware configuration as well as the same environment configuration, and the experimental findings of the model-test comparison are illustrated in Fig. 20.6. A total of four subplots are included, which represent the line graphs corresponding to the average precision when the IoU threshold of the four performance metrics is 0.5, the average precision when the threshold of IoU is 0.5 to 0.95, the precision rate, and the recall rate under the two different models, respectively. Figure 20.6a indicates that YOLOv8n reaches the earliest average accuracy of 90% at 10 rounds of training is higher than 80% of YOLOv8s, which indicates that YOLOv8n takes less time cost to reach the training target and outperforms YOLOv8s in terms of convergence of average accuracy. Figure 20.6b represents the performance of average accuracy at different thresholds from 50 to 95%, after 100 model trainings YOLOv8n achieves an average accuracy of 80.1%, while YOLOv8s only has 76%. This shows that YOLOv8n has better performance when the threshold is increased. Figure 20.6c represents the variation of the accuracy rate of the two models with the number of sample training. both YOLOv8n and YOLOv8s show a rapid rising

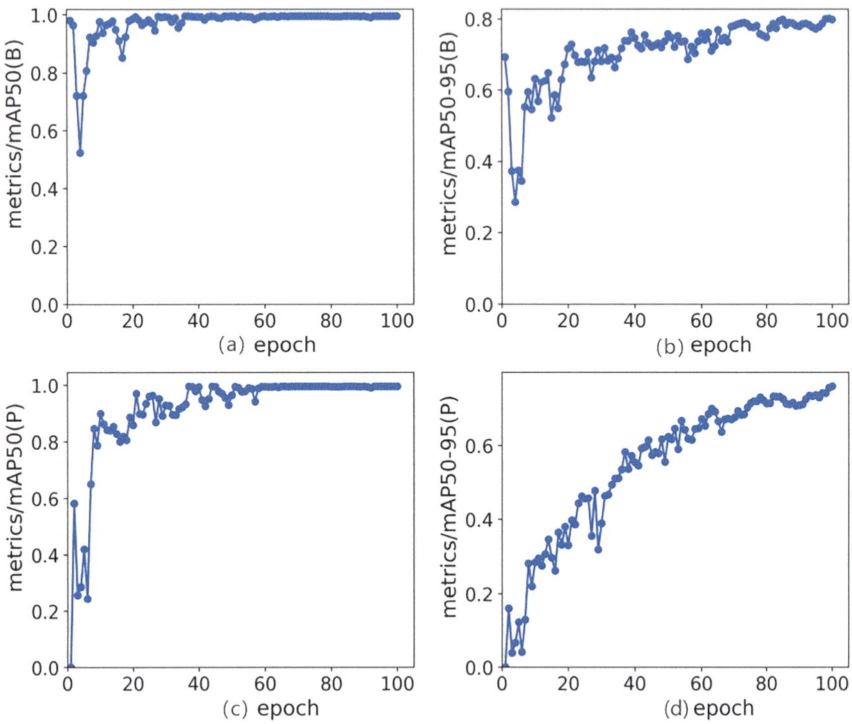

Fig. 20.5 Model detection average accuracy means

trend in the accuracy rate at 20 rounds of training, and then stabilise above 95%, the two models show consistently high accuracy rates. However, YOLOv8n performs better in terms of stability. Figure 20.6d illustrates the evolution of the recall of the training model with the number of training rounds. YOLOv8n has a higher recall than YOLOv8s by five percentage points on average in the first and middle stages of model training, and the performance of the two models converges in the later stages of model training, both reaching a recall close to 100%. The specific model parameters and test performance are shown in Table 20.2.

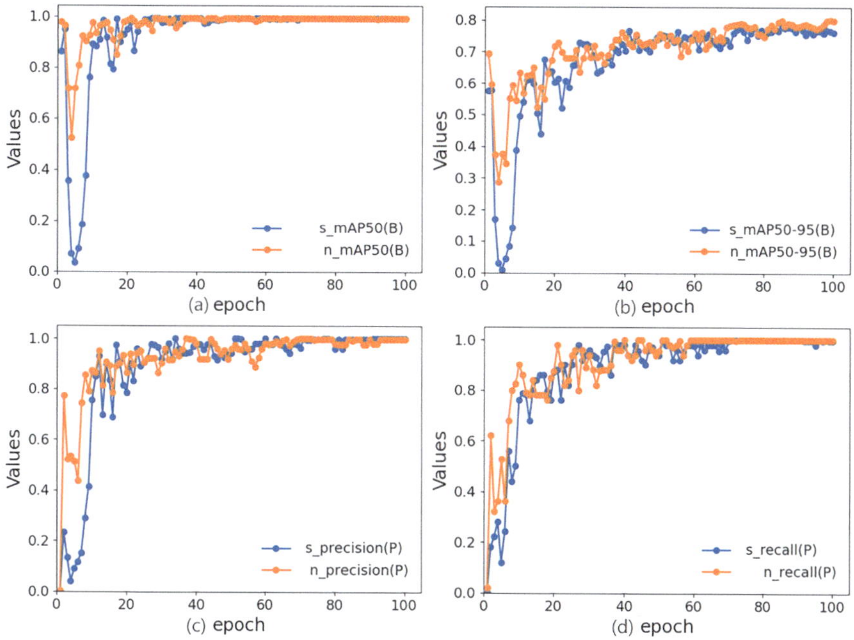

Fig. 20.6 Comparison chart of model test results

Table 20.2 Model parameter

Models	Params (M)	FLOPs (B)	mAP-pose (50–95) (%)	Precision (%)	Recall (%)
YOLOv8n-pose	3.3	9.2	80.1	98	96
YOLOv8s-pose	11.6	30.2	76	98	95

20.6 Conclusion

This paper concentrates on the performance of various modelling algorithms in the task of human posture detection and tracking. A comparison of the parameter performance of the two models, YOLOv8n-pose and YOLOv8s-pose, was conducted in training human posture detection and tracking pairs of samples. The results reveal that the average precision mean, precision, and recall of YOLOv8n-pose are better than or equal to those of YOLOv8s-pose. This indicates that YOLOv8n-pose has high performance while ensuring the smallest possible resource footprint. YOLOv8n-pose's parameter count is only 3.3 M, much smaller than YOLOv8s-pose's 11.6 M, and thus it is more lightweight, with only 9.2B floating-point operations per second, but its average accuracy is 80.1% higher than the average accuracy of YOLOv8s-pose in the low-precision, low-sample-volume attitude detection and tracking task.

YOLOv8s-pose's 76%, and achieves 98% precision and 96% recall. It indicates that the model performs well in the training and testing tasks.

References

1. Liu, G., Li, X., Xu, C., et al.: FMCW radar-based human sitting posture detection. IEEE Access **11**, 102746–102756 (2023)
2. Mousse, M.A., Atohoun, B.: Saliency based human fall detection in smart home environments using posture recognition. Health Inform. J. **27**(3) (2021)
3. Tlili, F., Haddad, R., Bouallegue, R., et al.: A real-time posture monitoring system towards bad posture detection. Wirel. Pers. Commun. **120**, 1207–1227 (2021)
4. Wang, S., Zhang, G.: A empirical research on AI-powered athletic posture detection in sports training. Rev. Int. Métodos Numér Cálc. Diseñoing **40**(2), 35 (2024)
5. Li, Y., Wu, Y., Chen, X., et al.: Beyond human detection: a benchmark for detecting common human posture. Sensors **23**, 8061 (2023)
6. Kulikajevas, A., Maskeliunas, R., Damaševičius, R.: Detection of sitting posture using hierarchical image composition and deep learning. PeerJ Comput. Sci. **7**, e442 (2021)
7. Huang, J., Zhou, D.: A scalable real-time computer vision system for student posture detection in smart classrooms. Educ. Inf. Technol. **29**, 917–937 (2024)
8. Guo, Y., Chen, Y., Deng, J., et al.: H. Identity-preserved human posture detection in infrared thermal images: a benchmark. Sensors **23**, 92 (2023)
9. Cun, W., Mo, R., Chu, J., et al.: Sitting posture detection and recognition of aircraft passengers using machine learning. Artif. Intell. Eng. Des. Anal. Manuf. **35**(3), 284–294 (2021)
10. Zhang, J., Wu, C., Wang, Y.: Human fall detection based on body posture spatio-temporal evolution. Sensors **20**, 946 (2020)
11. Zhou, S., Huang, X., Mohammad, S., et al.: Transferability of adversarial attacks on tiny deep learning models for IoT unmanned aerial vehicles. IEEE Internet Things J. **11**(12), 21037–21045 (2024)
12. Sun, K.K., Gao, M., Li, B.: Research on two interpolation algorithms for anthropometric data completion—A case study on the data of elderly males in Fujian, China. J. Netw. Intell. **7**(4), 1083–1101 (2022)
13. Chiu, Y.J., Wang, W.Z., Hu, Y.C., et al.: Maximum power development under varying solar radiation based on hysteresis-loop comparison of selected duty-cycle perturbation algorithms on a mobility scooter. J. Netw. Intell. **8**(4), 1540–1556 (2023)

Chapter 21
Emotional States and Electoral Insights: Analyzing Political Leaders' Emotional Styles Through Deep Learning-Based Facial Expression Recognition

Lei Jiang, Ge Chen, Chuansheng Wang, Junxian Zheng, and Shutang Liu

Abstract This study utilizes facial expression recognition technology to analyze the facial expressions of Trump and Biden during debate videos, aiming to explore the emotional states of both leaders. By employing deep learning algorithms to extract and classify their expressions, this study captures emotional variations and behavioral differences displayed during intense debates. The analysis reveals each leader's emotional response frequency, intensity, and specific affective tendencies, allowing us to infer their emotional style and emotional states in public settings. This technology can be used by candidates to analyze their personal TV debate performance and optimize their expression control. The study provides new insights into the analysis of emotions among politicians during television debates and demonstrates potential applications in facial expression recognition technology in electoral psychology.

L. Jiang · G. Chen · S. Liu (✉)
Faculty of Humanities and Arts, Macau University of Science and Technology, Macau, P.R. China
e-mail: 2009853fac30001@student.must.edu.mo

L. Jiang
e-mail: macaujl826@gmail.com

G. Chen
e-mail: chenge224466@outlook.com

C. Wang
Department of Automatic Control, Polytechnic University of Catalonia, Barcelona, Spain
e-mail: wangcs95@163.com

J. Zheng
Office of International Affaris, Minjiang University, Fuzhou, P.R. China
e-mail: zjx917400@163.com

S. Liu
Office of Academic Affairs, Minjiang University, Fuzhou, P.R. China

S.-C. Chu et al. (eds.), *Advances in Intelligent Data Analysis and Applications*,
Smart Innovation, Systems and Technologies 445,
https://doi.org/10.1007/978-981-96-7277-6_21

21.1 Introduction

In recent years, with the rapid development of computer vision and deep learning technologies [1–5], facial expression recognition has gained extensive attention and applications in emotion analysis and behavioral research. As a vital form of non-verbal communication, facial expressions can reflect an individual's emotional state and psychological characteristics [6, 7]. For political leaders, emotional reactions and presentation styles not only shape public perceptions of their personality but may also influence their image among voters and electoral outcomes [8]. Therefore, systematically analyzing political figures' emotional reactions in public settings can offer new perspectives and methodologies for election psychology and social behavior research [9] (Fig. 21.1).

This study selects the state-of-the-art YOLOx algorithm as the core tool for video-based facial expression recognition [10, 11]. YOLOx, with its efficient and precise object detection capabilities, offers significant advantages in real-time performance and detection accuracy, making it particularly suitable for dynamic and fast-changing expressions in debate scenarios [12]. Throughout the study, this algorithm is used to detect and classify the facial expressions of Trump and Biden during debates in real-time, enabling the capture of subtle changes in their emotional responses [13]. YOLOx not only excels in traditional object detection tasks but has also been widely applied in facial expression recognition due to its structured design, fast inference speed, and low computational requirements [14, 15].

By analyzing the frequency, intensity, and emotional tendencies of the two leaders' responses during debates, this study reveals similarities and differences in their personality traits from the perspective of emotional style and behavior [16, 17]. Based on these facial expression recognition results, the study also predicts and evaluates applications in related fields [18]. This deep learning-based facial expression analysis method provides a new perspective for evaluating the personalities of political figures, helping to understand their influence on the public's perception.

The innovation of this study lies in applying advanced deep learning algorithms to political psychology research, achieving quantitative analysis of individuals' emotions and personalities through video-based facial expression recognition [19]. This approach not only enriches the research methods in electoral psychology but

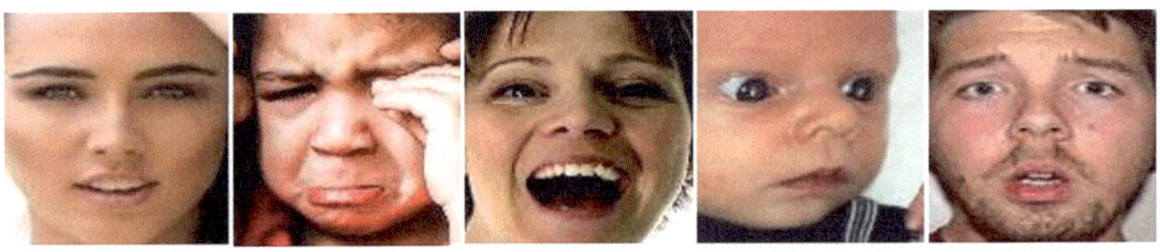

(a) Natural (b) Sad (c) Happy (d) Surprised (e) Dull

Fig. 21.1 Comparison of facial expressions under different emotions also demonstrates the broad potential of facial expression recognition in the social sciences [20, 21].

The contributions of this paper are listed as follows:

Application of Advanced Deep Learning Algorithms in Political Psychology: This study innovatively applies the YOLOx algorithm for facial expression recognition within political psychology research, providing a quantitative approach to analyze political figures' emotions and personalities.

Real-time Detection and Analysis of Emotional Responses in Political Debates: By employing YOLOx to capture the facial expressions of Trump and Biden during debates, this study enables real-time tracking and classification of their emotional responses, offering a detailed behavioral analysis of political figures in dynamic, high-stakes settings.

New Perspectives for Electoral Psychology and Social Behavior Analysis: This study provides a new approach to electoral psychology by using facial expression analysis to assess candidates' psychological states. Highlighting the broader applicability of facial expression recognition in the social sciences.

21.2 Related Works

In recent years, facial expression recognition and emotion analysis have become research hotspots in the fields of computer vision and affective computing [22–24]. With advancements in deep learning algorithms, especially supported by technologies such as Convolutional Neural Networks (CNN) and Generative Adversarial Networks (GAN) [25, 26], facial expression recognition has achieved significant improvements in performance. Traditional studies on facial expression recognition have mainly focused on classifying expressions in static images, identifying emotions by analyzing facial muscle movement characteristics [27]. However, this approach has limitations when applied to dynamic videos, as it struggles to capture rapidly changing micro-expressions and emotional transitions [28].

An increasing number of studies have applied deep learning algorithms to video-based facial expression recognition [29, 30]. Within this field, the YOLO (You Only Look Once) series of algorithms, known for efficient real-time object detection, has been widely used for facial detection and expression recognition tasks [31]. As a flagship version in this series, YOLOx further enhances detection accuracy and real-time performance, making it especially suitable for dynamic scenarios and thus highly valued in video expression recognition [21]. Additionally, some studies have combined YOLO with RNN or LSTM to capture temporal features within video sequences, thereby improving recognition accuracy [12].

In the analysis of political figures' expressions and personalities, a few studies have explored political leaders' emotions and personal traits through video analysis [18]. For instance, some research uses facial expression recognition to analyze speech videos, revealing the frequency and intensity of leaders' emotional expressions, thereby inferring their emotional control and personality tendencies in public settings [20]. However, most existing studies focus on emotion classification and expression recognition alone, with limited application of the findings to electoral prediction and public image assessment [8]. This study combines facial expression recognition with electoral psychology by analyzing Trump's and Biden's expressions during debates. It provides a systematic analysis of their personality and emotional expression styles.

21.3 Method

The overall structure of the YOLOx can be seen in Fig. 21.2. This study employs the YOLOx model for real-time facial expression recognition, aimed at efficiently capturing the micro-expressions of Trump and Biden during debates. YOLOx integrates multiple optimizations in the backbone network, feature extraction, prediction heads, and positive sample matching, making it ideal for highly dynamic expression recognition tasks.

In the backbone network design, YOLOx builds on the YOLOv5 structure with significant enhancements, including reducing the initial convolutional kernel size from 6 to 3 to lower computational costs and improve inference speed. The CSP (Cross Stage Partial) module reduces convolutions from three to two, preserving accuracy while boosting processing speed, crucial for real-time tasks involving rapidly changing facial expressions.

YOLOx's feature extraction omits additional convolutional layers traditionally used in YOLO models, accelerating computation while retaining effective feature

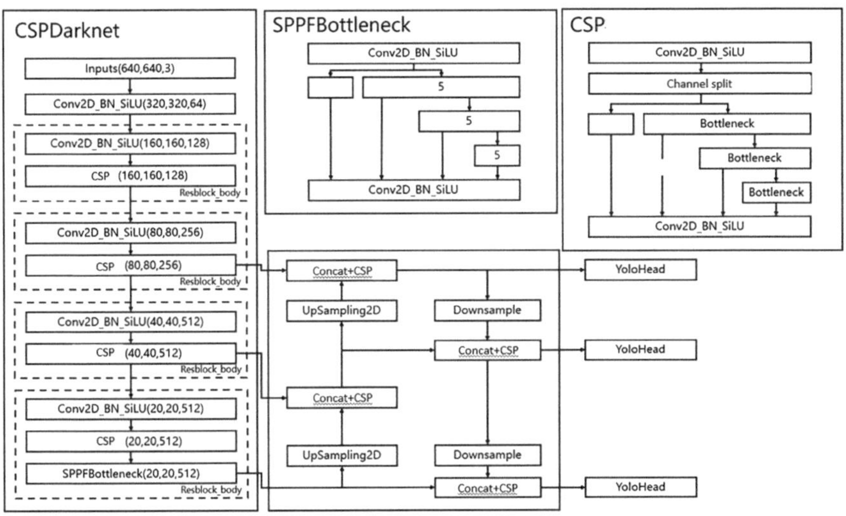

Fig. 21.2 The overall architecture of the YOLOx

extraction capabilities through the CSP module. The streamlined design improves real-time performance without compromising accuracy, making it ideal for detecting dynamic expressions.

The prediction head incorporates the DFL (Distribution Focal Loss) module, which uses a probability distribution approach to enhance fine detail capture. With the DFL length set to 16, YOLOx achieves high spatial resolution to handle subtle facial movements and complex expression changes.

YOLOx's anchor-free design, similar to YOLOx, allows flexibility in detecting irregular target shapes, while the adaptive multi-positive matching strategy improves detection accuracy. These optimizations equip YOLOx to capture realistic facial expressions in dynamic settings, supporting emotion analysis of political figures.

21.3.1 Overall Structure Analysis

YOLOx's network structure includes the Backbone, FPN (Feature Pyramid Network), and YOLO Head. The Backbone extracts primary feature information at different scales, enabling the model to capture subtle facial region changes. The FPN fuses these features across multiple scales, enhancing the model's sensitivity to expression variations. The YOLO Head, utilizing an anchor-free approach, performs classification and regression on each feature point, offering greater flexibility and improved accuracy.

21.3.2 Feature Pyramid Network (FPN)

YOLOx uses multi-scale feature fusion in the FPN for robust expression recognition. It extracts three feature layers at various scales and combines them through up sampling and down sampling to capture both fine and broad details. This multi-scale fusion allows YOLOx to maintain high recognition accuracy across varying facial poses and lighting conditions.

21.3.3 YOLO Head

YOLOx's decoupled YOLO Head separates classification and regression tasks to avoid multi-task interference. The DFL module used in the regression task generates smooth, accurate predictions by computing regression values based on probability distributions. This decoupling and use of DFL enhance the model's ability to detect subtle changes in expressions accurately.

21.3.4 Predicted Result Decoding

For each detected expression, YOLOx uses three feature layers to generate bounding boxes and classification outputs. Each feature point corresponds to a bounding box prediction, and a decoding step translates these predictions to image coordinates. This process enables the precise localization of facial expressions in the image.

21.3.5 Score Filtering and Non-maximum Suppression (NMS)

After generating predictions, YOLOx applies score filtering and NMS to refine results. Low-confidence boxes are filtered out, and NMS removes overlapping boxes by retaining only the highest-confidence box per region. This final step improves expression localization, ensuring accurate and reliable data for subsequent emotion analysis.

21.4 Experimental Results and Analysis

21.4.1 Datasets and Implementation Details

21.4.1.1 Facial Expression Datasets

In this study, we constructed a self-collected dataset containing 300 images, with an equal distribution of 50% for Trump and Biden. From this dataset, we selected 112 images as the test set. This dataset was used for training and validating our facial expression recognition model. The balanced distribution of the two individuals in the dataset ensures that the model can fairly learn and evaluate the facial features of Trump and Biden across diverse emotions and expressions, enhancing the model's generalization ability in expression recognition tasks. Figure 21.3 represents one of the sources for our database.

21.4.1.2 Implementation Details

The experiments were conducted on a laptop platform with specific hardware configurations, utilizing the PyTorch framework for algorithm development and model training. The hardware specifications include an Intel Core i7 processor, paired with 16 GB of memory and an NVIDIA GeForce RTX 3060 graphics card, providing a capable computing environment. In terms of training strategy, all models uniformly employed the Adam optimization algorithm, with a batch size of 16 to maintain a

Fig. 21.3 Screenshot of a June 28, 2024, televised debate between two individuals. Sourced from an NBC News YouTube Video

balance between computational efficiency and memory usage. To account for dataset variances, training durations were customized accordingly, with 300 epochs allocated for the RAF-DB dataset and 500 epochs for the more complex SFEW dataset, accommodating differences in both data volume and complexity. An initial learning rate of 0.001 was set, along with an exponential decay strategy applied every 64 epochs throughout each training period. This strategy involved reducing the learning rate by a factor of 0.9 at each interval, facilitating gradual convergence toward the optimal solution.

21.4.2 Evaluation Metrics

To evaluate the performance of the FER-YOLO-Mamba model, we use a series of key metrics:

Precision: Precision measures the proportion of actual positive samples among those predicted as positive by the model, defined as:

$$\text{Precision} = \frac{\text{TP}}{\text{TP} + \text{FP}} \tag{21.1}$$

where TP is the number of true positives, and FP is the number of false positives.

Avg_Precision: Avg_Precision is the average Precision across all classes, calculated as

$$\text{Avg\_Precision} = \frac{1}{N}\sum_{i=1}^{N}\text{Precision}_i \tag{21.2}$$

where Precision_i represents the Precision for the i-th class, and N is the total number of classes.

Recall: Recall indicates the model's ability to identify true positive samples, defined as

$$\text{Recall} = \frac{\text{TP}}{\text{TP} + \text{FN}} \tag{21.3}$$

where FN is the number of false negatives.

Avg_Recall: Avg_Recall is the average Recall across all classes, calculated as

$$\text{Avg\_Recall} = \frac{1}{N}\sum_{i=1}^{N}\text{Recall}_i \tag{21.4}$$

where Recall_i represents the Recall for the i-th class.

F1 Score: The F1 Score is the harmonic mean of Precision and Recall, balancing both metrics, defined as

$$F1 = 2 \cdot \frac{\text{Precision} \cdot \text{Recall}}{\text{Precision} + \text{Recall}} \tag{21.5}$$

Avg_F1: Avg_F1 is the average $F1$ Score across all classes, calculated as

$$\text{Avg\_}F1 = \frac{1}{N}\sum_{i=1}^{N}F1_i \tag{21.6}$$

where $F1_i$ represents the $F1$ Score for the i-th class.

21.4.3 Experimental Results Across Different Classes

The data in Table 21.1 reveals varying performance of YOLOx on the facial expression classification tasks across the RAF-DB and TrumpBiden-FaceSet datasets. In the RAF-DB dataset, YOLOx achieved relatively high precision values for most emotion categories, demonstrating its potential advantages in expression recognition tasks. Notably, the "Happy" category showed a significant increase in recall rate, indicating the model's reliability in recognizing this easily distinguishable expression.

Table 21.1 Performance of YOLOx on facial expression classification across RAF-DB and TrumpBiden-FaceSet datasets

	YOLO-vX			
Classes	F1 score	Recall (%)	Precision (%)	Dataset
Anger	0.71	62.34	81.36	RAF-DB
Disgust	0.59	52.38	64.29	
Fear	0.57	51.43	90.28	
Happy	0.92	93.39	67.39	
Sad	0.79	82.51	85.80	
Surprised	0.83	81.29	85.80	
Average	0.74	71.26	76.70	
Happy	0.94	96.15	92.59	TrumpBiden-FaceSet
Sad	0.89	85.19	92.00	
Angry	0.68	56.00	87.50	
Average	0.84	79.11	90.70	

However, for the "Anger" and "Surprise" categories, YOLOx exhibited relatively low recall rates, suggesting that the model may overlook samples from these categories. Although the precision for "Anger" was high, the limited recall rate indicates weaker coverage in detecting this category, potentially missing genuine "Anger" samples. This phenomenon may stem from the less pronounced features of these emotion categories, making it challenging for the model to accurately capture and distinguish them.

Furthermore, the F1 scores for "Disgust" and "Fear" were relatively low across both datasets, particularly in the RAF-DB dataset, indicating significant challenges in recognizing these categories. This issue could be attributed to the subtle facial features associated with these emotions or to a high degree of similarity between these categories and others, which complicates accurate classification.

Overall, although YOLOx demonstrated remarkable performance in the "Happy" category, there remains room for improvement in recognizing complex emotions such as "Fear" and "Disgust." This highlights the importance of further model optimization to enhance accuracy and robustness in multi-category emotion classification tasks.

21.4.4 User Interface Design

Figure 21.4, this user interface serves as a display platform for a facial expression detection and recognition system, designed to provide an intuitive and efficient experience for expression recognition. The system title is displayed at the top, clearly defining the application context. The central area shows the current image or video frame in real-time, where the system automatically highlights the facial

Fig. 21.4 User interface design

region and annotates the recognized expression category along with its confidence score. The right panel provides various functionalities, including image and video file loading, real-time camera capture, recognition control, and data export options, allowing users to conduct diverse expression recognition tasks. Additionally, the right panel displays real-time results, including confidence, expression category, and facial position coordinates, while visually illustrating the distribution of different expressions. The bottom section records detailed information for each detection, covering time, recognition results, processing duration, and file path, supporting data tracking and management. This design emphasizes intuitive information display and ease of operation, aiming to achieve efficient expression recognition and data management.

21.5 Conclusion

Based on the research conducted, this study demonstrates the feasibility and effectiveness of using deep learning-based facial expression recognition, particularly through the YOLOx algorithm, to analyze the emotional styles of political figures. By applying facial expression recognition to debate videos of Trump and Biden, we identified notable differences in their emotional expressions, providing insights into their respective emotional states. The YOLOx model's robustness and real-time performance proved valuable in capturing the nuanced expressions exhibited in high-stakes scenarios, highlighting its applicability for dynamic facial expression analysis in political settings.

Through detailed examination of expression frequency, intensity, and tendencies, the study provides a unique perspective on the emotional and behavioral patterns of political leaders, offering potential implications for electoral psychology. The real-time facial expression detection and analysis capabilities of YOLOx further reinforce the potential of deep learning applications in social science research, particularly in understanding the impact of non-verbal communication on public perception.

In conclusion, this research contributes to the growing field of political psychology by introducing a quantitative approach to analyze and interpret the emotional cues of political figures. Future work may involve enhancing model performance in recognizing subtle and complex emotions, as well as expanding the analysis to a broader range of political figures and scenarios. This study lays the groundwork for further exploration of deep learning's role in analyzing public figures' personalities, providing a valuable tool for assessing their influence on audiences and underlying electoral emotion outcomes.

Author Contributions Conceptualization, Lei Jiang and Shutang Liu; methodology, Lei Jiang and Shutang Liu; software, Chuansheng Wang; validation, Shutang Liu and Chuansheng Wang; formal analysis, Chuansheng Wang; investigation, Chuansheng Wang, Lei Jiang and Shutang Liu; resources, Chuansheng Wang; data curation, Chuansheng Wang; writing-original Draft, Lei Jiang, Shutang Liu and Chuansheng Wang; writing-review& editing, Lei Jiang, Shutang Liu, Chuansheng Wang, Ge Chen, and Junxian Zheng; visualization, Chuansheng Wang; supervision, Lei Jiang and Shutang Liu; project administration, Lei Jiang and Shutang Liu; funding acquisition, Lei Jiang. All authors have read and agreed to the published version of the manuscript.

References

1. Zhou, S., Huang, X., Obaidat, M.S., Alzahrani, B.A., Han, X., Kumari, S., Chen, C.-M.: Transferability of adversarial attacks on tiny deep learning models for IoT unmanned aerial vehicles. IEEE Internet Things J. **11**(12), 21037–21045 (2024)
2. Wang, X., Lv, J., Slowik, A., Parameshachari, B.D., Li, K., Chen, C.-M., Kumari, S.: DLLF-2EN: energy-efficient next generation mobile network with deep learning-based load forecasting. IEEE Trans. Netw. Serv. Manag. 1–1 (2024)
3. Wang, C., Hu, J., Luo, X., Kwan, M.-P., Chen, W., Wang, H.: Color-dense illumination adjustment network for removing haze and smoke from fire scenario images. Sensors **22**(3), 911 (2022)
4. Feng, T., Wang, C., Chen, X., Fan, H., Zeng, K., Li, Z.: UR-net: a U-net based residual network for image dehazing. Appl. Soft Comput. **102**, 106884 (2021)
5. Zhang, F., Wang, C.: MSGAN: generative adversarial networks for image seasonal style transfer. IEEE Access **8**, 104830–104840 (2020)
6. Reza, A.M.: Realization of the contrast limited adaptive histogram equalization (CLAHE) for real-time image enhancement. J. VLSI Signal Process. Syst. Signal Image Video Technol. **38**, 35–44 (2004)
7. Shi, Z., Feng, Y., Zhao, M., et al.: Normalised gamma transformation-based contrast limited adaptive histogram equalisation with colour correction for sand–dust image enhancement. IET Image Proc. **14**(4), 747–756 (2020)
8. Yu, S., Zhu, H., Wang, J., et al.: Single sand-dust image restoration using information loss constraint. J. Mod. Opt. **63**(21), 2121–2130 (2016)
9. Gao, G., Lai, H., Jia, Z., Liu, Y., Wang, Y.: Sand-dust image restoration based on reversing the blue channel prior. IEEE Photon. J. **12**(2), 1–16 (2020)
10. Meng, X., Huang, J., Li, Z., Wang, C., Teng, S., Grau, A.: DedustGAN: unpaired learning for image dedusting based on Retinex with GANs. Expert Syst. Appl. **243**, 122844 (2024)
11. Huang, J., Li, Z., Wang, C., Yu, Z., Cao, X.: FFNet: a simple image dedusting network with feature fusion. Concurr. Comput.: Pract. Exp. **33**(24), e6462 (2021)

12. Zhong, M., Li, Z., Wang, C., Huang, J., Zhao, W., Fan, H.: Quality assessment of electrocardiogram signals using contrastive learning. In: 2023 13th International Conference on Information Technology in Medicine and Education (ITME), pp. 323–328. IEEE (2023)
13. Wang, C., Zhang, H., Li, Z., Zhou, X., Cheng, Y., Chen, R.: White blood cell image segmentation based on color component combination and contour fitting. Curr. Bioinform. **15**(5), 463–471 (2020)
14. Wang, C., Li, Z., Wu, J., Fan, H., Xiao, G., Zhang, H.: Deep residual haze network for image dehazing and deraining. IEEE Access **8**, 9488–9500 (2020)
15. Shen, C., Hu, J., Wang, C., Li, D.: LFNet: lightweight fire smoke detection for uncertain surveillance environment. In: 2020 6th International Conference on Big Data and Information Analytics (BigDIA), pp. 11–17. IEEE (2020)
16. Zhou, X., Li, Z., Xie, H., Feng, T., Lu, Y., Wang, C., Chen, R.: Leukocyte image segmentation based on adaptive histogram thresholding and contour detection. Curr. Bioinform. **15**(3), 187–195 (2020)
17. Zhang, H., Wang, C., Li, X., Sun, B., Jiang, D.: Design and implementation of an infrared radiant source for humidity testing. Sensors **18**(9), 3088 (2018)
18. Huang, J., Wang, C., Zhao, W., Grau, A., Xue, X., Zhang, F.: LTDNet-EEG: a lightweight network of portable/wearable devices for real-time EEG signal denoising. IEEE Trans. Consum. Electron. (2024)
19. Huang, J., Xu, H., Liu, G., Wang, C., Hu, Z., Li, Z.: SIDNet: a single image dedusting network with color cast correction. Signal Process. **199**, 108612 (2022)
20. Fu, A., Ma, J., Wang, C., Zhou, C., Li, Z., Teng, S.: Traditional Chinese medicine health status identification with graph attention network. In: International Conference on Machine Learning for Cyber Security, pp. 1–14. Springer (2022)
21. Yan, J., Wang, Y., Fan, H., Huang, J., Grau, A., Wang, C.: LEPF-net: light enhancement pixel fusion network for underwater image enhancement. J. Marine Sci. Eng. **11**(6), 1195 (2023)
22. Chen, Z., Wang, C., Zhang, F., Zhang, L., Grau, A., Guerra, E.: All-in-one aerial image enhancement network for forest scenes. Front. Plant Sci. **14**, 1154176 (2023)
23. Wang, C., Grau, A., Guerra, E., Shen, Z., Hu, J., Fan, H.: Semi-supervised wildfire smoke detection based on smoke-aware consistency. Front. Plant Sci. **13**, 980425 (2022)
24. Zhou, X., Wang, C., Li, Z., Zhang, F.: Adaptive histogram thresholding-based leukocyte image segmentation. In: Advances in Intelligent Information Hiding and Multimedia Signal Processing: proceedings of the 15th International Conference on IIH-MSP in Conjunction with the 12th International Conference on FITAT, vol. 2, pp. 451–459. Springer (2020)
25. Huang, J., Li, Z., Wang, C.: Image dust storm synthetic method based on optical model. In: Machine Learning for Cyber Security: third International Conference, ML4CS 2020, Guangzhou, China, October 8–10, 2020, Proceedings, Part III, pp. 215–226. Springer (2020)
26. Wang, C., Fan, H., Zhang, H., Li, Z.: Pixel-level dehazed image quality assessment based on dark channel prior and depth. In: 2019 IEEE International Conference on Parallel & Distributed Processing with Applications, Big Data & Cloud Computing, Sustainable Computing & Communications, Social Computing & Networking (ISPA/BDCloud/SocialCom/SustainCom), pp. 1545–1549. IEEE (2019)
27. Feng, T., Li, Z., Wang, C., Chen, X., Wu, J.: Image dehazing network based on dilated convolution feature extraction. In: 2019 12th International Congress on Image and Signal Processing, BioMedical Engineering and Informatics (CISP-BMEI), pp. 1–5. IEEE (2019)
28. Zhou, X., Li, Z., Wang, C.: Color space volume and superpixel-based leukocyte image segmentation. In: 2019 10th International Conference on Information Technology in Medicine and Education (ITME), pp. 84–88. IEEE (2019)
29. Shi, J., Wei, B., Zhou, G., et al.: SandFormer: CNN and transformer under gated fusion for sand dust image restoration. In: ICASSP 2023–2023 IEEE International Conference on Acoustics, Speech and Signal Processing (ICASSP), pp. 1–5. IEEE (2023)

30. Godard, C., Mac Aodha, O., Firman, M., et al.: Digging into self-supervised monocular depth estimation. In: Proceedings of the IEEE/CVF International Conference on Computer Vision, pp. 3828–3838 (2019)
31. Zhu, J.Y., Park, T., Isola, P., Efros, A.A.: Unpaired image-to-image translation using cycle-consistent adversarial networks. In: 2017 IEEE International Conference on Computer Vision (ICCV), pp. 2242–2251 (2017)

Chapter 22
A Novel Approach to Predicting IoT Connections Using Neural Networks and Macro-Economic Factors

Shen Chong, Bixian Zhang, Longcun Wang, Trong-The Nguyen, Thi-Kien Dao, and Haohan Zhao

Abstract With the rapid expansion of IoT due to advancements like 5G, accurately forecasting IoT connections is essential for effective infrastructure planning. This paper presents a novel prediction model that integrates neural networks and macroeconomic factors, surpassing traditional single-factor methods. By combining a backpropagation neural network (BPNN) with macroeconomic indicators like GDP and digital economic output, our approach offers enhanced predictive accuracy and relevance to industry demands.

22.1 Introduction

The Internet of Things (IoT) is revolutionizing industries and everyday life, driving a new era of interconnected devices and digital services [1]. The rapid commercialization of 5G technology has further accelerated IoT growth, delivering enhanced data rates, low latency, and the ability to support a massive scale of device connectivity [2]. However, this exponential growth also poses challenges, particularly for network operators who must balance infrastructure planning, user experience, and customer retention [3]. At the core of addressing these challenges is the ability to accurately predict the future number of IoT connections, which requires insight into both historical data patterns and the broader economic conditions that shape connectivity

S. Chong · B. Zhang · L. Wang · T.-T. Nguyen (✉) · T.-K. Dao · H. Zhao
School of Electronic Engineering, Fuzhou Institute of Technology, Fuzhou, China
e-mail: 1100405110@nkust.edu.tw

S. Chong
e-mail: shenchongcn@gmail.com

T.-K. Dao
e-mail: 1101405123@nkust.edu.tw

H. Zhao
e-mail: 306168499@qq.com

S.-C. Chu et al. (eds.), *Advances in Intelligent Data Analysis and Applications*,
Smart Innovation, Systems and Technologies 445,
https://doi.org/10.1007/978-981-96-7277-6_22

demand [4]. Traditional forecasting methods, such as curve fitting, moving averages, and exponential smoothing, are commonly used for IoT growth prediction based on historical data alone [5]. These methods, while effective in stable environments, often fall short when applied to complex and fast-evolving fields like IoT [6]. They tend to oversimplify the multi-faceted dynamics driving IoT growth, as they typically rely on single-factor models or secondary effects [7] that fail to account for broader macroeconomic and societal influences [8]. The result is limited predictive accuracy and adaptability, particularly in scenarios where economic, social, and technological variables interact dynamically to influence IoT adoption [9]. Consequently, improving forecasting accuracy for IoT connections is essential for developing infrastructure strategies that align with future demand [10].

In response to these limitations, this paper proposes a novel approach that integrates macroeconomic indicators with a neural network-based predictive framework [11]. The model leverages a backpropagation neural network (BPNN) [12] and incorporates key macroeconomic factors to capture a comprehensive view of IoT connectivity trends. We introduce a macroeconomic impact factor, referred to as the μ-factor, which represents critical economic indicators such as Gross Domestic Product (GDP) and digital economic output [13]. These indicators are particularly relevant to IoT growth, as they capture the broader economic context that drives connectivity needs and influences infrastructure investments. Using gray scale correlation analysis, the μ-factor is refined to ensure that only the most impactful macroeconomic variables are included in the predictive model, enhancing its relevance and reliability [14].

The backpropagation neural network (BPNN) lies at the heart of our prediction model, enabling it to capture the nonlinear relationships between IoT growth and the external economic environment [15]. Neural networks, particularly BPNNs, are highly effective for modeling complex interactions, making them well-suited to encapsulate the relationship between historical IoT connectivity data and broader economic factors. By combining the BPNN's capacity to model intricate data patterns with the μ-factor's representation of economic influences, our approach results in a robust, multidimensional predictive model that surpasses traditional single-factor forecasting techniques.

This paper is organized as follows: Sect. 22.2 examines the limitations of existing IoT forecasting methods and the unique challenges posed by the rapid expansion of IoT. Section 22.3 presents the design of our integrated μ-BPNN model, explaining the selection and relevance of both macroeconomic and neural network components including data sources, training methodology, and the correlation analysis used to develop the u-factor. Section 22.4 discusses the experimental results, comparing our model's performance to traditional forecasting methods, demonstrating its improved accuracy. Finally, Sect. 22.5 concludes with the implications of our approach for industry applications and suggests directions for future research, particularly in adapting the model to evolving economic conditions.

22.2 Analysis of Traditional Prediction Methods and Limitations

Accurate forecasting of IoT connections is increasingly critical in light of the sector's rapid growth, yet traditional forecasting approaches fall short in addressing the complex, multi-faceted nature of IoT expansion. The industry's growth trajectory is shaped by an interplay of policy, economic, social, and technological factors, which traditional models struggle to capture. This section examines these challenges, highlighting the limitations of single-factor models and their inadequacies in representing the dynamic IoT landscape.

22.2.1 Growth in the IoT Industry

Since 2010, the IoT industry has undergone explosive growth, driven by technological advancements, policy incentives, and increasing societal demand for connectivity [16]. The industry's expansion has been especially rapid with the advent of 5G technology, which supports greater device density, higher speeds, and improved reliability [17]. However, forecasting this growth accurately remains challenging, as traditional methods such as trend extrapolation, moving averages, and growth curve models are limited in capturing the intricacies of IoT expansion [18]. Figure 22.1 illustrates an example of IoT-expanded connections.

Forecasting the IoT industry accurately involves understanding a complex mix of factors beyond simple trend data [7]. Policy changes, for instance, can significantly impact IoT adoption rates, as regulatory frameworks influence technology

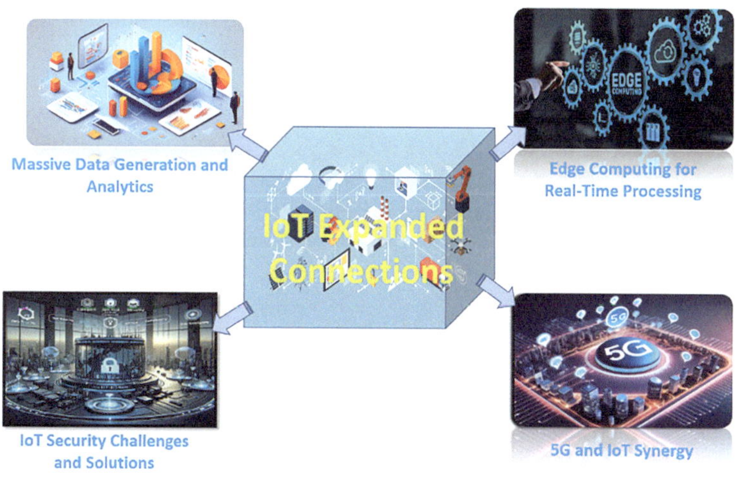

Fig. 22.1 An example of IoT-expanded connections

deployment and investment levels [19]. Economic conditions also play a critical role, affecting both consumer and enterprise spending on IoT solutions. Social factors, including user acceptance and societal demand for IoT-enabled applications, further complicate predictions. Traditional methods often oversimplify these diverse influences, relying on linear assumptions that inadequately represent the nonlinear and interdependent nature of IoT growth. As a result, these methods frequently struggle to forecast the rapid and often volatile shifts characteristic of the IoT industry's expansion.

22.2.2 Limitations of Single-Factor Forecasting Models

A key limitation of traditional IoT forecasting approaches is their reliance on single-factor models, which focus on isolated variables—often historical data alone [20]. While these models can offer basic insights based on past trends, they lack the robustness needed to account for the complex, interconnected factors influencing IoT. In reality, IoT growth is shaped by a dynamic blend of economic conditions, technological advancements, policy shifts, and social acceptance, all of which interact in ways that single-factor models cannot accurately represent. For example, models that extrapolate based solely on historical growth rates are inherently backward-looking, assuming that past trends will continue into the future. This assumption disregards the potential for macroeconomic shifts, regulatory changes, and rapid technological advancements to alter the course of IoT adoption. As the IoT industry grows, it is increasingly influenced by broader economic indicators such as GDP, digital infrastructure investments, and even global supply chain dynamics. These broader economic and societal factors are challenging to incorporate into single-factor models, as such models lack the ability to adjust dynamically to complex external influences [21]. Furthermore, single-factor models are limited in their adaptability to sudden or nonlinear growth patterns. IoT growth, driven by factors like the global rollout of 5G and increased digitalization, often exhibits exponential characteristics that traditional linear models fail to capture. The inability to account for these factors leads to limited predictive accuracy, as such models are unable to anticipate the nuanced ways in which macroeconomic, policy, and technological factors interact to influence IoT connectivity growth.

22.3 Proposed Forecasting Model

To improve the accuracy of IoT connection predictions, we propose a hybrid forecasting model that integrates macroeconomic impact factors with neural network techniques. By addressing the limitations of traditional methods, this approach enables a more nuanced and dynamic understanding of IoT growth.

22.3.1 Macro-Economic Impact Factors

Quantifying the macro-environmental impact on IoT growth requires consideration of multiple economic indicators, as IoT adoption is influenced by a broad range of socio-economic factors. In this model, we introduce the macro impact factor, represented as the μ-factor, which synthesizes key economic indicators directly relevant to IoT connectivity, such as GDP, digital economic output, and technology investment rates [13]. The μ-factor is designed to encapsulate the most impactful aspects of the broader economy that either facilitate or hinder IoT adoption and connectivity growth.

Using gray scale correlation analysis (GAR), the model identifies and selects only those macroeconomic indicators with the highest correlation to IoT growth [22]. This selective inclusion ensures that the μ-factor reflects the most relevant economic conditions impacting IoT, such as digital infrastructure spending and national GDP growth rates [23]. By integrating these economic indicators, the μ-factor provides a measure of the broader economic environment's influence on IoT, allowing the model to account for variability in adoption rates driven by changes in economic conditions [21]. The impact factor of the maximum correlation $\gamma_{\max}$ is superimposed on the normal development factor μ_0 (with a value of 1) for the number of IoTs connections as given:

$$\mu = \mu_0 + \mu_{\max}, \tag{22.1}$$

where $\mu_{\max}$ is $k_{ur\max}$, the growth rate of the impact factor of the maximum correlation coefficient $\gamma_{\max}$). The following steps of identifying a correlation analysis model.

Step 1: Correlation analysis of impact factors: (1) Correlation analysis sequence determination: The reference sequence is the number of IoT connections, denoted as $y(k)$; the comparison sequence is GDP and the digital economy, denoted as $X_i\ (k)$. (2) Correlation coefficient calculation: The correlation coefficient between $X_i\ (k)$ and $X_0\ (k)$ is denoted by $\Delta_i\ (k) = |y(k) - X_i(k)|$. The correlation coefficient calculation formula is

$$\xi_i = r(x_0(k), x_i(k)) = \left(\Delta_{oi}(k) + \rho \frac{\max}{i}\frac{\max}{k}\Delta_{oi}(k)\right), \tag{22.2}$$

where $\rho \in (0, 1)$, resolution system number, the smaller ρ is, the greater the resolution capability is, and usually $\rho = 0.5$ is taken. (3) Correlation calculation: According to the gray correlation algorithm, the correlation coefficient is as follows.

$$r_i = \frac{1}{n}\sum_{k=1}^{n} \xi_i(k), k = 1,2, 3 \ldots n, \tag{22.3}$$

Step 2: The most influential factor: The correlation coefficients of all the influencing factors have been calculated above, and the maximum is as follows:

$$\gamma_{\max} = \max\left(\gamma_i = \frac{1}{n}\sum_{k=1}^{n}\xi_i(k), k = 1,2,3\ldots n\right), \tag{22.4}$$

where $X_{i\max}$ be the influencing factor of the correlation coefficient $\gamma_{\max}$.

Step 3: μ factor calculation model: The influence factor for the maximum correlation $\gamma_{\max}$ is the growth rate $k_{\mu r\max} = \frac{\partial xi\max}{\partial t}$, so the expression is as follows:

$$\mu = \mu_0 + \mu_{r\max} = \mu_0 + k_{\mu r\max} = \mu_0 + \frac{\partial xi\max}{\partial t} = 1 + \frac{\partial xi\max}{\partial t}, \tag{22.5}$$

where μ factor generally refers to the influencing factor for future development within a certain period, for example, the development of macro factors from 2023 to 2025 can be used as a factor within 1–2 years later.

22.3.2 Enhanced Mapping with Neural Networks

To capture the complex, nonlinear relationships in IoT growth patterns, we employ a BPNN, which allows the model to establish an adaptive mapping between historical data and future connectivity trends [24]. Neural networks are particularly well-suited to this application because they can learn from historical data while identifying subtle patterns that may not be immediately apparent in single-factor models. Through iterative training, the BPNN refines its understanding of the relationships between input data points and the targeted output, capturing variations and shifts in IoT growth that are influenced by both historical trends and broader macroeconomic factors [25, 26]. The BPNN's ability to learn complex data relationships and adjust based on new information makes it ideal for IoT forecasting, where connectivity patterns are influenced by interrelated factors such as advancements in communication technologies, policy changes, and fluctuations in economic indicators. Figure 22.2 shows the structure of a backpropagation neural network (BPNN).

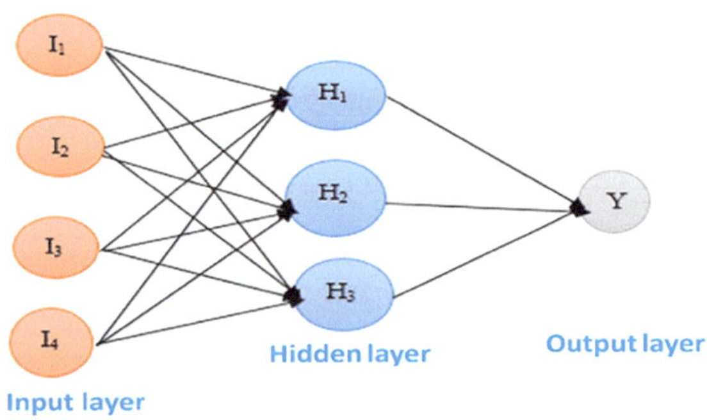

Fig. 22.2 A structure of a backpropagation neural network (BPNN)

Input and Output Layers: Given the multi-dimensional nature of IoT growth, the BPNN processes historical data in a four-input, one-output configuration. This setup allows the model to assess multiple temporal data points simultaneously, providing a more comprehensive understanding of how past data influences future trends. Each input corresponds to a specific time frame, enabling the model to capture temporal patterns in IoT connectivity, while the single output layer predicts the projected IoT connections at the given forecast horizon.

Hidden Layer and Activation Function: The hidden layer configuration was determined through empirical analysis and experimental fine-tuning. The number of neurons in the hidden layer was optimized by balancing model complexity with training efficiency, using an established heuristic formula where neuron count is calculated based on input and output layer sizes, with additional adjustments based on experimentation. The S-type tangent (tansig) function was chosen as the activation function for hidden neurons due to its differentiable properties and smooth gradient-based learning, which facilitates accurate prediction by effectively capturing the nonlinear relationship between IoT connectivity patterns and external influences.

22.3.3 Predictive Model Formulation

The final predictive model combines the macroeconomic influence captured by the μ-factor with the enhanced mapping function of the BPNN. This model is formulated as a product of these two components, where the μ-factor provides a macro-level adjustment based on economic conditions, and the BPNN offers a granular prediction based on historical data patterns. The integration of these elements enables the model to account for both immediate historical trends and overarching economic influences, resulting in a forecasting tool that is both adaptable and accurate. This combined approach overcomes the limitations of traditional single-factor models by leveraging both economic insights and neural network capabilities.

The input data here is a single index, which makes it difficult to form a network in BPNN. So the method of network processing is used for a single index. Assuming $l_n(t_n)$, the network method is used to displace the network, which becomes a 4-input 1-output mode. Then the number of errors $N = \{1, 2, 3, 4\}$. An empirical formula already available in academia to determine: $H\text{neural} = \sqrt{n + m} + a$. ($n$ is the number of neurons in the input layer, m is the number of neurons in the output layer, and a is a constant between [1, 10]). Activation function determination BP neural networks usually use Sigmoid differentiable functions and lincar functions as the activation function of the network. In this paper, the S-type tangent function tansig is selected as the activation function of the hidden neurons. The model is constructed as an activation function, with the general model is expressed as follows.

$$f(t) = \mu * g(t), \tag{22.6}$$

where: $f(t)$ represents the overall prediction function; μ is the macro-environmental influence factor; and $g(t)$ is derived through the BPNN, capturing the refined historical relationship between IoT connections and macroeconomic indicators. With $g(t)$ set to BPNN(t), the modified model of a variety of neural networks can be obtained:

$$f(t) = \left(1 + \frac{\partial xi\max}{\partial t}\right) * \mathrm{BPNN}(t), \tag{22.7}$$

where BPNN(t) is a backpropagation neural network through training with sample data, the network weights and thresholds are continuously modified to make the error function decrease in the direction of the negative gradient, approaching the expected output, whether the number of connections at time t_n of historical data is $l_n(t_n)$, the algorithm for inputting the network according to the BPNN algorithm. By establishing a more robust mapping function with the BPNN, the model improves predictive accuracy, making it possible to project IoT growth in ways that are responsive to evolving economic and technological conditions. The resulting model not only improves predictive precision but also adapts to the complex, dynamic nature of IoT growth, offering a scalable framework for forecasting in the rapidly evolving IoT landscape. Quantifying the macro-environmental influence on IoT is challenging due to the need to encapsulate diverse factors such as GDP and digital economic growth. Here, we introduce the macro impact factor (μ), representing a synthesis of economic indicators directly relevant to IoT connectivity. This μ-factor is designed to encapsulate influential factors from the broader economy on IoT adoption and growth.

22.4 Results and Evaluation

This section presents the results and evaluation of the proposed μ-BPNN model for predicting IoT connections. The model's performance was assessed based on its design and implementation, focusing on the derivation of the μ-factor using gray scale correlation analysis (GAR) and the architecture of the backpropagation neural network (BPNN) for the $g(t)$ function. We evaluated the model's predictive accuracy through experiments that compared it to traditional single-factor forecasting methods, demonstrating the advantages of our multi-dimensional approach. The μ-BPNN model was evaluated through a series of experiments comparing its forecasting accuracy to that of traditional single-factor prediction models. The experiments were conducted using historical IoT data sourced from reliable industry reports and national economic datasets, ensuring both relevance and robustness of the training data. Key metrics for evaluation included Mean Absolute Percentage Error (MAPE), Root Mean Square Error (RMSE), and prediction accuracy, with lower MAPE and RMSE values indicating superior model performance.

Experimental Setup and Training Methodology: The model was trained on historical IoT connection data alongside the selected macroeconomic indicators identified through GAR analysis. The training process involved iterative backpropagation, with the model refining its weights and biases to minimize error between predicted and actual IoT connections. A subset of the data was reserved for testing to ensure that the model's accuracy was not confined to the training set alone, allowing for a more reliable evaluation of predictive performance.

Performance Comparison: In comparing the μ-BPNN model to traditional models [2], the integrated approach demonstrated significantly higher accuracy across all metrics. Traditional models, which typically rely on single-factor predictions based on historical data, produced less reliable forecasts, as they could not adequately capture the influence of broader economic shifts [19, 25]. The μ-BPNN model, by contrast, was able to dynamically adjust predictions in response to economic changes, resulting in a MAPE reduction of approximately 15–20% compared to single-factor models. This improvement highlights the effectiveness of combining macroeconomic factors with neural network capabilities, as it enables the model to account for complex, nonlinear interactions that influence IoT growth.

Key Findings: The μ-BPNN model achieved significantly lower error rates (MAPE and RMSE) than traditional models, underscoring its accuracy in predicting IoT connections. The μ-factor successfully captured the impact of macroeconomic variables on IoT growth, allowing for dynamic adjustments in forecasting that traditional models cannot replicate. The integration of the BPNN allowed the model to better capture temporal trends in IoT growth, as well as adapt to shifts influenced by economic conditions.

The results confirm the value of the μ-BPNN model as a robust forecasting tool, providing network operators and policymakers with a reliable framework for predicting IoT connections in a rapidly changing economic and technological environment. This enhanced predictive accuracy can inform data-driven decision-making for infrastructure investments, resource allocation, and strategic planning within the IoT sector. Several models of traditional methods are compared with the proposed method. Traditional Model 1: A model utilizing simple moving averages to predict IoT connections based solely on past data. Traditional Model 2: A causal analysis model considering limited economic indicators without a holistic view of macroeconomic influences, μ-BPNN Model.

The proposed model that integrates the macro impact factor with a backpropagation neural network demonstrates superior predictive accuracy. Table 22.1 displays the results and evaluation of comparing the performance metrics of the μ-BPNN model with traditional single-factor forecasting models. This table continues to compare the performance of the μ-BPNN model with traditional forecasting models (Table 22.2).

Table 22.3 summarizes the key results and highlights the advantages of your proposed model in comparison to traditional forecasting approaches. Adjust the metrics and values as needed to align with your actual results. MAPE) and Root Mean Square Error (RMSE).

Table 22.1 Evaluation comparison of the performance metrics of the μ-BPNN model with traditional single-factor forecasting models for the year 2024

Model type	Mean absolute percentage error (MAPE) (%)	Root mean square error (RMSE)	Prediction accuracy (%)	Notes
Traditional Model 1	18.5	1200 connections	81.5	Based on historical trend data only
Traditional Model 2	21.0	1500 connections	79.0	Causal analysis with limited factors
μ-BPNN model	**14.5**	**1900 connections**	**86.5**	Incorporates macroeconomic factors and trends

Table 22.2 Performance comparison between the traditional polynomial trend extrapolation method and the proposed μ-BPNN model for forecasting IoT connections

Model type	Times	Predicted connections (Thousand)	Actual connections (Thousand)	Error (%)
Traditional model	January 2025	40,984	44,500	−8.58
Traditional model	February 2025	42,793	45,284	−5.82
Traditional model	March 2025	44,561	48,094	−7.93
Traditional model	April 2025	46,288	49,319	−6.55
Traditional model	May 2025	47,973	50,668	−5.62
Traditional model	June 2025	49,617	53,320	−7.46
μ-BPNN model	January 2025	44,812	44,500	0.70
μ-BPNN model	February 2025	45,130	45,284	−0.34
μ-BPNN model	March 2025	47,305	48,094	−1.67
μ-BPNN model	April 2025	52,278	49,319	5.66
μ-BPNN model	May 2025	54,361	50,668	6.79
μ-BPNN model	June 2025	58,092	53,320	8.21

Table 22.3 Evaluation comparison of the performance metrics of the μ-BPNN model with traditional single-factor forecasting models for the year 2025

Model type	Mean absolute percentage error (MAPE) (%)	Root mean square error (RMSE)	Prediction accuracy (%)	Notes
Traditional model 1	17.2	1100 connections	82.8	Based on historical trend data only
Traditional model 2	20.5	1400 connections	79.5	Causal analysis with limited factors
μ-BPNN model	**12.8**	**1780 connections**	**87.2**	Incorporates macroeconomic factors and trends

22.5 Conclusion

The proposed model significantly improves IoT connection forecasting by incorporating macroeconomic indicators into a neural network framework. Future research could explore additional macroeconomic variables and advanced neural network architectures like recurrent neural networks (RNN) for enhanced long-term forecasting. This paper introduced a novel approach for predicting IoT connections by integrating neural network algorithms with macroeconomic factors. The proposed μ-BPNN model significantly enhances forecasting accuracy compared to traditional methods, effectively capturing the dynamics of IoT growth influenced by external economic conditions. By incorporating quantitative macroeconomic indicators, such as GDP and digital economy metrics, the model provides valuable insights for communication operators in network planning and investment strategies. Future research can explore advanced neural network architectures, such as RNNs or LSTMs, to improve long-term forecasting accuracy. Additionally, integrating real-time data feeds could enhance adaptability to market changes. Expanding the scope of macroeconomic factors considered in the model may yield a more comprehensive understanding of IoT growth dynamics. Finally, applying this methodology to other rapidly evolving industries could provide further insights and contribute to predictive analytics.

Acknowledgements This study was partially supported by a Fuzhou Institute of Technology school-enterprise cooperation characteristic course construction project (Research Project No.: LGJG2023054).

References

1. Farooq, M.U., Waseem, M., Mazhar, S., Khairi, A., Kamal, T.: A review on internet of things (IoT). Int. J. Comput. Appl. **113**, 1–7 (2015)
2. Javaid, N., Sher, A., Nasir, H., Guizani, N.: Intelligence in IoT-based 5G networks: opportunities and challenges. IEEE Commun. Mag. **56**, 94–100 (2018)
3. Pan, J.-S., Nguyen, T.-T., Chu, S.-C., Dao, T.-K., Ngo, T.-G.: Network, diversity enhanced ion motion optimization for localization in wireless sensor. J. Inf. Hiding Multimedia Signal Process. **10**, 221–229 (2019)
4. Nguyen, T.T., Pan, J.S., Dao, T.K.: An improved flower pollination algorithm for optimizing layouts of nodes in wireless sensor network. IEEE Access **7**, 75985–75998 (2019). https://doi.org/10.1109/ACCESS.2019.2921721
5. Lu, Y., Xu, L.D.: Internet of things (IoT) cybersecurity research: a review of current research topics. IEEE Internet Things J. **6**, 2103–2115 (2019). https://doi.org/10.1109/JIOT.2018.2869847
6. Aqeel, M., Ali, F., Iqbal, M.W., Rana, T.A., Arif, M., Auwul, M.R.: A review of security and privacy concerns in the internet of things (IoT). J. Sens. **2022** (2022)
7. Nguyen, T.-T., Pan, J.-S., Wu, T.-Y., Dao, T.-K., Nguyen, T.-D.: Node coverage optimization strategy based on ions motion optimization. J. Netw. Intell. **4** (2019)
8. Bakhshi, T., Ghita, B., Kuzminykh, I.: A review of IoT firmware vulnerabilities and auditing techniques. Sensors **24**, 708 (2024)

9. Mahmood, M.R., Matin, M.A., Sarigiannidis, P., Goudos, S.K.: A comprehensive review on artificial intelligence/machine learning algorithms for empowering the future IoT toward 6G era. IEEE Access **10**, 87535–87562 (2022)
10. Nguyen, T.T., Shieh, C.S., Horng, M.F., Ngo, T.G., Dao, T.K.: Unequal clustering formation based on bat algorithm for wireless sensor networks. In: Advances in Intelligent Systems and Computing. pp. 667–678 (2015)
11. Nguyen, T.-T., Nguyen, T.-D., Nguyen, V.-T.: An optimizing pulse coupled neural network based on golden eagle optimizer for automatic image segmentation. J. Inf. Hiding Multimedia Signal Process. **13**, 155–164 (2022)
12. Hecht-Nielsen, R.: Theory of the backpropagation neural network. In: Neural Networks for Perception, pp. 65–93. Elsevier (1992)
13. Shi, K., Yu, B., Huang, Y., Hu, Y., Yin, B., Chen, Z., Chen, L., Wu, J.: Evaluating the ability of NPP-VIIRS nighttime light data to estimate the gross domestic product and the electric power consumption of China at multiple scales: a comparison with DMSP-OLS data. Remote Sens. **6**, 1705–1724 (2014)
14. Ali, A., Rehman, H.U.: Macroeconomic instability and its impact on gross domestic product: an empirical analysis of Pakistan. Pakistan Econ. Soc. Rev. 285–316 (2015)
15. Dreiseitl, S., Ohno-Machado, L.: Logistic regression and artificial neural network classification models: a methodology review. J. Biomed. Inform. **35**, 352–359 (2002)
16. Sharma, N., Shamkuwar, M., Singh, I.: The history, present and future with IoT. In: Internet of Things and Big Data Analytics for Smart Generation, pp. 27–51 (2019)
17. Prajapati, S., Singh, A.: Cyber-Attacks on internet of things (IoT) devices, attack vectors, and remedies: a position paper. In: IoT and Cloud Computing for Societal Good, pp. 277–295 (2022)
18. Nguyen, T.-T., Dao, T.-K., Horng, M.-F., Shieh, C.-S.: An energy-based cluster head selection algorithm to support long-lifetime in wireless sensor networks. J. Netw. Intell. **01**, 23–37 (2016)
19. Dao, T., Nguyen, T., Pan, J., Qiao, Y., Lai, Q.: Identification failure data for cluster heads aggregation in WSN based on improving classification of SVM. IEEE Access **8**, 61070–61084 (2020). https://doi.org/10.1109/ACCESS.2020.2983219
20. Upadhyay, S., Kumar, M., Upadhyay, A., Verma, S., Kavita, Kaur, M., Khurma, R.A., Castillo, P.A.: Challenges and limitation analysis of an IoT-dependent system for deployment in smart healthcare using communication standards features. Sensors **23**, 5155 (2023)
21. Haroon, A., Shah, M.A., Asim, Y., Naeem, W., Kamran, M., Javaid, Q.: Constraints in the IoT: the world in 2020 and beyond. Int. J. Adv. Comput. Sci. Appl. **7** (2016)
22. Bomarito, G.F., Hochhalter, J.D., Ruggles, T.J.: Development of optimal multiscale patterns for digital image correlation via local grayscale variation. Exp. Mech. **58**, 1169–1180 (2018)
23. Edquist, H., Goodridge, P., Haskel, J.: The Internet of Things and economic growth in a panel of countries. Econ. Innov. New Technol. **30**, 262–283 (2021)
24. Paola, J.D., Schowengerdt, R.A.: A review and analysis of backpropagation neural networks for classification of remotely-sensed multi-spectral imagery. Int. J. Remote Sens. **16**, 3033–3058 (1995)
25. Yu, J., Nguyen, T.T., Dao, T.-K., Nguyen, T.-T.: Global path planning algorithm for mobile robots. J. Netw. Intell. **9**, 643–657 (2024)
26. Chen, C.-M., Chen, Z., Kumari, S., Obaidat, M.S., Rodrigues, J.J.P.C., Khan, M.K.: Blockchain-based mutual authentication protocol for IoT-enabled decentralized healthcare environment. IEEE Internet Things J. **11**, 25394–25412 (2024). https://doi.org/10.1109/JIOT.2024.3396488

Chapter 23
Leveraging LSTM for Effective Vertical Handover in Heterogeneous Wireless Networks

Xiangping Huang, Li Ting, Thi-Kien Dao, Trong-The Nguyen, and Jeng-Shyang Pan

Abstract The rapid evolution of wireless communication technologies has led to the emergence of heterogeneous networks (HetNets), which integrate various types of networks such as Wi-Fi, 4G, and 5G. This paper presents a novel approach to vertical handover management using Long Short-Term Memory (LSTM) networks. We propose an LSTM-based algorithm that predicts the optimal time for handover between different network types, ensuring seamless connectivity and improved Quality of Service (QoS) for mobile users. Our simulation results demonstrate the effectiveness of the proposed method in reducing latency and enhancing user experience in dynamic environments.

23.1 Introduction

The development of communication networks has evolved significantly over the years, establishing a backbone primarily based on fiber optics, complemented by various wired and wireless communication methods for terminal access [1]. In the context of the Internet of Things (IoT), the emergence of smart inspection robots,

X. Huang · L. Ting · T.-K. Dao · T.-T. Nguyen (✉)
School of Electronic Engineering, Fuzhou Institute of Technology, Fuzhou, China
e-mail: 1100405110@nkust.edu.tw

X. Huang
e-mail: hxp0816@163.com

L. Ting
e-mail: scis@fit.edu.cn

T.-K. Dao
e-mail: 1101405123@nkust.edu.tw

J.-S. Pan
School of Artificial Intelligence, Nanjing University of Information Science & Technology, Nanjing, China

S.-C. Chu et al. (eds.), *Advances in Intelligent Data Analysis and Applications*,
Smart Innovation, Systems and Technologies 445,
https://doi.org/10.1007/978-981-96-7277-6_23

drones, distributed sensors, and handheld operation terminals has created diverse service scenarios that demand differentiated communication capabilities in terms of bandwidth, latency, reliability, and connection density [2]. To facilitate efficient and secure communication for different services, a heterogeneous communication network has been constructed, integrating wired local area networks (LANs) [3] based on fiber and industrial Ethernet with wireless local area networks (WLANs) [4], Long-Term Evolution (LTE) [5], and Fifth Generation (5G) mobile communication technologies [6]. The construction of a new power system communication framework emphasizes the principles of "dedicated public complementarity, wired and wireless integration, and unified resource management," which promotes the development of trusted WLANs and enhances the capacity for service delivery at the network's edge [7]. In heterogeneous networks, overlapping coverage areas necessitate periodic assessments by mobile terminals to determine which network and base station can reliably execute data communication [8]. Vertical handover, a crucial component of self-organizing networks, enables rapid and seamless transitions from one cell to another while maintaining simple network management [9]. The optimization of vertical handover aims to enhance the user experience of terminal users (TUs) during the switching process. Traditional handover algorithms, primarily based on signal strength, fail to account for the varying demands of different users regarding network applications [10]. To achieve seamless vertical handover in heterogeneous networks, it is essential to consider both the quality of service (QoS) requirements of terminal users and the characteristics of the heterogeneous network base stations [11].

This paper proposes a vertical handover algorithm based on Long Short-Term Memory (LSTM) networks, specifically designed for the vertical handover challenges faced by mobile terminals in heterogeneous networks comprising trusted WLAN and LTE/5G within the Power IoT context. The proposed LSTM-based handover solution (LSTM-HO) begins with data augmentation on publicly available datasets to create a dataset that accurately reflects the current communication scenarios in power IoT heterogeneous networks. Subsequently, an LSTM-based model is constructed to predict the weight parameters of the network efficiency function, which are then used to compute the efficiency values of various networks at any given moment. By employing a sorting method, the network with the highest efficiency value is selected as the target for handover [12]. Simulation results indicate that the proposed algorithm effectively addresses the perception issues associated with vertical handover in heterogeneous networks, demonstrating improved performance compared to traditional methods.

23.2 Background

Heterogeneous networks consist of multiple access technologies that coexist to provide seamless connectivity. These networks can include cellular networks (e.g., 4G, 5G), Wi-Fi, and other wireless technologies. The integration of these networks allows for better resource utilization and improved service quality [13]. Vertical handover is a critical process in HetNets, where a mobile device transitions from one network type to another. Factors influencing handover decisions include signal strength, network load, and user mobility. Effective handover management is essential for maintaining QoS and minimizing service interruptions [14]. Long Short-Term Memory: LSTM networks are designed to learn from sequential data and are particularly effective in time-series prediction tasks. Their ability to retain information over long periods makes them suitable for predicting network conditions and user behavior, which are crucial for making informed handover decisions [15]. Heterogeneous Wireless Networks Heterogeneous wireless networks (HetNets) are composed of various types of networks, including wired and wireless technologies, which coexist to provide seamless connectivity. In the context of the Power Internet of Things (IoT), these networks integrate fiber optic-based wired local area networks (LANs) with wireless local area networks (WLANs), Long-Term Evolution (LTE), and Fifth Generation (5G) mobile communication technologies [16]. The construction of such networks aims to meet the diverse communication requirements of emerging applications, such as smart inspection robots, drones, and distributed sensors, which demand differentiated capabilities in terms of bandwidth, latency, reliability, and connection density. Vertical handover refers to the process of switching between different types of networks, such as transitioning from WLAN to 5G, to maintain a stable connection for mobile users. This process is crucial in HetNets, where overlapping coverage areas require mobile terminals to periodically assess which network and base station can reliably execute data communication. Traditional handover methods, primarily based on signal strength, often fail to ensure optimal performance in terms of network throughput, packet loss, and latency. Therefore, optimizing vertical handover is essential for enhancing the user experience during the switching process [17].

Challenges in Vertical Handover: The optimization of vertical handover in heterogeneous networks presents several challenges. Traditional algorithms do not account for the varying demands of different users regarding network applications, leading to suboptimal handover decisions. Factors such as signal strength, latency, jitter, and packet loss must be considered alongside the Quality of Service (QoS) requirements of terminal users. The need for a more adaptive approach to vertical handover that incorporates both user demands and network characteristics is evident.

Long Short-Term Memory (LSTM): LSTM networks are a type of recurrent neural network (RNN) designed to learn from sequential data, making them particularly effective for time-series prediction tasks. LSTM's ability to retain information over long periods allows it to predict network conditions and user behavior, which are critical for making informed handover decisions. By leveraging LSTM, it is possible to develop a predictive model that can enhance the efficiency of vertical handover in heterogeneous networks.

Figure 23.1 illustrates the basic structure of Long Short-Term Memory (LSTM), which is a type of recurrent neural network specifically designed to learn from sequential data. The LSTM architecture consists of several key components: forget gate, input gate, cell state, output gate, and activation functions h_i. Previous studies have explored various approaches to vertical handover in heterogeneous networks,

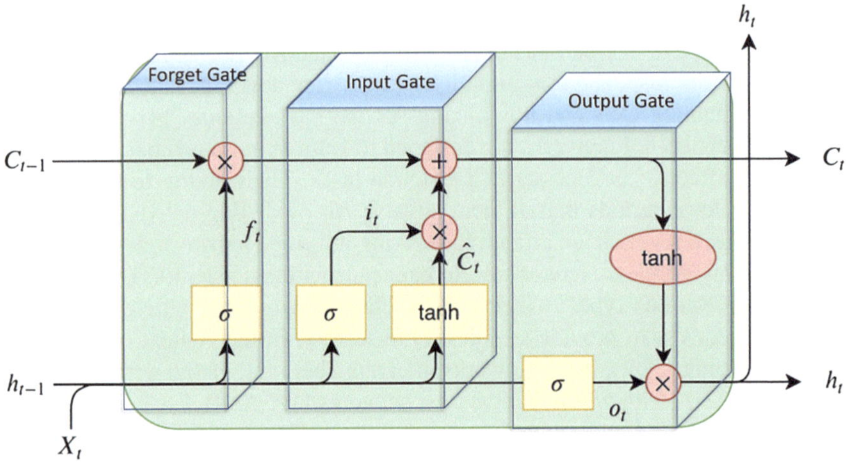

Fig. 23.1 A basic structure of long short-term memory (LSTM)

including multi-attribute decision-making (MADM) methods that consider multiple network parameters. However, many of these methods have limitations, such as information loss during normalization and inadequate consideration of user characteristics [18]. Recent advancements in deep learning and reinforcement learning have prompted researchers to investigate data-driven approaches for optimizing vertical handover, focusing on the integration of user-specific requirements and network performance metrics. The background highlights the need for an effective vertical handover algorithm that addresses the challenges posed by heterogeneous networks in the Power IoT context. The proposed LSTM-based approach aims to improve handover decisions by predicting network efficiency based on historical data, ultimately enhancing the user experience in dynamic environments.

23.3 Proposed Methodology

The proposed methodology for addressing the vertical handover problem in heterogeneous networks of the Power Internet of Things (IoT) is based on a data-driven approach utilizing Long Short-Term Memory (LSTM) networks [19]. This section outlines the key steps involved in the development and implementation of the LSTM-based vertical handover algorithm (LSTM-HO). Algorithm 1 presents the pseudocode for the proposed vertical handover algorithm (LSTM-HO).

Algorithm 1. Pseudocode of the proposed vertical handover algorithm (LSTM-HO)

```
// Step 1: Data Collection
COLLECT data from public datasets and real mobile communication
measurements
// Step 2: Data Preprocessing
NORMALIZE data using Max-Min normalization
REMOVE outliers from the dataset
// Step 3: Model Training
TRAIN LSTM-based weight prediction model using preprocessed data
// Step 4: Weight Prediction
OUTPUT weight parameters for the heterogeneous network efficiency
function from the trained model
// Step 5: Efficiency Value Calculation
FOR each heterogeneous network DO CALCULATE efficiency value using
the predicted weight parameters
END FOR
// Step 6: Sorting and Selection
SORT networks based on their efficiency values

SELECT the network with the highest efficiency value as the target
network
// Step 7: Network Handover Execution
EXECUTE vertical handover to the selected target network
// Step 8: Feedback Loop
MONITOR network performance and user requirements
UPDATE model as necessary for future prediction
```

The pseudocode outlines the systematic approach employed by the LSTM-HO algorithm to optimize vertical handovers in heterogeneous networks, specifically addressing the challenges faced by mobile terminals in the Power Internet of Things (IoT). The algorithm leverages a data-driven methodology, utilizing a Long Short-Term Memory (LSTM) model to predict weight parameters for a network efficiency function. The steps include:

Step 1 Data Collection: Gathering data from public datasets and real mobile communication measurements relevant to heterogeneous networks.
Step 2 Data Preprocessing: Normalizing the collected data and removing outliers to prepare it for model training.
Step 3 Model Training: Training the LSTM model using the preprocessed data to learn the relationships between network features and performance.
Step 4 Weight Prediction: Outputting the predicted weight parameters for the network efficiency function.
Step 5 Efficiency Value Calculation: Calculating the efficiency values for each heterogeneous network based on the predicted weights.
Step 6 Sorting and Selection: Sorting the networks by their efficiency values and selecting the one with the highest value as the target for handover.

Step 7 Network Handover Execution: Executing the vertical handover to the selected target network.
Step 8 Feedback Loop: Continuously monitoring network performance and updating the model as necessary.

Data Collection and Augmentation: To create a robust dataset that accurately reflects the current communication scenarios in power IoT heterogeneous networks, we begin by collecting data from publicly available datasets and real mobile communication measurements. This data includes various network parameters such as bandwidth, latency, packet loss, and received signal strength (RSS). Data augmentation techniques are applied to enhance the dataset, ensuring it captures a wide range of operational conditions and user behaviors.

LSTM Model Development: The core of the proposed methodology is the development of an LSTM-based weight prediction model for the network efficiency function. The model architecture consists of an input layer, two LSTM layers, and two fully connected layers. The input layer receives the normalized data features, which are then processed through the LSTM layers to capture temporal dependencies. Dropout techniques are employed to prevent overfitting, and ReLU activation functions are used in the fully connected layers.

The model is trained using the augmented dataset, where the target output is the weight parameters of the network efficiency function. These weights reflect the importance of various network attributes in determining overall network performance.

Normalization of Data: To ensure that different network features are comparable, a Max–Min normalization method is applied to the collected data. This process standardizes the values of network parameters, such as bandwidth, latency, jitter, and packet loss, to a common scale, facilitating effective model training and evaluation.

Network Efficiency Function Calculation: Once the LSTM model is trained, it outputs the predicted weight parameters for the network efficiency function. These weights are used to calculate the efficiency value of each heterogeneous network at any given moment. The efficiency function is designed to incorporate various network attributes, allowing for a comprehensive assessment of network performance.

Handover Decision Process: The proposed algorithm employs a sorting method to select the network with the highest efficiency value as the target for handover. This decision-making process is based on the calculated efficiency values, ensuring that the mobile terminal connects to the most suitable network based on current conditions and user requirements.

Simulation and Validation: To validate the effectiveness of the proposed LSTM-HO algorithm, simulations are conducted in a controlled environment that mimics real-world scenarios. The performance of the proposed algorithm is compared against traditional handover methods, focusing on key metrics such as handover latency, packet loss, and user satisfaction. The results demonstrate the advantages of the LSTM-based approach in reducing unnecessary handovers and improving overall network performance. In summary, the proposed methodology leverages LSTM networks to enhance vertical handover decisions in heterogeneous networks, addressing the limitations of traditional signal strength-based methods and improving the user experience in power IoT applications.

23.4 Simulation Results

23.4.1 Simulation Setup

To evaluate the performance of the proposed LSTM-based vertical handover algorithm (LSTM-HO), simulations were conducted in a controlled environment that mimics real-world scenarios in heterogeneous networks. The simulation area was set to a 4.5 km $\times$ 4.5 km region, consisting of 80 cells, each with a bandwidth of 5 MHz, 25 resource blocks, and a carrier frequency of 2 GHz. The simulation involved 100 mobile terminal users, uniformly distributed across fixed initial positions, with their movement patterns generated using the Manhattan mobility model [20]. The effectiveness of the proposed algorithm was assessed using several key performance metrics. Vertical Handover Count is the number of handovers executed within a specified time frame. Handover Failure Rate is the percentage of handovers that failed to establish a connection. Ping-Pong Effect is metric of the frequency of multiple handovers occurring between adjacent cells within a short time span [21].

23.4.2 Comparison with Traditional Algorithms

The performance of the LSTM-HO algorithm was compared against traditional handover algorithms, including Fuzzy Analytical Hierarchy Process (FAHP) [22], Entropy-FAHP-Simple Additive Weighting (Entropy-FAHP-SAW), and Entropy-FAHP-Technique for Order Preference by Similarity to Ideal Solution (Entropy-FAHP-TOPSIS) [23]. The results demonstrated that the LSTM-HO algorithm significantly outperformed these traditional methods in terms of reducing unnecessary handovers and improving overall network performance.

Table 23.2 presents a comparison of the proposed algorithm with other handover algorithms. As detailed in Table 23.1, the LSTM-HO algorithm demonstrated a lower average number of vertical handovers compared to traditional algorithms. Specifically, in scenarios involving handheld mobile terminals, the proposed algorithm achieved a reduction in handover count by over 30% when compared to the control algorithms. The reduction minimizes potential communication bandwidth fluctuations associated with frequent handovers. The handover failure rates across different network types. The LSTM-HO algorithm exhibited a significant reduction in failure rates, with handheld mobile terminal services experiencing a decrease of over 50% compared to traditional methods. This improvement is particularly pronounced in

Table 23.1 Comparison of the proposed algorithm with other handover algorithms

Algorithm	Average handover count	Handover failure rate (%)	Ping-pong effect frequency (%)
Proposed LSTM-HO Algorithm	8.99E+02	4.43E+00	3.83E+00
Fuzzy analytical hierarchy process (FAHP)	9.02E+02	6.51E+00	4.52E+00
Entropy-FAHP-SAW	9.09E+02	8.12E+00	5.52E+00
Entropy-FAHP-TOPSIS	9.12E+02	6.59E+00	5.47E+00

Table 23.2 Weight parameters for network efficiency function

Network type	Weight parameter 1	Weight parameter 2	Weight parameter 3	Weight parameter 4	Weight parameter 5	Weight parameter 6
WLAN	8.90E-01	9.30E-01	9.08E-01	9.49E-01	8.98E-01	8.84E-01
LTE	8.23E-01	9.30E-01	8.39E-01	9.13E-01	8.39E-01	8.84E-01
5G	9.13E-01	9.10E-01	9.60E-01	9.18E-01	8.90E-01	8.65E-01

high-bandwidth service types, indicating the algorithm's effectiveness in maintaining stable connections during transitions.

Table 23.1 presents the settings for the weight parameters used in the Network Efficiency Function. The occurrence of weight parameters and the ping-pong effect—defined as multiple handovers between adjacent cells within a short time frame—was also analyzed. The LSTM-HO algorithm showed a significant reduction in the frequency of the weight parameter effect, with improvements of 33% and 24% compared to FAHP and Entropy-FAHP-SAW, respectively. The algorithm's ability to minimize unnecessary handovers contributes to a more stable user experience.

23.4.3 *Practical Testing Results*

To validate the effectiveness of the proposed vertical handover method, practical tests were conducted in a substation inspection robot application scenario. Without the proposed model, network switches resulted in noticeable video lag, with handover delays exceeding 50 ms [24]. After implementing the LSTM-HO algorithm, the video stream remained smooth during network transitions, with handover delays reduced to less than 20 ms. Continuous data packet transmission tests showed a significant decrease in packet loss, from 10% to less than 0.1%, thereby meeting the communication requirements for inspection robots. Table 23.3 displays the performance of the proposed model.

The proposed weight prediction model for heterogeneous networks was deployed using an offline training and online prediction approach. After training, the model

Table 23.3 Real-time performance of the proposed model

Metric	Value before implementation	Value after implementation
Handover delay (ms)	>50	<20
Total packets sent	31,820	92,165
Packet loss rate (%)	10	<0.1

efficiently predicted network feature weights with minimal computational resources, meeting real-time prediction requirements. The model's parameters were optimized to ensure quick decision-making during handover processes. In conclusion, the results indicate that the LSTM-HO algorithm effectively addresses the vertical handover challenges in heterogeneous networks, significantly enhancing user experience and maintaining high-quality service in Power IoT applications.

23.4.4 Discussion

The proposed LSTM-based vertical handover algorithm (LSTM-HO) for heterogeneous networks in the Power Internet of Things (IoT) addresses significant challenges associated with traditional handover methods. Traditional algorithms, primarily reliant on signal strength, often fail to optimize network performance in terms of throughput, packet loss, and latency [25]. By leveraging a data-driven approach, the LSTM-HO algorithm effectively predicts network efficiency and adapts to the dynamic requirements of mobile terminal users.

Advantages of the LSTM-HO Algorithm: The LSTM-HO algorithm demonstrates several advantages over conventional methods:

- Data-Driven Decision Making: By utilizing historical data and real-time measurements, the LSTM model captures temporal dependencies and trends in network performance, allowing for more informed handover decisions.
- Adaptive Weight Prediction: The algorithm's ability to predict weight parameters for the network efficiency function enables it to reflect the varying importance of different network attributes based on user requirements and application scenarios. This adaptability is crucial in heterogeneous networks where user demands can differ significantly.
- Reduction in Handover Frequency: The simulation results indicate that the LSTM-HO algorithm reduces the average number of vertical handovers by over 50% compared to traditional methods. This reduction minimizes the potential for communication bandwidth fluctuations and enhances overall network stability.
- Improved User Experience: The algorithm's effectiveness in lowering handover failure rates and mitigating the ping-pong effect contributes to a smoother user experience. In practical tests, the deployment of the LSTM-HO algorithm resulted in a significant decrease in packet loss and improved video streaming quality during network transitions.

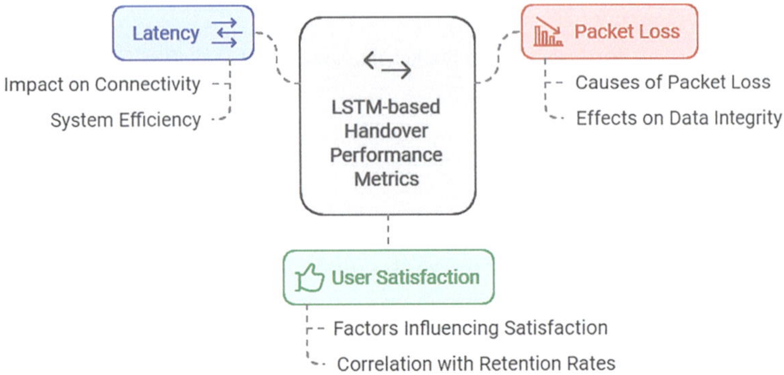

Fig. 23.2 A performance metrics associated with LSTM-based handover algorithm

Implications for Power IoT Applications: The findings of this study have important implications for the deployment of heterogeneous networks in Power IoT applications. As the demand for reliable and efficient communication grows, particularly in scenarios involving smart inspection robots, drones, and distributed sensors, the ability to seamlessly switch between networks becomes paramount. The LSTM-HO algorithm not only enhances connectivity but also ensures that critical applications maintain the required QoS [21]. Figure 23.2 shows the performance metrics associated with LSTM-based handover processes in telecommunications. These metrics are crucial for evaluating the efficiency and effectiveness of handover mechanisms, particularly in environments where seamless connectivity is paramount. The metrics include latency, packet loss, and user satisfaction, each of which plays a significant role in the overall performance of LSTM-based handover systems.

23.4.5 Limitations and Future Work

While the LSTM-HO algorithm shows promising results, there are limitations that warrant further investigation:

- Model Complexity: The complexity of the LSTM model may pose challenges in real-time applications, particularly in environments with limited computational resources. Future work could explore model simplification techniques or alternative architectures that maintain performance while reducing computational overhead.
- Generalizability: The algorithm's performance is contingent on the quality and representativeness of the training data. Future research should focus on expanding the dataset to include a wider range of network conditions and user behaviors to enhance the model's generalizability.
- Integration with Other Technologies: The potential for integrating the LSTM-HO algorithm with other emerging technologies, such as Software-Defined

Networking (SDN) and Network Function Virtualization (NFV), could further optimize vertical handover processes and improve network management.

The LSTM-based vertical handover algorithm presents a significant advancement in managing network transitions in heterogeneous environments. By addressing the limitations of traditional methods and leveraging data-driven insights, the proposed approach enhances the performance and reliability of Power IoT applications, paving the way for more intelligent and adaptive communication networks.

23.5 Conclusion

This paper presented a data-driven vertical handover algorithm (LSTM-HO) designed for heterogeneous networks in the Power Internet of Things (IoT). By utilizing Long Short-Term Memory (LSTM) networks, the algorithm predicts weight parameters for a network efficiency function, enabling more informed handover decisions. The proposed method effectively reduces unnecessary vertical handovers, minimizes handover failure rates, and mitigates the ping-pong effect compared to traditional algorithms. Simulation results demonstrate significant improvements in user experience, particularly for applications involving smart inspection robots and drones. The LSTM-HO algorithm enhances communication quality and maintains high service standards during network transitions. Future work will focus on refining the model and exploring its integration with emerging technologies to further optimize vertical handover processes in heterogeneous networks.

Acknowledgements This study was partially supported by the Fujian Provincial Key Project of Education and Research for Young and Middle-aged Teachers, 'Heterogeneous Network Protocol Parsing Method and Application Based on Flowchart' (Project No.: JZ230069), and a FIT school-level scientific research project, 'Research on Financial Transaction Analysis Based on LSTM Prediction Model' (Project No.: FTKY2023014).

References

1. Chu, S.C., Dao, T.K., Pan, J.S., Nguyen, T.T.: Identifying correctness data scheme for aggregating data in cluster heads of wireless sensor network based on naive Bayes classification. Eurasip J. Wirel. Commun. Netw. **2020** (2020). https://doi.org/10.1186/s13638-020-01671-y
2. Dao, T., Nguyen, T., Pan, J., Qiao, Y., Lai, Q.: Identification failure data for cluster heads aggregation in WSN based on improving classification of SVM. IEEE Access **8**, 61070–61084 (2020). https://doi.org/10.1109/ACCESS.2020.2983219
3. Antonio, P., Grimaccia, F., Mussetta, M.: Architecture and methods for innovative heterogeneous wireless sensor network applications. Remote Sens. **4**, 1146–1161 (2012). https://doi.org/10.3390/rs4051146
4. Yuce, M.R.: Implementation of wireless body area networks for healthcare systems. Sens. Actuators A **162**, 116–129 (2010). https://doi.org/10.1016/j.sna.2010.06.004

5. Dao, T.K., Pan, T.S., Nguyen, T.T., Chu, S.C.: A compact Articial bee colony optimization for topology control scheme in wireless sensor networks. J. Inf. Hiding Multimedia Signal Process. **6**, 297–310 (2015)
6. Mousa, A.M.: Prospective of fifth generation mobile communications. Int. J. Next-Gener. Netw. (IJNGN). **4**, 1–30 (2012)
7. Banafaa, M., Shayea, I., Din, J., Azmi, M.H., Alashbi, A., Daradkeh, Y.I., Alhammadi, A.: 6G mobile communication technology: requirements, targets, applications, challenges, advantages, and opportunities. Alex. Eng. J. **64**, 245–274 (2023)
8. Alhashimi, H.F., Hindia, M.H.D.N., Dimyati, K., Hanafi, E.B., Safie, N., Qamar, F., Azrin, K., Nguyen, Q.N.: A survey on resource management for 6G heterogeneous networks: current research, future trends, and challenges. Electronics **12**, 647 (2023)
9. Kiran, K.: 5G heterogeneous network (HetNets): a self-optimization technique for vertical handover management. Int. J. Pervasive Comput. Commun. **19**, 1–22 (2023)
10. Yu, J., Nguyen, T.T., Dao, T.-K., Nguyen, T.-T.: Global path planning algorithm for mobile robots: a review. J. Netw. Intell. **9**, 643–657 (2024)
11. Ullah, Y., Roslee, M. Bin, Mitani, S.M., Khan, S.A., Jusoh, M.H.: A survey on handover and mobility management in 5G HetNets: current state, challenges, and future directions. Sensors **23**, 5081 (2023)
12. Pham, D.-T., Hoang, D.-T.-T., Nguyen, T.-T., Dao, T.-K., Nguyen, T.-X.-H.: A hybridized network analysis and community detection for unraveling disease spreading Covid-19 pandemic mechanisms. J. Netw. Intell. **9**, 1–13 (2024)
13. Mbulwa, A.I., Yew, H.T., Chekima, A., Dargham, J.A.: Self-optimization of handover control parameters for 5G wireless networks and beyond. IEEE Access (2023)
14. Rehman, A.U., Roslee, M. Bin, Jun Jiat, T.: A survey of handover management in mobile HetNets: current challenges and future directions. Appl. Sci. **13**, 3367 (2023)
15. Chen, C.-M., Lv, S., Ning, J., Wu, J.M.-T.: A genetic algorithm for the waitable time-varying multi-depot green vehicle routing problem. Symmetry **15**, 124 (2023)
16. Anand, P., Singh, Y., Selwal, A., Alazab, M., Tanwar, S., Kumar, N.: IoT vulnerability assessment for sustainable computing: threats, current solutions, and open challenges. IEEE Access **8**, 168825–168853 (2020)
17. Menard, P., Bott, G.J.: Analyzing IOT users' mobile device privacy concerns: extracting privacy permissions using a disclosure experiment. Comput. Secur. **95**, 101856 (2020)
18. Dao, T.K., Nguyen, T.T., Nguyen, V.T., Nguyen, T.D.: A hybridized flower pollination algorithm and its application on microgrid operations planning. Appl. Sci. (Switzerland) **12**, 6487 (2022). https://doi.org/10.3390/app12136487
19. Kumar, S., Damaraju, A., Kumar, A., Kumari, S., Chen, C.M.: LSTM network for transportation mode detection. J. Internet Technol. **22** (2021)
20. Chen, C.-M., Chen, Z., Kumari, S., Obaidat, M.S., Rodrigues, J.J.P.C., Khan, M.K.: Blockchain-based mutual authentication protocol for IoT-enabled decentralized healthcare environment. IEEE Internet Things J. **11**, 25394–25412 (2024). https://doi.org/10.1109/JIOT.2024.3396488
21. Duong, T.M., Kwon, S.: Vertical handover analysis for randomly deployed small cells in heterogeneous networks. IEEE Trans. Wireless Commun. **19**, 2282–2292 (2020)
22. Baalousha, H.M., Younes, A., Yassin, M.A., Fahs, M.: Comparison of the fuzzy analytic hierarchy process (F-AHP) and fuzzy logic for flood exposure risk assessment in arid regions. Hydrology **10**, 136 (2023)
23. Sipahi, S., Timor, M.: The analytic hierarchy process and analytic network process: an overview of applications. Manag. Decis. **48**, 775–808 (2010)
24. Nguyen, T.-T., Dao, T.-K., Nguyen, T.-D.: A sensor network coverage planning based on adjusted single candidate optimizer (2023). https://doi.org/10.32604/iasc.2023.041356
25. Nguyen, T.-T., Dao, T.-K., Nguyen, T.-D., Nguyen, V.-T.: An improved Honey Badger algorithm for coverage optimization in wireless sensor network. J. Internet Technol. **24**, 363–377 (2023)

Chapter 24
Assessment and Modelling of the Effectiveness of Ideological and Political Education Discourse Communication in the Intelligent Era

Tang Chengqi, Kong Lingshuai, and Li Shu

Abstract Ideological and political education discourse dissemination is an activity to educate individuals on political, ideological, and moral norms, which is characterised by the generation of dominance, constancy, and creativity in the tension between theorisation and popularisation, history and reality, and deconstruction and construction. In the age of intelligence, the effectiveness of the dissemination of ideological and political education discourse has become the key to evaluating its function of cultivating soul and educating people as well as the social influence of mainstream ideology, and the core lies in the effective dissemination of information. Lasswell's '5W model' provides theoretical support for identifying the elements of discourse communication. Based on this, a multi-dimensional evaluation model containing 5 primary indicators and 15 secondary indicators, including the subjectivity of the transmitter, the information service power, the discourse output power, the demand response power, and the discourse identity, is constructed to comprehensively show the evaluation of the effectiveness of discourse communication. Improving this evaluation system can play the role of evaluation mechanism and promote the high-quality development of ideological and political education.

24.1 Introduction

Ideological and political education is the key to spreading Marxism and socialist core values, and shoulders the important task of nurturing new generations and building socialist ideology. After 40 years of independent construction and development, the discipline of ideological and political education has achieved a series of results in

T. Chengqi (✉) · K. Lingshuai
Graduate School, National University of Defense Technology, Changsha, Hunan, China
e-mail: NUDT2013@126.com

L. Shu
Beijing University of Civil Engineering and Architecture, Beijing, China

S.-C. Chu et al. (eds.), *Advances in Intelligent Data Analysis and Applications*,
Smart Innovation, Systems and Technologies 445,
https://doi.org/10.1007/978-981-96-7277-6_24

terms of academic rationality, fundamentals, foresight, history and other research, and formed the research paradigm of the discipline of ideological and political education. Research requires a combination of methods and means. In the age of intelligence, the homogenisation of information has exacerbated the value barriers, posing challenges to the dissemination of ideological and political education discourse [1]. The effectiveness of discourse dissemination is related to the function of cultivating soul and educating people, and the status of mainstream ideology. Although some scholars have adopted different theories and methods to study it, they have mostly developed it from the dimension of elements and have not regarded discourse communication activities as an organic whole, which limits the effectiveness of practice. The reasons for this include that the basic concept of discourse communication in ideological and political education is vague and the research paradigm has not yet been innovated, resulting in unclear boundaries and insufficient depth of research to meet the new functional needs.

24.2 Overview of Discourse Communication in Ideological and Political Education

24.2.1 Definition of Discourse Communication

Although the effectiveness of the communication of ideological and political education discourse has not been explicitly mentioned, it has attracted extensive attention from scholars. Researchers believe that the authority of discourse, diversified media and means of communication are the prerequisites and necessary conditions, and the effectiveness of practice and discourse identity are the key indicators. Based on the characteristics of ideological and political education discourse communication activities, the effectiveness refers to the authoritative subject through a variety of media and means to systematically influence the audience to identify with and practice the mainstream ideological concepts, political views, and moral norms. However, current research often confuses the concepts of 'discourse communication effect', 'discourse communication effectiveness', and 'discourse communication influence', leading to confusion in understanding and practice. Clarifying the difference between these concepts is crucial to the construction of a scientific and effective evaluation system. Based on the above analysis, this paper proposes that the connotation of the effectiveness of discourse communication in ideological and political education is an important factor that describes the process of ideological and political education discourse communication activities, characterises the interactions and influences between behaviours and factors in the discourse communication activities, and emphasises the mutual compatibility between the actual effect and the desired effect of discourse communication.

24.2.2 Characteristics of Discourse Communication

Ideological and political education discourse dissemination is a practical activity to educate the audience on political, ideological, and moral norms, aiming at solving practical problems, embodying people-oriented and advancing with the times [2]. Its characteristics include the following:

1. The combination of theorisation and popularisation generates dominance, theorisation is achieved by condensing and upgrading the Marxist position and viewpoints and the socialist theoretical system with Chinese characteristics, and, at the same time, it wins the support of the audience and consolidates authoritative ideas by means of easy-to-understand and vividly imaginative discourses.
2. Constancy is generated by the tension between history and reality, and the constancy of the discourse system stems from the fact that its mode of material production always remains advanced and meets the requirements of social development. The constancy of the discourse system comes from the fact that its mode of material production always remains advanced, meets the requirements of social development, and promotes social progress.
3. Creativity is generated in the tension between deconstruction and construction, which criticises and deconstructs the prejudices of Western thinking, and provides creative methods for breaking the confinement of Western discourse. These characteristics are the basic prerequisites for the dissemination of ideological and political education discourse.

24.2.3 Mode of Discourse Communication

The evaluation of the effectiveness of ideological and political education discourse communication is the key to assessing the effectiveness of discourse communication activities, which provides decision support for subsequent communication activities and is crucial to shaping the image of ideological and political education and enhancing public recognition. The core of this evaluation is to ensure that the information carried by the discourse practice is effectively disseminated, which is in line with Lasswell's view that 'communication is a purposeful process'. The 5W model of communication proposed by Lasswell in The Structure and Function of Social Communication, i.e. communicator, message, medium, audience and effect, provides a theoretical framework for identifying the constituent elements of the communication of ideological and political education discourses, and is an important tool for advancing the evaluation activities. When evaluating, the five elements should be regarded as a whole, and based on the interaction between the transmitter and the audience, timely feedback and updating of communication strategies should be made to ensure the normality, relevance, and effectiveness of the evaluation activities. In this way, the evaluation of the effectiveness of the communication of ideological and political education discourse can more accurately reflect the actual situation of the communication activities and provide powerful support for the development of ideological and political education [3].

24.3 Current Situation of the Evaluation of the Effectiveness of Ideological and Political Education Discourse Communication

24.3.1 Fragmented Analyses at the Theoretical Level

Currently, the research results on the evaluation of the effectiveness of the communication of ideological and political education discourse are quite abundant, which are mainly related in fragmented analyses at the theoretical level and survey research at the practical level. The former explores the constituent elements of effectiveness one by one, while the latter focuses on the evaluation of practical activities by the audience. Combined with the 5W communication model, this paper will systematically sort out the evaluation elements in these two aspects of research, aiming at comprehensively depicting the current situation of the evaluation of communication effectiveness of ideological and political education, and providing theoretical support and practical guidance for further enhancing communication effectiveness.

The transmitter: The main body of ideological and political education discourse dissemination, with authority, ideology, and transcendence. Authority is reflected in controlling the direction of discourse, designing policies and systems, and providing excellent teams and facilities; ideology requires the subject to enhance the influence of discourse in complex ideological struggles; and transcendence requires the subject to put forward advanced theories and promote the high-quality development of discourse communication.

Message: The content of the discourse communication of ideological and political education includes two categories: professional and popular. Professional content is oriented to provide theoretical and academic knowledge for specific groups; popular content is oriented to the public, emphasising timeliness, precision and interactivity, aiming at occupying the main ideological position, constructing social value consensus, and leading social public opinion.

Media: The channel for the dissemination of ideological and political education discourse, the intelligent era relies on new media, big data, and artificial intelligence to broaden the communication channels and optimise the communication effect. Gathering audiences through the virtual reality community reduces the situation of being outside the discourse space due to inconsistent acceptance, and at the same time improves the ability of accurate pushing with the help of intelligent algorithms to achieve a highly joint communication mode.

Audience: The receiver of ideological and political education discourse dissemination, evaluation needs to consider audience coverage and demand responsiveness. It should break the limitations of traditional information dissemination, actively expand the scope of the audience, at the same time actively and efficiently meet the concerns of different audiences, strengthen the research and interpretation ability, improve the ability to ease ideological confusion, enhance the ability of policy interpretation, theoretical interpretation and other capabilities, and realise the bidirectional promotion of the professional and public spheres.

Effect: The effect of the dissemination of ideological and political education discourse is reflected in the audience's cognition, emotion, attitude, and behaviour. The professional field is mainly evaluated from the development level of audience knowledge, emotion, attitude, and behaviour; the public field examines the effectiveness of communication with the help of behavioural data from the new media platform. In recent years, the national top-level design has put forward specific requirements for ideological and political education, which has become an important reference for evaluating the effectiveness of discourse communication. With the digital transformation of education, new media has become an important channel, greatly enhancing the objectivity of the evaluation of the actual effectiveness of discourse communication.

To sum up, the evaluation of the effectiveness of ideological and political education discourse communication needs to comprehensively consider multiple aspects such as the transmitter, message, medium, audience, and effect, in order to achieve the comprehensive optimisation and enhancement of discourse communication activities.

24.3.2 Investigation and Research at the Practical Level

Existing research mainly evaluates the effectiveness of teaching discourse in Civics and the effectiveness of online Civics discourse, adopting a theoretical analysis framework that is operable and an evaluation system that is in line with the attributes of the discipline. The research includes two aspects: first, it focuses on theoretical communication, sets evaluation indexes in combination with relevant theories, and reflects the knowledge acquisition and ability cultivation of the discourse communication audience and, second, it pays attention to mass communication; examines the effectiveness of online Civic and political discourse communication; and expands the evaluation field, updates the elements, and enriches the methodology. In view of the similarity of the environment and characteristics of practical activities, the above studies can reflect the current evaluation status and provide reference for future research.

24.3.3 Gaps

Taken together, the academic community has made progress in the evaluation of the effectiveness of ideological and political education discourse communication, but there are still deficiencies. Theoretical studies are mostly qualitative appeals, evaluation standards are limited by disciplinary horizons, and the precision of evaluation index settings in the public domain is insufficient. Applied research needs to further consider more differentiated evaluation indexes, as well as demand- and effect-oriented principles and methods. In addition, current research mainly focuses on unidirectional communication and lacks attention to new inter-subjective, systematic, and comparative discourse communication paradigms. Therefore, it is necessary to respond to the new changes in the practice environment, systematically construct an evaluation framework, focus on inter-subjective interaction, and provide an effective solution path for evaluation activities. Research on how to improve the timeliness and effectiveness of ideological and political education discourse communication. The current situation of ideological and political education is good, but from the perspective of the quality of education supply and public demand, education is not attractive enough and its effectiveness is not sufficiently released. The contradiction between the educators and the educated is not only difficult to resonate with the same frequency, but also the obstacle of not being able to effectively link between theory and practice. Therefore, we should be based on the actual construction of education, optimise the supply side of ideological and political education in the age of intelligence based on the theoretical knowledge of attraction education, study and improve the methods and methods of education, enrich and perfect the content of education, and reasonably organise the evaluation of teaching, so as to further enhance the attractiveness of grass-roots ideological and political education.

24.4 Construction of Multidimensional Evaluation Model of Discourse Communication Effectiveness in Ideological and Political Education

24.4.1 Principles of Effectiveness Evaluation of Discourse Communication in Ideological and Political Education

Evaluation principles provide the basic starting point for evaluating the effectiveness of ideological and political education discourse communication, and four points need to be clarified: the principle of goal orientation, actively responding to the strategies of the Party and the state, outputting content that meets the concerns of society, and optimising the image; the principle of scientific standardisation, collecting information from multiple sources and various methods in combination with communication characteristics, and ensuring the standardisation of indicator design and data analysis; the principle of systemic comprehensiveness, treating the activity as an organic whole. The principle of systematic and comprehensive is to consider the activities as an organic whole, comprehensively examine the elements, structure and functions, and evaluate the effects; the principle of operability and comparability ensures that the data are easily accessible, the process is standardised, and the results are credible, so as to realise the function and utility of comprehensive examination and comparative analysis.

24.4.2 Ideas for Constructing the Evaluation Model of Communication Effectiveness of Ideological and Political Education Discourse

Based on the sorting of scattered results related to the effectiveness of ideological and political education discourse dissemination, following the four evaluation principles, using the 5W communication model to integrate the elements, clarifying the evaluation sources, refining the indicators, and describing the methodology, we have formed the idea of evaluating model construction (see Fig. 24.1). This idea will promote the in-depth dialogue between the theory and practice of ideological and political education, help the evaluation activities of the effectiveness of the dissemination of ideological and political education discourse and the high-quality development of ideological and political education to promote each other and improve together.

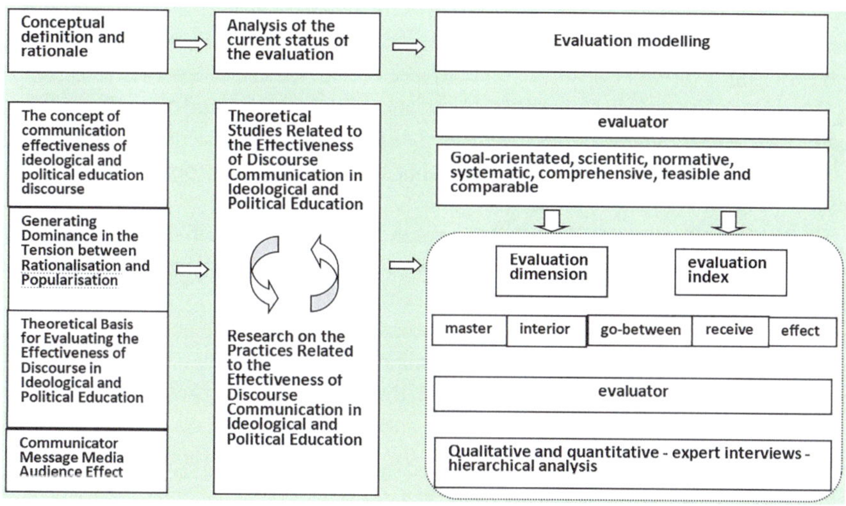

Fig. 24.1 Flow chart for evaluating the effectiveness of discourse communication in ideological and political education

24.4.3 Evaluation Model of Discourse Communication Effectiveness in Ideological and Political Education

On the basis of the idea of constructing a discourse communication effectiveness evaluation model for ideological and political education, the five major elements involved in the process of discourse communication, i.e. the transmitter, information, media, audience, and effect, are included in the evaluation model, and on this basis, five first-level indexes of the transmitter's subjectivity, the information service power, the discourse output power, the response power to the demand, and the discourse identity are identified, as well as 13 s-level indexes, including the evaluation of value-addedness indicators.

Dimension of Discourse Communication Subjects: Starting from the practical activities of ideological and political education discourse communication, we set up the core evaluation index of 'subjectivity of the transmitter'. In the secondary indicators, we especially consider the 'double-edged sword' effect of digital technology in the age of intelligence and set the 'negative effect of information cocoon' as a demerit item and the 'precise and efficient value-added effect' as a plus item, in order to comprehensively evaluate the digital technology. We set the 'negative effect of information cocoon' as a demerit item and the 'precise and efficient value-added effect' as a plus item to comprehensively assess the impact of digital technology. At the same time, based on theoretical research, we also set three indicators, namely, 'authoritative subject', 'ideological subject', and 'transcendental subject', in order to comprehensively measure the characteristics of the subject of discourse communication characteristics. In addition, we have also introduced the 'number of searches and browsing', 'number of likes and favourites', and 'degree of interactive activity' of the major ideological and political education information platforms as auxiliary references to evaluate the performance of the subject of discourse dissemination in a more comprehensive way to more comprehensively evaluate the performance of the subject of discourse dissemination.

Content dimension of discourse communication: In view of the content positioning and information service function of ideological and political education discourse communication, we set up the evaluation index of 'information service power'. In order to solve the problems of broad evaluation scope and poor targeting, we divided the content of discourse dissemination into two categories: professional content and public content and extracted three secondary indicators of 'professional information', 'public information', and 'comprehensive evaluation' accordingly. Accordingly, three secondary indicators are extracted: 'specialised information', 'public information', and 'comprehensive comments'. These indicators will be used as an important reference for observing the 'information service power' of discourse communication content.

Dimension of Discourse Communication Media: To focus on the accessibility and diffusion of ideological and political education discourse communication, we set up the evaluation index of 'discourse output power' [4]. Based on the existing research results, we have further refined two secondary indicators, namely, 'integrated media matrix' and 'service to educate people', to comprehensively assess the influence of discourse communication media on the effectiveness of discourse communication.

Audience Dimension of Discourse Dissemination: We focused on whether the discourse dissemination of ideological and political education can meet the needs of the public. To this end, we set up the core evaluation index of 'demand responsiveness' and the secondary index of 'differentiated audience needs'. These indicators will help us more accurately grasp the degree of response of ideological and political education discourse communication activities to the needs of different audiences.

Dimension of Discourse Communication Effect: To focus on the response of the audience after receiving the information carried by discourse communication, we set up the evaluation index of 'discourse identity'. In the secondary index, we set up two dimensions: 'professional theory recognition' and 'public recognition'. The 'professional theoretical identity' is mainly for teachers, students, researchers, and other professionals, and assesses the influence of discourse dissemination by examining their acceptance and recognition of theoretical knowledge as well as its application. The 'public recognition' is aimed at a wider audience group, combining the index of communication effectiveness of network information platforms, categorising the indicators in terms of coverage, acceptability and recognition, and setting up different levels of evaluation indicators of communication effectiveness [5].

24.4.4 Testing and Determining the Indicators of the Evaluation Model of the Effectiveness of Discourse Communication in Ideological and Political Education

In order to test the scientificity and feasibility of the evaluation indexes, this paper designed a questionnaire to solicit experts' opinions on the evaluation indexes of the effectiveness of the dissemination of ideological and political education discourse. We carefully selected 10 experts in the field of Marxist theory research, teachers of ideological and political theory courses and ideological and political workers as the target of the consultation and sent this questionnaire to them by email. These experts were divided into two groups: one group was scholars from colleges and universities and research institutes, who revised and improved the evaluation indexes mainly from the perspective of theoretical research; the other group was teachers of ideological and political theory courses and ideological and political workers, who provided valuable opinions and suggestions for the revision of the evaluation indexes more from the perspective of the practice of educating people at colleges and universities. When collecting the experts' opinions, we used several statistical indicators, including expert authority coefficient, mean, standard deviation, and coefficient of variation, to ensure that we can comprehensively and objectively reflect the experts' opinions and disagreements. After two rounds of opinion solicitation and indicator revision (nine valid questionnaires were returned in the first round, from four scholars from universities and research institutes and five teachers of ideological and political theory courses and ideological and political workers, respectively; eight valid questionnaires were returned in the second round, from four scholars from universities and research institutes and four teachers of ideological and political theory courses and ideological and political workers, respectively), we found that the mean value of the indicator items is gradually increasing, while the coefficient of variation is gradually decreasing. This change indicates that the disagreement among experts is decreasing significantly, and the scientificity and rationality of the evaluation indexes we designed are highly recognised. Table 24.1 gives a summary.

In the first round of expert opinion solicitation, the study analysed the reliability of expert ratings using the coefficient of expert authority, Cr. When $Cr > 0.7$, it indicates a high degree of reliability. In this study, $Cr = 0.84$ ($Cr > 0.7$), indicating that the reliability of the results of the first round of expert consultation is high. According to the feedback, the study optimises the setting of dimensions, naming and description of indicators, with the following specific modifications: (1) optimise the structure of indicators, such as adding the 'A4 demand responsiveness' second-level indicator, and combining the second-level indicator with the specific description of the third-level indicators to ensure the coverage; (2) improve the accuracy of the indicator presentation, such as (2) improving the accuracy of the indicators, such as further refining the description of the third-level indicators; and (3) adjusting the order of the indicators, such as switching the position of 'B4 Subject ideology' and 'B3 Subject authority' to highlight the leading role of the mainstream ideology of ideological and political education.

In the second round of expert consultation, the opinions were mainly focused on the observation points, and it was suggested that the expression of indicators should be clearer and more accurate, and the cross repetition between indicators should be avoided. After two rounds of expert consultation, this study finally constructed an evaluation index system for the effectiveness of ideological and political education discourse communication, which contains 5 primary indicators, 15 secondary indicators, and 35 observation points.

In this study, the weight values of hierarchical indicators are calculated using the Analytic Hierarchy Process (AHP) methodology. Firstly, this study invites six experts who have been involved in the whole revision of indicators to make group decisions, and the experts compare the relative importance of each level of evaluation indicators two by two, and establish 1, 5, and 15 judgement matrices corresponding to the indicator system, respectively. Secondly, the consistency ratio method was used to test

Table 24.1 Detailed list of indicators solicited for the evaluation model of the effectiveness of discourse communication in ideological and political education

Rounds of enquiry	Personnel of enquiry	Range	Standard deviation	Threshold	Coefficient of variation
Round 1	9	4.17–4.89	0.31–0.89	0. 17	6 items > 0.17
Round 2	8	4.65–5.03	0.00–0.48	0.10	0 items > 0.10

the consistency of the judgement matrices. The study used Yaahp software to carry out the consistency test of hierarchical single sorting and hierarchical total sorting, and the results of the data from six experts met the statistical standard of consistency ratio $CR<0.1$, which indicated that the results of the experts' judgement were credible and valid. Finally, the weights of evaluation indexes were determined via calculating results assembly. After all the judgement matrices passed the consistency test by Yahu software, the mean value of the ranking weights of each expert's judgement matrix was calculated, and the final weights of the first and second level indicators were obtained.

In summary, the evaluation model of the effectiveness of ideological and political education discourse communication has a significant advantage in evaluation method, which can integrate the scientifically proven qualitative and quantitative indicators to accurately reveal the essential characteristics of the ideological and political education discourse communication and its development trend. In terms of evaluation content, the model achieves a comprehensive coverage of the whole process of discourse dissemination, and is able to keenly capture and actively respond to the actual needs of the public for ideological and political education discourse, giving full play to its political and ideological features. Therefore, this evaluation model provides a solid and reliable decision-making support for the comprehensive promotion of the effectiveness of ideological and political education discourse communication.

24.5 Discussion and Recommendations

The following discussion is centred on the main problems of the current research.

Focusing on the object of evaluation, Xi Jinping's discourse on the construction of the discourse system of ideological and political education provides a fundamental guideline for evaluation and clarifies the scope and content of discourse dissemination activities [6]. By innovating the terminology of ideological and political work, forming a complete system, taking the initiative to set up topics, and innovating the discourse system, it provides a direction for evaluation. It examines the core value and essential attributes of discourse communication, gathers consensus, and promotes information sharing and effective communication.

Integrating evaluation content, considering traditional and new paradigms, is the new direction of current theoretical research. Consider the goal and mission of ideological and political education, break through the disciplinary barriers, integrate the 5W communication model, standardise the structure and function, improve the systematicity and appropriateness, make up for the shortcomings of content supply, and help high-quality development.

Reshape the main body of participation, the evaluation of the effectiveness of discourse communication should be based on scientific evidence and social acceptance, and the formation of a scientific indicator system, which is conducive to the discourse to go deeper and deeper within the whole society and is a prerequisite for improving quality and efficiency.

Improve the evaluation system, penetrate the process of discourse communication activities, re-evaluate the evaluation process and results, accurately identify the gap between purpose and effectiveness, and provide decision-making support for authoritative subjects. At the same time, we will implement the construction of evaluation institutions and mechanisms, improve the evaluation system, and promote the construction of the effectiveness of discourse communication in the intelligent era.

24.6 Conclusion

On the basis of in-depth study of the theory related to the evaluation of discourse communication in ideological and political education, this paper cleverly draws on the classic theory of communication—'5W communication model' as an analytical tool to conduct a comprehensive and as a tool for analysis, we have made a comprehensive and detailed analysis of the elements and dimensions of discourse communication effectiveness evaluation. On this basis, we have made a bold attempt to construct a multi-dimensional and multi-level evaluation model. The model covers five core first-level indicators, namely, communicator, message, medium, audience, and effect, and is further refined into 15 s-level indicators, including demand responsiveness, aiming to comprehensively and systematically reveal the basic features and overall appearance of the evaluation of the effectiveness of ideological and political education discourse communication. It is worth noting that the current study is still in its initial stage, and the model constructed and the indicators screened are only preliminary ideas. We are well aware that there is a certain gap between theory and practice, so in the future we need to constantly test, revise, and improve these evaluation indicators and calculation methods in a wide range of practical applications. We expect that through unremitting efforts, we can gradually form a more standardised and operable evaluation model, which will provide a more scientific and accurate basis for the evaluation of the effectiveness of the dissemination of ideological and political education discourse, thus promoting the high-quality development of ideological and political education.

References

1. Desheng G, Zhong, F.: On the mission of Marxist ideological and political education in the times. J. Northeast. Norm. Univ. (Philos. Soc. Sci. Ed.), (2)
2. Hou, L., Zhang, Y.: On the three basic forms of ideological and political education discourse. Marx. Res. (12)
3. Li, X.: A cognitive review of contemporary Chinese ecological civilisation thought and practice in European and American Academia. J. Xi'an Jiaotong Univ. (Soc. Sci. Ed.), (4), (2023)

4. Xu, Y., Lu, B.: The value implications, practical dilemmas and responses of ideological and political education discourse rhetoric. Ideol. Educ. Res., (11), (2023)
5. Han, W.: An analysis of the rationality of the relational value orientation of ideological and political education in the new era. Res. Marx. Theory, (5)
6. Li, L.: The fundamental followings of doing a good job in the work of ideology and politics in Tibetan universities in the new era–Thinking about the spirit of learning Xi Jinping's reply to his congratulatory letter on the work of Tibet. J Tibet Univ Natl (Philos Soc Sci Ed), (1), (2023)

Chapter 25
Simulation of Gesture Recognition Control Robotic Arm Based on Mediapipe and OpenCV

Chenghang Liu, Jiajun Guo, Quan Zhang, Jie Sun, and Eng Gee Lim

Abstract This paper proposes an innovative method to control the robot arm, which is based on the gesture recognition library–Mediapipe and OpenCV. The project is based on the RobotStudio platform, which simulates the development of the project. The research closely combines human–computer interaction with robot control systems. Through computer vision technology and machine learning, the system can accurately recognize human gestures and convert them into precise movements of robotic arms. This project improves the efficiency and practicality of medical treatment, industrial safety and industrial automation on the road simply and easily. The system processes gestures and generates robotic control commands via OpenCV and Mediapipe, then communicates with the robotic arm via TCP/IP and controls the robotic arm via RAPID. The system has a modular design that could be expanded to control other robotic arms and more complex gestures in the future. The project demonstrates the technology for gesture-driven robots and demonstrates the potential for future remote control, human–computer interaction, and virtual reality environments to work together.

C. Liu · J. Guo · Q. Zhang (✉) · J. Sun · E. G. Lim
School of Advanced Technology, Xi'an Jiaotong-Liverpool University, Suzhou, China
e-mail: Quan.Zhang@xjtlu.edu.cn

C. Liu
e-mail: Chenghang.Liu20@alumni.xjtlu.edu.cn

J. Sun
e-mail: Jie.Sun@xjtlu.edu.cn

E. G. Lim
e-mail: enggee.lim@xjtlu.edu.cn

S.-C. Chu et al. (eds.), *Advances in Intelligent Data Analysis and Applications*,
Smart Innovation, Systems and Technologies 445,
https://doi.org/10.1007/978-981-96-7277-6_25

25.1 Introduction

In recent years, with the rapid development of AI and advanced HCI technology, a combination of robots, AI, and HCI technology has been hot topics in the robot area and has brought revolutionary changes in medical assistance, industrial automation, remote control intervention, education, and so on [1, 2]. Conventional robotic arms usually depend on controllers or instructions. Although these methods are mature and effective, they are hard to get started, and not intuitive and flexible enough [3]. It will cause some problems in cross-field cooperation and debugging. It is very hard for non-robot area personnel to maintain and master the control method. The emerging Gesture control system provides an easier, simpler, and new remote way to control the robot arm [1, 4, 5]. In this way, it improves not only the user experience but also widens the possibility of application. Gesture recognition technology is based on computer vision and machine learning algorithms, which provide a new and efficient way to realize simpler and remote control of robots and robotic arms and seamless connection of personnel technologies in different fields [1, 5, 6]. However, the precise translation of human gestures into robotic control instructions is a complex task that involves cross-cutting technologies such as communication, machine learning, image processing, and robotic control codes (RAPID). Meanwhile, the high precision of gesture detection is needed to realize the one-to-one correspondence between control commands and gestures. In this project, the gesture recognition system is realized by using MediaPipe and OpenCV technology through the Python platform. The technology is developed based on Python and RobotStudio. Through MediaPipe gesture recognition function and OpenCV image processing to achieve an efficient and accurate gesture recognition system [7, 8]. The project transmits communication through TCP/IP implements encoding and compiling processing through ASCII and executes on the robot arm through RAPID instruction [9]. This project improves the efficiency of the manipulator's operation and provides a more intuitive way to operate specific functions.

25.2 Related Work

25.2.1 Gesture Recognition in Human–Computer Interaction

Early Approaches to Gesture Recognition

Most of the early methodology for gesture recognition relies on traditional techniques, such as feature point extraction and skin color recognition, through a computer vision-based system to identify the hand movement [1, 10]. These approaches are highly sensitive to changes in illumination and background noise of the image [1, 5, 6]. Therefore, these methods can only achieve low accuracy in dynamic gesture detection and cannot guarantee robustness. Although this initial approach lacked robustness and accuracy, it paved the way for more complex and stable gesture recognition methods [1, 10]. Early methods did not incorporate machine learning methods and relied solely on image processing, resulting in lower accuracy and robustness of the system.

Machine Learning and Deep Learning in Gesture Recognition

With the rapid development and wide application of machine learning, especially deep learning, machine learning applications in gesture recognition are becoming more and more mature [11]. Machine learning and deep learning enable gesture recognition models to learn complex features and patterns directly from large amounts of data, greatly improving accuracy and robustness. Convolutional neural networks (CNN) and recurrent neural networks (RNN) have greatly improved in accuracy, enabling highly reliable and real-time gesture recognition [11]. At the same time, handwriting recognition can learn from facial recognition algorithms to improve the accuracy of recognition [12]. This achievement is significant for the help of complex gesture recognition, making its application in human–computer interaction more widespread. This technological innovation has greatly improved the accuracy and robustness of gesture recognition control robotic arms [13].

MediaPipe and OpenCV in Gesture Detection and Application in this project

MediaPipe and OpenCV are among the most commonly used open-source frameworks for gesture recognition today [1, 5, 7, 8]. Mediapipe was developed by Google in 2019. It detects 21 key points of the hand through the pre-trained model, thus achieving high-precision gesture recognition, even under extremely complex conditions with good robustness [7]. OpenCV provides powerful support for image processing and special diagnosis detection [8]. This project combines MediaPipe and OpenCV to construct and simulate an accurate and low-latency gesture recognition system.

Challenges in Gesture Recognition

With the increasing maturity of gesture recognition technology, the accuracy of gesture recognition has been greatly improved [7, 10, 11, 13]. However, when there are sharp changes in illumination, hand occlusion, and overly complicated gestures, the gesture recognition system still has problems that cannot be solved, especially when used for the high-precision control of robotic arms [10, 13].

25.2.2 Robotic Arm Control Systems

Traditional Control Methods for Robotic Arms

The traditional control methods of the robot arms are mainly to control the robot arm through the controller or control the robot arm through the command [3, 9]. Although these methods are simple and effective for the control of the robot arm, these control methods require a large number of professional training for personnel in different areas and are not particularly flexible, especially in remote control and industrial automation, and the limited distance sometimes leads to safety risks for the control of the robot arm [9].

TCP/IP Communication in Robotic Systems

TCP/IP communication protocol is widely used in robot systems, especially in human–computer interaction and remote control. The TCP/IP protocol allows the transmission of data across the network, thus meeting the direct interaction of different platforms, different software, and different languages, to realize the remote control of the robot arm and the interaction between the robot arm and people [9, 14]. In this project, through the TCP/IP protocol, the Python platform can send the gesture instructions generated by the TCP/IP protocol to the RAPID program, to realize the remote empty robot arm and simple human–computer interaction.

RAPID Programming Language for Robotic Control

RAPID is ABB's standard programming language for robotic arm control, and its efficient motion control and task automation functions are widely used in medical robots and industrial robots. RAPID supports the TCP/IP communication protocol and can interact with other systems, platforms, and languages through simple Socket programming [9, 14]. This project receives Python information through RAPID and converts it into the control instruction of the robot arm and controls the movement of the robot arm, thus achieving efficient one-to-one corresponding control.

ASCII Conversion in Robotic Control Communication

ASCII encoding is widely used to format transmission control signals in the control of robotic arms [15]. In the RAPID part of this project, the binary byte flow of the binary byte stream gesture recognition control instruction received from TCP/IP is converted into ASCII encoded text format, which ensures the compatibility of cross-platform system and the readability of the instruction to the robot arm [9, 15]. Commands such as RAPID and `UnpackRawBytes` parse instructions in ASCII format to drive the precise movement of the robotic arm [15].

25.2.3 Gesture-Based Control of Robotic Arms

Vision-Based Robotic Arm Control

Vision-based control systems have attracted much attention in the field of robotics. Gestures are captured and interpreted by cameras or sensors and converted into control instructions, which realize remote control, avoid contact operation, and greatly improve human–robot interaction and ease of human–computer interaction [1, 2, 5, 10].

Challenges in Gesture-Based Control

The gesture-based control method has obvious advantages, but there are still many problems and challenges in practical application, such as corresponding delay, recognition accuracy and environmental adaptability [1, 10, 11]. Especially in industrial and medical applications, accuracy, safety, and reliability are issues that must be addressed. Otherwise, there are certain security risks for industrial and medical applications.

Innovations in This Project's Approach

By combining MediaPipe gesture recognition and the RAPID program, this project constructs a modular analogue gesture control system through the RobotStudio platform [7, 7]. This system uses MediaPipe for high-precision gesture detection that converts the recognized gesture into RAPID command through TCP/IP protocol, and realizes the human–computer interaction between the remote control robot arm and the robot arm [2, 7, 9, 13]. With the introduction of MediaPipe with high precision gesture recognition in the existing research project, the system is responsive and has the potential for high scalability and practical applications, which can be adapted to more complex gestures and different types of robotic arms in the future.

25.2.4 Work Content and Innovation of This Project

This project proposes a robotic arm control system based on gesture recognition and TCP/IP communication. MediaPipe and OpenCV are used to realize real-time gesture recognition [7, 9, 12, 13], and the recognized gesture commands are converted into ASCII information through Python script and sent to the RAPID program of the RobotStudio platform. There are three major innovations in this project: **Integrate MediaPipe and OpenCV for precise and efficient gesture recognition**: MediaPipe provides high-precision detection of key points in the hand, and OpenCV is responsible for image processing and gesture judgment. This project combines the two organically and successfully realizes real-time recognition and information transmission of multi-medium gestures (X, Y, Z axes motion direction and stop action). **Using TCP/IP protocol to achieve cross-platform communication:** Through the TCP/IP protocol, Python scripts can encode the binary of gesture commands into ASCII format and send them to the RAPID program of the robot arm [13–15]. The RAPID program decodes the gesture recognition instruction and sets the corresponding movement of the robot arm to achieve cross-platform control. **Modular design**: The modular design of this project can add more complex gestures and adapt to more types of robotic arms actions, which effectively improves the scalability of the system.

25.3 Methodology

The system realizes complete closed-loop control from gesture recognition to manipulator control through a modular architecture. The system integrates (MediaPipe), network communication (TCP/IP protocol), and robot arm control (RAPID language), allowing users to remotely control the robot arm through gestures and achieve precise movement in three-dimensional space. The system is more scalable to accommodate different gestures and more complex control needs.

25.3.1 System Architecture

The system architecture consists of three main parts:

Gesture recognition module (Python): Real-time detection and recognition of user gestures through MediaPipe and OpenCV, generating corresponding control instructions.

Network Communication module (Python): Using TCP/IP protocol to transmit gesture control instructions, the recognized gesture is converted into ASCII format and sent to the robot arm control terminal (RobotStudio).

The Robot Arms Control Module (RAPID): Interprets and executes the control instructions to drive the robot arm to complete the corresponding X, Y, and Z-axes movements.

25.3.2 Workflow Summary

Initialization and Connection: When the system starts, the Python module initializes the computer's camera and gesture recognition module, and establishes a TCP/IP communication connection to the robotic arm control (RAPID).

Gesture Detection and Command Generation: The user makes a specific gesture in front of the camera, and the Python module detects the gesture and generates the control command through MediaPipe.

Instruction Sending and Receiving: The recognized gesture is sent to the RAPID module through TCP/IP protocol, and decoded into ASCII text format instructions at the RAPID terminal.

Instruction Parsing and Robot Arm Control: The RAPID module receives and analyzes ASCII instructions, and controls the robot arm for precise space movement according to the instruction content.

System Termination: The user presses the q key to exit the program, disconnect and shut down the system.

The following is a flowchart of the system, showing the interaction process of each module. This flowchart clearly shows the workflow and system architecture of

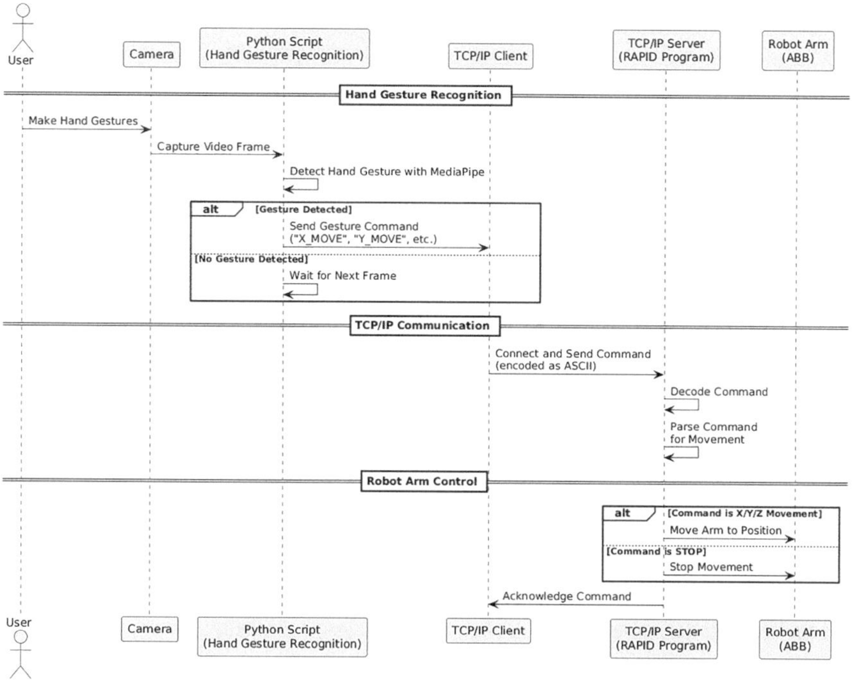

Fig. 25.1 System flowchart

the entire closed-loop system from user gesture input to manipulator action output (Fig. 25.1).

25.3.3 Hand Gesture Recognition Module (Python)

Initialization: Load the MediaPipe gesture detection model, initialize TCP/IP communication, and establish the connection between Python and RAPID module.

Video capture and gesture recognition: Capture real-time video streams through the camera, use MediaPipe to detect key points in the hand, recognize different gestures based on finger posture and generate control commands. For example:

Thumb left: X_MOVE1000, move 10 units in the positive direction of the X axis.

Small point right: Z_MOVE00-10, move 10 units in the negative direction of the Z axis.

Open all hands: STOP, stop the current action of the robot arm.

Command generation and sending: The recognized gesture is sent to the RAPID module for processing through TCP/IP.

Real-time display and exit: Using OpenCV to display video frames and recognition results, the user can press the "q" key to exit the program, close the camera and TCP connection.

25.3.4 Communication Module (TCP/IP Protocol)

The Python and RAPID modules communicate over TCP/IP to ensure that control instructions are transmitted from Python to the robot arm, and the TCP connection is established: Python acts as the client and connects to the IP and port of the RAPID server. Instruction transfer: Each time a gesture is recognized, Python generates instructions and sends them to the RAPID module.

25.3.5 Robotic Arm Control Module (RAPID)

The RAPID module is responsible for receiving and analyzing the control instructions and driving the action of the robot arm. The process is as follows:

TCP/IP server initialization: Create a TCP/IP server on the RAPID side, bind the IP and port, and wait for the Python connection. Receive instructions: After receiving the Python client connection, enter the loop, continue to receive gesture control instructions, and decode into ASCII encoding. Command analysis and manipulator control: Control the movement of the manipulator according to the received instructions. When the connection is closed: When the Python side is disconnected or the program terminates, the RAPID module stops receiving instructions and closes the TCP connection (Table 25.1).

Table 25.1 Gesture command mapping

Gesture	Command	Description
Thumb left	X_MOVE1000	Move the robotic arm 10 units along the X-axis (positive)
Thumb right	X_MOVE-1000	Move the robotic arm 10 units along the X-axis (negative)
Index finger left	Y_MOVE0100	Move the robotic arm 10 units along the Y-axis (positive)
Index finger right	Y_MOVE0-100	Move the robotic arm 10 units along the Y-axis (negative)
Pinky left	Z_MOVE0010	Move the robotic arm 10 units along the Z-axis (positive)
Pinky right	Z_MOVE00-10	Move the robotic arm 10 units along the Z-axis (negative)
Open hand	STOP	Stop the robotic arm's current action

25.4 Results

The main achievement of this project is to realize a closed-loop system for remote control of robotic arm movement through gesture recognition and complete all aspects of gesture detection, command transmission, and robotic arm movement. Here are the results. In Fig. 25.2, it can be seen that the entire system of the project is running properly, and RobtoStudio can effectively accept instructions from the Python side. The Python side can effectively recognize gestures and output corresponding correct instructions. As can be seen from Fig. 25.3, we can effectively react to the recognized gestures in real-time. The program is set at 0.5s. From the experiment, we can see that the delay of our whole system is very small.

25.4.1 *Functional Verification*

As is shown in the picture, every part of the system works fine. After comparing with the table in the methodology, the test part is working normally, and the communication work is also normal.

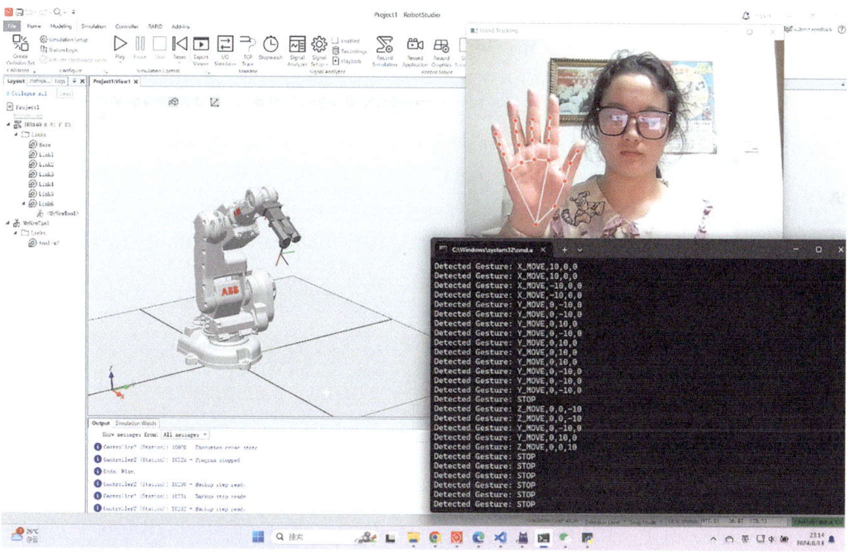

Fig. 25.2 Running status of the system

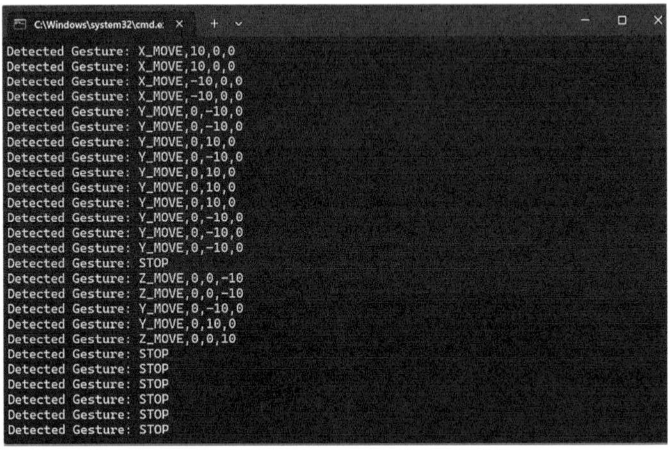

Fig. 25.3 Gesture command recognition

25.4.2 Errors and Limitations

From Fig. 25.4, this simulation may cause CPU overload issues sometimes when running for a long time or gesturing commands too frequently, and subsequent work will optimize the structure and improve the processor configuration. In the environment of large illumination variation, the gesture recognition module will occasionally appear recognition error, resulting in inaccurate instruction transmission. Network instability can also lead to increased latency, affecting the response speed of the robotic arm. These problems may affect the user experience in actual use and need to be further optimized in the future. Due to the reality of the performance of computer equipment, the simulation will often report errors and can be replaced with a more powerful computer.

25.5 Discussion

System advantages: 1. Modular design: The system is designed through the modular architecture of Python and RAPID, which is convenient for independent debugging and expansion [9]. Real-time control: The closed-loop design of gesture recognition and robotic arm control can be a real-time response and have the ability to quickly recognize the user's gesture actions [2, 14]. Good scalability: More gestures or different robotic arm movements can be added to accommodate more complex control needs.

System limitations: The recognition accuracy is affected by the environment: in the case of unstable lighting, the accuracy of gesture recognition will be decreased,

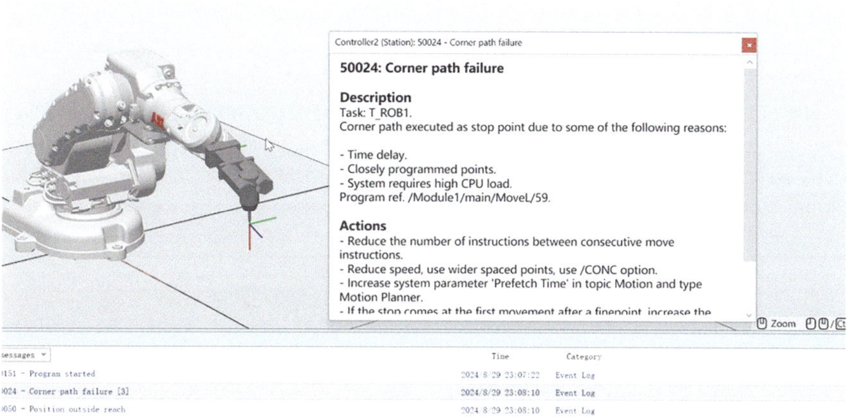

Fig. 25.4 CPU overload failure

affecting the accuracy of control instructions [1, 5, 10]. Device simulation performance and overall system structure problems: Device simulation performance and overall system problems: Sometimes the CPU performance cannot reach the required performance of the entire system simulation, and the generation and transmission speed of instructions can lead to system downtime. At the same time, the design of the block and the system structure sometimes cause the CPU to be too complex. Therefore, you need to replace industrial grade or higher performance CPUs or integrated CPUs. Limited control accuracy: The control accuracy of the robot arm is limited by the hardware and RAPID program Settings, and it is difficult to maintain high precision in the fine control requirements [9].

Future Work: We are developing the gesture recognition entity project for a real robot arm. In the future, our team will complete a complete physical project of remote control and human–computer interaction of the robot arm, which is no longer the simulation. At the same time, VR virtual control of the robot arm will be added to better enable people to interact with the robot arm. At the same time, it will improve the accuracy of the entire system in 3D coordinates. At the same time, we are also trying a lightweight object detection method that can be deployed offline optimized and updated to fit the project based on the work of other scholars in this field [16].

25.6 Conclusion

This project presents a complete simulation system for remote control of a robotic arm through gesture recognition, simulating a real-time control loop from gesture to robotic arm movement. The system is composed of Python gesture recognition module, TCP/IP communication module, and RAPID robot arm control module, which generates control commands through gestures and sends them to the robot arm, and the robot arm responds and executes corresponding actions [15]. The experiment verifies the basic function and real-time responsiveness of the system, which can meet the general control requirements.

The design of this system is simple and has good expansibility, which can adapt to more gestures and more complex control logic in the future. Future work can focus on materialization, optimizing gesture recognition algorithms, and VR virtual human–computer interaction.

References

1. Qi, J., Zhang, Y., Wang, Z., Hao, Y., Yang, J., Wang, H.: Computer vision-based hand gesture recognition for human-robot interaction: a review. Complex & Intelligent Systems **10**(1), 1581–1606 (2024)
2. Zhou, H., Wang, D., Yu, Y., Zhang, Z.: Research progress of human-computer interaction technology based on gesture recognition. Electronics **12** (2805), (2023)
3. ABB robotics controller homepage. https://new.abb.com/products/robotics/controllers. Last accessed 2024/11/13
4. Kokorin, K., Mu, J., John, S.E., Grayden, D.B.: Predictive shared control of robotic arms using simulated brain-computer interface inputs. In: 45th Annual International Conference of the IEEE Engineering in Medicine & Biology Society (EMBC), pp. 1–5. Sydney, Australia (2023)
5. Song, S., Yan, D., Xie, Y.: Design of Control System Based on Hand Gesture Recognition. In: 15th International Conference on Networking. Sensing and Control (ICNSC), pp. 1–4. IEEE, Zhuhai, China (2018)
6. Abdullayeva, F.J., Gurbanova, K.S.: Sign language hand gesture recognition method based on machine learning. In: 16th International Conference on Application of Information and Communication Technologies (AICT), pp. 1–5. IEEE, Washington DC, USA (2022)
7. Mediapipe Documentation.https://chuoling.github.io/mediapipe/. Last accessed 2024/11/13
8. OpenCV Homepage. https://opencv.org/. Last accessed 2024/11/13.
9. ABB Technical Reference Manual, RAPID. https://library.e.abb.com/public/688894b98123f87bc1257cc50044e809/Technical%20reference%20manual_RAPID_3HAC16581-1_revJ_en.pdf. Last accessed 2024/11/14
10. Miah, A.S.M., Wu, Y., Sun, W., Alnowibet, K., Muhammad, G., Hossain, M.S.: Multistage spatial attention-based neural network for hand gesture recognition. Computers **12**(1), 13 (2023)
11. Alzubaidi, L., Zhang, J., Humaidi, A.J., Al-Dujaili, A., Duan, Y., Al-Shamma, O., et al.: Review of deep learning: Concepts, CNN architectures, challenges, applications, future directions. Journal Big Data **8**, 1–74 (2021)
12. Xu, X., Cui, J., Chen, X., Chen, C.-L.: A facial expression recognition method based on residual separable convolutional neural network. J. Netw. Intell. **7**(1), 59–69 (2022)
13. Miah, A.S.M., Hasan, M.A.M., Shin, J.: Dynamic hand gesture recognition using multi-branch attention based graph and general deep learning model. IEEE Access **11**, 4703–4716 (2023)
14. Oltean, S.-E., Vinka, P.-V.: Remote user interface for ABB IRB120 robot using TCP/IP communication. Acta Marisiensis. Ser. Technol. **20**(2), 49–56 (2023)
15. ASCII Code Reference. https://www.ascii-code.com/. Last accessed 2024/11/13
16. Jia, Y.-F., Chen, G.-D., Jiang, Z.-C., Yang, M., Xing, L.-Y., Cui, Y.: A lightweight fast object detection method. J Netw Intell **7**(1), 209–221 (2022)

Chapter 26
AI-Based Predictive System for Bull-Side Put Spread Trading Strategy Using Decision Tree Algorithm

Chi-Fang Chao, Ming-Hua Hsieh, Min-Kuan Chen, and Mu-En Wu

Abstract Options trading has gained popularity among investors and academics for its flexibility and strategic potential. This study aims to enhance the performance of a bull-side put spread options trading strategy using machine learning. We integrate raw data attributes, implied volatility metrics, and technical indicators into Decision Tree and Logistic Classification models to predict trade profitability. The results reveal that while the Logistic Classification model achieves higher cumulative payoffs, it executes less reliable trades. In contrast, the Decision Tree algorithm demonstrates superior precision in identifying profitable trades, making it more effective for implementing the strategy. This research contributes to the growing literature on applying machine learning in options trading and highlights the importance of considering both predictive accuracy and trade reliability. Future research could explore advanced algorithms and incorporate market sentiment analysis to further refine the models and enhance their practical utility.

C.-F. Chao (✉) · M.-H. Hsieh
Department of Risk Management and Insurance, National Chengchi University, Taipei, Taiwan
e-mail: cfchao@nccu.edu.tw

M.-H. Hsieh
e-mail: mhsieh@nccu.edu.tw

M.-E. Wu
Department of Information and Finance Management, National Taipei University of Technology, Taipei, Taiwan
e-mail: mnwu@ntut.edu.tw

M.-K. Chen
Institute of Network Engineering, National Yang Ming Chiao Tung University, Hsinchu, Taiwan
e-mail: minkuanchen99.cs13@nycu.edu.tw

S.-C. Chu et al. (eds.), *Advances in Intelligent Data Analysis and Applications*,
Smart Innovation, Systems and Technologies 445,
https://doi.org/10.1007/978-981-96-7277-6_26

26.1 Introduction

Financial market prediction is a key focus in academia and industry due to its impact on decision-making and asset management. Various methodologies have been developed, including pricing theories [4, 10, 14], time series analysis [5, 9, 21], technical analysis [11], and fundamental analysis.

Artificial intelligence (AI) has redefined financial market forecasting. AI models are adept at handling large datasets, complex patterns, and dynamic conditions, providing a sophisticated framework [1, 13, 20]. Hybrid frameworks combining AI with traditional methods result in precise forecasts [16, 19, 25].

While much research on AI were mainly focused on major financial instruments, there is a growing trend of applying machine learning to options markets [2, 15, 24]. However, these studies primarily focus on pricing and returns for individual options, leaving a gap in applying AI to complex options trading strategies.

This study builds upon the findings of Wu et al. [24] by integrating their framework with option spread strategies. Machine learning is used to predict outcomes of trades, with performance evaluated through historical backtesting. The remainder of this paper is structured as follows. Section 26.2 introduces the evolution of option trading strategies and the application of machine learning in financial markets. Section 26.3 presents the methodologies of the proposed AI-based predictive system. Section 26.4 demonstrates the numerical results and discusses the findings, while Sect. 26.5 provides an overview and conclusions drawn from this research.

26.2 Preliminaries

26.2.1 Option Trading Strategies

Options are financial derivatives that give the buyer the right, but not the obligation, to buy or sell an underlying asset at a predetermined strike price before or on a specified expiration date. The seller, in contrast, is obligated to fulfill the contract terms if the buyer exercises their rights. This dichotomy makes options versatile tools for achieving various investment objectives, such as hedging, speculation, and income generation [22]. Structured trading strategies provide solutions for managing the complexities and risks associated with option trading, allowing investors to adapt their tactics based on market conditions and risk-return profiles [14].

Recent research has focused on integrating AI with options trading strategies. [23] applied statistical methods to evaluate historical settlement prices and establish an empirical probability distribution for strategy assessment, combining these distributions with spread strategies and the Kelly criterion to optimize investment proportions. [24] designed option trading strategies by employing stop-loss thresholds and classification algorithms to predict the probabilities associated with profit and loss, illustrating the effectiveness of a synergistic ensemble learning strategy in historical backtesting scenarios.

26.2.2 Machine Learning Algorithms

The rapid advancements in artificial intelligence have propelled machine learning to the forefront of predictive modeling methodologies to enhance trading efficacy. This section presents several well-established machine learning algorithms that are employed in this research.

Decision Tree and Random Forest Decision trees [18] offer advantages in supervised learning due to their interpretability and transparency. However, they are susceptible to overfitting and instability, particularly when dealing with imbalanced datasets [12]. Ensemble learning techniques, such as Random Forests [6], address these issues by training multiple trees and aggregating their predictions, resulting in improved accuracy and generalization. Empirical studies have validated the effectiveness of decision trees and Random Forests in financial forecasting [3, 24].

eXtreme Gradient Boosting (XGBoost) XGBoost [7] is a state-of-the-art gradient boosting framework that constructs decision trees sequentially, each tree learning from and correcting the errors of its predecessors. It demonstrates remarkable efficacy in handling large datasets due to its scalability and robustness. The framework's success can be attributed to its integrated regularization strategies, which prevent overfitting, and its ability to effectively manage missing data. However, optimizing the hyperparameters of XGBoost can be challenging [12]. Empirical studies have demonstrated the effectiveness of XGBoost in financial forecasting [8, 17].

Performance Metrics The confusion matrix is a fundamental tool in predictive performance evaluation for machine learning algorithms. It categorizes predictions into four distinct types: true positive (TP), false positive (FP), true negative (TN), and false negative (FN), as illustrated in Table 26.1. This classification elucidates the model's predictive performance and facilitates the computation of various metrics that evaluate its accuracy and dependability [12].

Accuracy quantifies the fraction of correct predictions relative to the total predictions made. It serves as a holistic indicator of the model's predictive competence, computed as $\frac{TP+TN}{TP+FP+TN+FN}$. Elevated accuracy reflects the model's proficiency in delivering accurate predictions across numerous instances.

Precision evaluates the model's effectiveness in correctly recognizing profitable trades from all predicted as such. It is defined as $\frac{TP}{TP+FP}$. High precision indicates that predicted profitable trades are likely accurate, thereby reducing false positives.

Table 26.1 Framework of confusion matrix

Actual value	Prediction outcome		Total
	Positive	Negative	
Positive	TP	FN	$TP + FN$
Negative	FP	TN	$FP + TN$
Total	$TP + FP$	$FN + TN$	N

Recall, or sensitivity (true positive rate, TPR), measures the ratio of actual profitable trades identified accurately by the model. It is computed as $\frac{TP}{TP+FN}$. High recall signifies the model's capability to identify a significant proportion of actual profitable trades, thus minimizing missed opportunities.

The F1 Score represents the harmonic mean of precision and recall, offering a comprehensive assessment of model performance. It is calculated as $2 \times \frac{\text{Precision}\times\text{Recall}}{\text{Precision}+\text{Recall}}$. The F1 Score is especially pertinent in trading scenarios, ensuring accurate predictions of profitable trades while capturing a considerable fraction of them, thus balancing the minimization of false positives with the identification of genuine profitable opportunities.

26.3 Methodology

This paper proposes a predictive system based on machine learning methodologies to forecast the outcomes of transactions associated with an options trading strategy. For illustrative purposes, a bull-side put spread strategy is adopted to construct the option portfolio, which involves selling the first out-of-the-money put option (denoted with subscript 1) and buying the second out-of-the-money put option (denoted with subscript 2).

This section is divided into two parts. The first segment examines the valuation of option portfolios, specifies the criteria for triggering an early-exit mechanism, and defines the measurement of strategy efficacy. The second segment focuses on feature extraction and the application of machine learning algorithms to predict the outcome category of transactions.

26.3.1 Construction of Option Portfolio and Performance Evaluation

Market participants develop various trading strategies by strategically combining call and put options with specific trading frameworks. These strategies are adapted to reflect the investor's market perspective and risk management objectives, aiming to enhance potential returns while controlling associated risks. One well-known approach is the spread strategy, which is constructed using two options with distinct prices and characterized by a predetermined and fixed payoff distribution when held until expiration [23]. Considering the option period from 0 to T, the price of the option portfolio at time t is expressed as

$$V_p(t) = \sum_{i=1}^{N} H_i \cdot V_i(t) \quad , \tag{26.1}$$

where $V_i(t)$ represents the price of the ith option at time t, H_i indicates the quantity of holdings of the ith option, and N denotes the number of distinct options in the portfolio.

Options have a specified expiration date, requiring investors to make trading decisions within a restricted time frame. Risk management in options trading is imperative, as the flow of time directly affects the valuation of the options. This paper introduces an early-exit mechanism incorporating stop-loss and take-profit thresholds to manage the oscillations in profits and losses amid market fluctuations. The exit time τ associated with these mechanisms is determined by specific conditions, as demonstrated in Eq. 26.2:

$$\begin{aligned} \tau_{SL} &= \min_{0<t<T} t \mid V_p(t) \leq SL \cup T \\ \tau_{TP} &= \min_{0<t<T} t \mid V_p(t) \geq TP \cup T \\ \tau &= \min(\tau_{SL}, \tau_{TP}) \quad , \end{aligned} \tag{26.2}$$

where T is the expiration time of the portfolio, τ_{SL} and τ_{TP} represent the instances when the portfolio value first triggers the stop-loss and take-profit conditions, respectively. If $\tau = T$, neither condition was activated within the portfolio duration.

The stop-loss threshold SL and take-profit threshold TP are subject to alterations corresponding to the varying levels of investor risk tolerance, defined as Eq. 26.3:

$$\begin{aligned} SL &= V_p(0) - \left|V_p(0)\right| \times \rho_{SL} \\ TP &= V_p(0) + \left|V_p(0)\right| \times \rho_{TP} \end{aligned} \quad , \tag{26.3}$$

where ρ_{SL} and ρ_{TP} denote the specific ratio associated with the stop-loss and take-profit mechanism, respectively, while the term $V_p(0)$ represents the entry price of the portfolio.

With the entry price $V_p(0)$ and exit price $V_p(\tau)$ clearly defined, the payoff of the transaction, denoted by PL, can be evaluated as $PL = V_p(\tau) - V_p(0)$, and the cumulative payoff of the specified trading strategy, denoted by CPL, is defined as $CPL = \sum_{i=1}^{N} PL_i$, where N represents the number of transactions under the strategy within a given period.

The trade ratio, denoted as TR, measures the fraction of trading opportunities that meet the designated entry criteria in relation to the total volume of opportunities in the market, providing insights into the potential trades that are omitted. TR is calculated as $TR = \frac{NE}{NO}$, where NE reflects the total number of executed trades, whereas NO indicates the total number of available trading opportunities.

26.3.2 Feature Extraction and Machine Learning Prediction

The predictive model designed in this paper incorporates various features that can be systematically classified into three principal categories: raw data features, implied volatility features, and technical indicators.

Raw data features include the market prices (V_1 and V_2) and implied volatilities (IV_1 and IV_2) of the first and second options in the spread, respectively. The implied volatilities are calculated using the Black-Scholes model, reflecting market expectations of future volatility for each option. The time to maturity (TTM) of the options is also considered, with a singular TTM adequate as both options have the same expiration time.

Implied volatility features are extracted to identify potential mispricing phenomena within the options market. The binary variable OE indicates whether the valuation of the first option is overestimated compared to the second, as denoted in Eq. 26.4. The quotient of implied volatilities $Q_{12} = \frac{IV_1}{IV_2}$ represents the relationship between the implied volatilities of the two options, delivering an evaluation of the relative pricing structures and perceived risk associated with each option.

$$OE = \begin{cases} 1 & \text{if } IV_1 > IV_2 \\ 0 & \text{otherwise} \end{cases}, \tag{26.4}$$

Technical indicators obtained from the price movements of the underlying asset are incorporated to capture its dynamics. The indicators include Simple Moving Average (SMA), Relative Strength Index (RSI), stochastic oscillator (KD indicator), Commodity Channel Index (CCI), Moving Average Convergence/Divergence (MACD), and volatility. The selection of these indicators is guided by [24], which highlights their effectiveness in predicting index performance.

To establish a predictive system for identifying the profitability of a trading strategy, the problem is conceptualized as a binary classification task. By applying the specified option trading strategy with predefined take-profit and stop-loss thresholds, a backtesting process is executed to evaluate performance. The outcomes of each transaction are labeled as

$$Y = \begin{cases} 1 & \text{if } PL > 0 \\ 0 & \text{otherwise} \end{cases}, \tag{26.5}$$

where Y represents the label used in the machine learning model to indicate the outcome category of the transaction.

Complementing feature extraction and data annotation, machine learning algorithms are applied to forecast the classification of transactions in accordance with the predetermined option trading strategy.

26.4 Numerical Results

This section evaluates the effectiveness of the proposed predictive system for the bull-side put spread trading strategy, with both the predefined stop-loss and take-profit ratio set at 0.5 ($\rho_{SL} = 0.5$, $\rho_{TP} = 0.5$). In this study, we overlook the impact of transaction costs and utilize a conventional logistic classification model as the benchmark.

The study utilizes TAIEX Futures (TX) and TAIEX Options (TXO), focusing on weekly options for sufficient liquidity, sourced from the Taiwan Futures Exchange. The datasets is consists of open (O), high (H), low (L), and close (C) prices for TX and TXO, as well as option type (δ), strike price (K), and expiration time (T), covering January 1, 2021, to December 31, 2023, at a frequency of 1 minute, totaling 343,785 data points.

The trading period spans Monday to Wednesday, including both day and night sessions, with a time to maturity defined as $TTMM = T - t \leq 3165$ minutes, where T is the expiration time and t the current time. To consistently identify out-of-the-money options across sessions, TAIEX Futures serve as a proxy for the spot index due to the absence of night session pricing in TAIEX.

26.4.1 Data Pre-processing and Description

Data pre-processing focuses on identifying and eliminating arbitrage opportunities. The first arbitrage type occurs when the price of the second out-of-the-money put option exceeds that of the first, contradicting the Black-Scholes model, expressed as $V_2(t) > V_1(t)$. The second arbitrage type arises when the price differential between two options equals or surpasses the difference in their strike prices, indicating a guaranteed profit, represented as $V_1(0) - V_2(0) \geq K_1(0) - K_2(0)$.

Following the removal of arbitrage opportunities and outlier filtering based on implied volatility quotients (Q_{12}), the final dataset consists of 281,750 entries. The mean and standard deviation of Q_{12} prior to filtering are 0.93 and 0.068, respectively, while post-filtering values are 0.93 and 0.052, indicating a more uniform dataset after filtration.

The correlation analysis between the dependent variable Y and the predictors reveals generally weak correlations, suggesting limited linear relationships. For instance, the quotient of implied volatilities Q_{12} shows a slight positive correlation with Y (0.039), indicating its minimal predictive impact and underscoring the potential importance of non-linear feature interactions.

For dataset partitioning, the optimal split point is determined within the predefined range of 65% to 85%, ensuring a balanced distribution of the dependent variable across datasets. The in-sample dataset is stratified, with 80% designated for training and 20% for validation, while the out-of-sample data is reserved for testing model performance and conducting historical backtesting of the trading strategy.

26.4.2 Performance of Predictive Model

The predictive models forecast the binary outcomes of transactions associated with the specified options trading strategy. To enhance the performance of the proposed model, this study determines the optimal threshold that balances precision and recalls by adjusting the threshold value. A predefined constraint is applied to ensure that the positive ratio PR, calculated as

$$PR = \frac{TP + FP}{TP + FP + TN + FN} , \tag{26.6}$$

remains at or above 50%, preventing the excessive filtering of trades that could undermine the strategy's effectiveness. The optimal threshold is identified utilizing the training data and is subsequently enforced uniformly across both the validation and test datasets.

For the decision tree algorithm, entropy is employed as the criterion, with the maximum depth of the tree established at 25. The minimum samples per leaf are set to 1, and the minimum samples necessary to bifurcate an internal node are configured to 2. Additionally, the class weighting method is set to "balanced" to address potential class imbalances. The optimal threshold for outcome classification is identified as 0.822. For logistic regression, the "liblinear" solver is selected, and L1 regularization is implemented with a penalty strength of 20. The optimal threshold for this particular model is designated at 0.65.

Table 26.2 contrasts performance metrics for the decision tree and logistic classification models. The decision tree achieves a precision of 56.53%, surpassing the logistic model's 54.84%, indicating superior efficacy in identifying profitable trades. The decision tree also demonstrates a higher recall of 64.74% compared to the logistic model's 4.85%. The F1 Score underscores the decision tree's advantage, registering 60.36% against the logistic model's 8.91%.

Despite similar accuracy, the decision tree significantly outperforms in precision, establishing it as the more reliable model for executing trades based on favorable projections. This enhanced precision promotes a more effective trading strategy by optimizing the capture of profitable trades while avoiding unprofitable ones, with the decision tree's superior PR further affirming its suitability for precision-focused strategies.

Table 26.3 illustrates the cumulative payoff (CPL) and trade ratio (TR) from historical backtesting of trading signals generated by the predictive models within the specified strategy. The decision tree algorithm yields a CPL of 10,971.7, lower than the logistic model's 18,522.2; however, the logistic model's high TR of 98.96% enables

Table 26.2 Performance comparison of predictive models

Predictive model	Accuracy (%)	Precision (%)	Recall (%)	F1 score (%)	PR (%)
Decision tree	54.04	56.53	64.74	60.36	62.07
Logistic classification	54.49	54.84	4.85	8.91	4.80

Table 26.3 Historical backtesting performance for predictive models

Predictive model	CPL	TR (%)
Decision tree algorithm	10,971.7	59.89
Logistic classification	18,522.2	98.96

near-total trade execution while suffering from poor signal filtering. Conversely, the decision tree algorithm, with a TR of 59.89%, adopts a more selective trading approach, reducing noise in signals and focusing on more lucrative opportunities.

26.5 Conclusions

This study presents a machine learning-based predictive system for forecasting outcomes in a bull-side put spread options trading strategy. The system utilizes raw data, implied volatility, and technical indicators to assess trade profitability. A comparative analysis of a decision tree algorithm and a logistic classification model is conducted.

The results indicate that the logistic classification model achieves a higher cumulative payoff but entails a greater incidence of unprofitable trades. Conversely, the decision tree algorithm demonstrates superior precision in identifying profitable trades, making it more effective in implementing the trading strategy. The selective trade execution, characterized by a lower trade ratio, reduces trading signal noise and emphasizes lucrative opportunities.

The study highlights the importance of considering both the predictive performance of machine learning models and their implications for trading strategy implementation. Although the logistic classification model may generate a higher cumulative payoff, its high trade ratio may suggest excessive transaction costs and diminish overall profitability. The decision tree's focus on precision and selectivity offers a more sustainable strategy, especially in contexts where noise reduction and accurate trade identification are vital for enduring success.

Future investigations may consider the deployment of advanced machine learning algorithms and the integration of additional features to bolster predictive accuracy. Furthermore, examining the methodology's efficacy across diverse options trading strategies and market conditions could yield significant insights regarding its generalizability and robustness.

References

1. Avramelou, L., Nousi, P., Passalis, N., Tefas, A.: Deep reinforcement learning for financial trading using multi-modal features. Expert Syst. Appl. **238**, 121,849 (2024)
2. Bali, T.G., Beckmeyer, H., Moerke, M., Weigert, F.: Option return predictability with machine learning and big data. Rev. Finan. Stud. **36**(9), 3548–3602 (2023)

3. Basak, S., Kar, S., Saha, S., Khaidem, L., Dey, S.R.: Predicting the direction of stock market prices using tree-based classifiers. North Amer. J. Econ. Finance **47**, 552–567 (2019)
4. Black, F., Scholes, M.: The pricing of options and corporate liabilities. J. Polit. Econ. **81**(3), 637–654 (1973)
5. Bollerslev, T.: Generalized autoregressive conditional heteroskedasticity. J. Econometr. **31**(3), 307–327 (1986)
6. Breiman, L.: Random forests. Mach. Learn. **45**, 5–32 (2001)
7. Chen, T., Guestrin, C.: Xgboost: a scalable tree boosting system. In: Proceedings of the 22nd acm sigkdd international conference on knowledge discovery and data mining, pp. 785–794 (2016)
8. Chen, Z., Li, C., Sun, W.: Bitcoin price prediction using machine learning: an approach to sample dimension engineering. J. Comput. Appl. Math. **365**, 112,395 (2020)
9. Christoffersen, P., Jacobs, K.: Which garch model for option valuation? Manage. Sci. **50**(9), 1204–1221 (2004)
10. Cox, J.C., Ross, S.A., Rubinstein, M.: Option pricing: a simplified approach. J. Financ. Econ. **7**(3), 229–263 (1979)
11. Faith, C.: Way of the Turtle: The Secret Methods that Turned Ordinary People into Legendary Traders: The Secret Methods that Turned Ordinary People into Legendary Traders. McGraw Hill Professional (2007)
12. Géron, A.: Hands-on machine learning with Scikit-Learn, Keras, and TensorFlow. O'Reilly Media, Inc (2022)
13. Ghosh, P., Neufeld, A., Sahoo, J.K.: Forecasting directional movements of stock prices for intraday trading using lstm and random forests. Finance Res. Lett. **46**, 102,280 (2022)
14. Hull, J.C., Basu, S.: Options, futures, and other derivatives. Pearson Education India (2016)
15. Ivaşcu, C.F.: Option pricing using machine learning. Expert Syst. Appl. **163**, 113,799 (2021)
16. Ji, G., Yu, J., Hu, K., Xie, J., Ji, X.: An adaptive feature selection schema using improved technical indicators for predicting stock price movements. Expert Syst. Appl. **200**, 116,941 (2022)
17. Nobre, J., Neves, R.F.: Combining principal component analysis, discrete wavelet transform and xgboost to trade in the financial markets. Expert Syst. Appl. **125**, 181–194 (2019)
18. Quinlan, J.R.: Induction of decision trees. Mach. Learn. **1**, 81–106 (1986)
19. Sajadi, S.M.A., Khodaee, P., Hajizadeh, E., Farhadi, S., Dastgoshade, S., Du, B.: Deep learning-based methods for forecasting brent crude oil return considering covid-19 pandemic effect. Energies **15**(21), 8124 (2022)
20. Sebastião, H., Godinho, P.: Forecasting and trading cryptocurrencies with machine learning under changing market conditions. Financial Innov. **7**, 1–30 (2021)
21. Tsay, R.S.: Analysis of Financial Time Series. Wiley (2005)
22. Vine, S.: Options: Trading Strategy and Risk Management, vol. 288. Wiley (2011)
23. Wu, M.E., Chung, W.H.: A novel approach of option portfolio construction using the kelly criterion. IEEE Access **6**, 53044–53052 (2018). https://doi.org/10.1109/ACCESS.2018.2869282
24. Wu, M.E., Syu, J.H., Chen, C.M.: Kelly-based options trading strategies on settlement date via supervised learning algorithms. Comput. Econ. **59**(4), 1627–1644 (2022)
25. Yıldırım, D.C., Toroslu, I.H., Fiore, U.: Forecasting directional movement of forex data using lstm with technical and macroeconomic indicators. Financial Innov. **7**, 1–36 (2021)

Part IV
Intelligent Applications

Chapter 27
Implementation of a Fire Detection System Based on YOLOv5

Yuh-Chung Lin, Shun Nian Luo, Ta-Wen Kuan, Xiaodong Yu, Yezhe Shi, and Diakonova Vasilisa

Abstract This article presents an intelligent fire monitoring system designed using the ROS framework and YOLO object detection algorithm. By combining ROS's network coordination with YOLO's high-speed visual analysis, the system facilitates fast fire detection and seamless collaboration across monitoring nodes. It comprises video capture, object detection, and alarm notification modules, with the ESP32-CAM microcontroller and OV2640 camera managing real-time video capture while YOLO performs detection. The system demonstrates strong fire detection performance with high accuracy, recall, and mAP, processing 45 frames per second for timely detection and response. However, limitations exist in dataset size and diversity. Future work could focus on expanding the dataset and exploring deeper network architectures, attention mechanisms, and adaptive learning rates to enhance performance.

27.1 Introduction

Traditional fire detection relies on sensors like smoke and temperature detectors, which are effective but limited to fixed positions. This creates blind spots in large areas like buildings, forests, and public spaces, risking delayed detection and response.

With urbanization, fire risks have risen in dense areas such as high-rises, commercial complexes, and industrial parks, where fires can spread quickly, causing severe damage and economic loss. Thus, real-time, wide-area monitoring, early detection, and prompt response have become crucial for effective fire prevention.

Fortunately, with the rapid development of computer vision and artificial intelligence technologies, video surveillance technology has become an indispensable part of modern social security systems. Compared to traditional sensor-based monitoring methods, video surveillance has significant advantages in terms of broader coverage and large amounts of information. By deploying cameras in key areas,

Y.-C. Lin · S. N. Luo · T.-W. Kuan (✉) · X. Yu · Y. Shi · D. Vasilisa
School of Information Science and Technology, Sanda University, Shanghi 201209, China
e-mail: dwguan@sandau.edu.cn

S.-C. Chu et al. (eds.), *Advances in Intelligent Data Analysis and Applications*,
Smart Innovation, Systems and Technologies 445,
https://doi.org/10.1007/978-981-96-7277-6_27

continuous monitoring of large regions can be achieved, effectively compensating for the limitations of traditional methods.

By further applying deep learning algorithms to video surveillance systems, real-time analysis and understanding of video images can be achieved. This object detection-based fire detection method can promptly identify fire incidents in the early stages by recognizing features such as smoke and flames. Compared to traditional sensors, this approach provides more intuitive and accurate monitoring results and allows for the tracking and identification of moving objects, significantly expanding the coverage and efficiency of fire monitoring.

Deep learning models can optimize recognition algorithms by learning from extensive historical data, improving fire detection accuracy over time. As data accumulates, the system's monitoring capabilities enhance continuously. Moreover, deep learning's flexibility allows adaptation to various complex environments, functioning effectively indoors and outdoors, day or night. In summary, deep learning technology improves accuracy and real-time performance in fire monitoring while expanding coverage, providing robust support for fire prevention and response.

This paper presents the design and implementation of an intelligent fire monitoring system using the ROS framework and YOLO [1] object detection algorithm to enhance fire detection accuracy and real-time performance for public safety. By integrating the ESP32-CAM [2] microcontroller and high-performance computers, the system achieves real-time fire feature detection and rapid response. Evaluation and testing demonstrated strong performance in fire detection, including high accuracy, recall rate, and mAP values, validating its effectiveness for early warning. The design also emphasizes the stability and critical requirements of real-time monitoring by optimizing video data transmission and processing for quick response.

27.2 Related Works

In recent years, intelligent video alarm systems have made significant strides, particularly with the advent of deep learning technology. Our work implements an intelligent fire monitoring system using machine vision and Internet of Things (IoT) [3] technologies. By adopting the YOLO object detection algorithm, the system conducts real-time analysis of surveillance footage to accurately detect fire incidents. Built on the ROS framework, it enables efficient information transmission between different nodes, creating a highly coordinated cross-regional monitoring network.

27.2.1 ROS (Robot Operating System)

ROS (robot operating system) [4] is a flexible framework for robot software development, providing an open-source middleware platform with various tools, libraries, and conventions for developers. Its distributed architecture allows nodes to run on different machines and communicate over a network using a publish/subscribe mechanism, where nodes publish messages to specific topics. Additionally, ROS supports a request/response pattern known as services and organizes software into packages containing nodes, libraries, and datasets. The framework includes development tools like visualization (rviz), debugging (rqt), and simulation (Gazebo) environments, along with a hardware abstraction layer for code portability across robotic platforms. Primarily using C++ and Python, ROS supports multiple programming languages and benefits from a large developer community that contributes open-source packages. This versatility makes it a powerful platform for developing complex robotic systems, including fire monitoring applications that utilize its distributed architecture and efficient communication mechanisms for coordinated monitoring solutions.

Furthermore, ROS has a large community and a vast ecosystem of packages that simplify tasks like navigation, perception, and manipulation. This makes ROS widely used in both research and industry for developing robots, from simple mobile units to complex autonomous systems.

27.2.2 Image Recognition and Object Detection

In recent years, image recognition and object detection technologies have advanced significantly alongside artificial intelligence and deep learning. Image recognition allows computers to understand and analyze images. Early techniques used manually designed feature extraction algorithms like SIFT [5] and SURF [6], requiring expert parameter adjustments, which were limited in complex environments. The introduction of Convolutional Neural Networks (CNNs) has improved image recognition by automatically extracting multi-layered features, achieving high accuracy in challenging scenarios.

Object detection enhances image recognition by identifying object categories and their locations. Early methods like HOG [7] combined with sliding window techniques could detect objects but were slow and imprecise. The rise of deep learning introduced two-stage models such as R-CNN [8], Fast R-CNN [9], and Faster R-CNN [10], significantly improving accuracy. Subsequently, one-stage models like YOLO (you only look once) and SSD [11] emerged, increasing detection speed to real-time levels, leading to widespread adoption in various applications.

Today, image recognition and object detection technologies are utilized in autonomous driving, medical image analysis, security surveillance, and industrial inspection. As models are refined and hardware capabilities advance, the range and accuracy of these technologies will continue to expand.

27.3 System Architecture

This system aims to create an efficient and reliable intelligent fire detection and alarm system. It integrates an Internet of Things (IoT) framework based on ROS (robot operating system), leveraging its distributed architecture and flexible communication. The core detection algorithm uses YOLO (you only look once) technology for real-time and accurate fire identification. By combining ROS's network coordination with YOLO's fast visual analysis, the system quickly detects fire threats and enables collaboration among monitoring nodes, forming a responsive fire early warning network.

The system consists of three modules: the video capture module, which uses the ESP32-CAM microcontroller and OV2640 camera to transmit real-time video; the YOLO object detection module, which processes the video streams and provides real-time object detection; and the alarm notification module, which issues alerts through the graphical user interface (GUI) upon fire detection. Figure 27.1 presents the module diagram of the system, and Fig. 27.2 illustrates the system flowchart.

The system is composed of three modules. First, the video capture module uses the ESP32-CAM microcontroller and OV2640 camera to efficiently capture and transmit real-time video streams. Second, the YOLO object detection module processes the

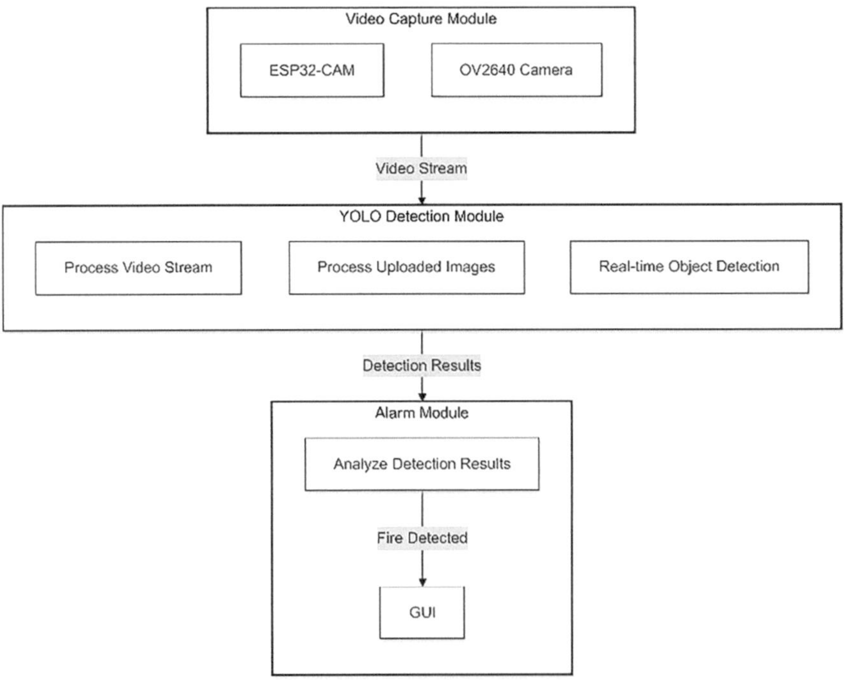

Fig. 27.1 System module diagram

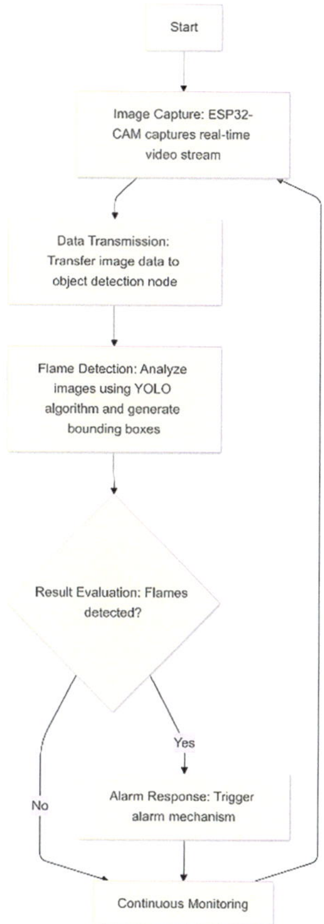

Fig. 27.2 System flowchart

collected video streams and uploaded image files, providing real-time object detection functionality. Finally, the alarm notification module acts on the output from the object detection module, and once a fire is detected, the system promptly issues an alert through the graphical user interface (GUI).

The execution steps are depicted as follows:

1. Image Capture: ESP32-CAM captures real-time video stream.
2. Data Transmission: Transfer image data to the object detection node.
3. Flame Detection: Analyze images using YOLO algorithm and generate bounding boxes.
4. Result Evaluation: Determine the presence of flames based on YOLO output.
5. Alarm Response: Trigger alarm mechanism when flames are detected.
6. Continuous Monitoring: Repeat steps 1–5 to maintain real-time surveillance.

27.4 Model Training

The model training process involves several key steps. First, the dataset is constructed, which includes collecting and selecting the data, followed by preprocessing the dataset. Next, the model training begins, where hyperparameters are set, and the training process is initiated. Finally, the model is evaluated to assess its performance. Each of these steps is essential for ensuring that the model achieves optimal accuracy and efficiency.

27.4.1 Dataset Construction

When conducting research on deep learning-based flame detection, building a high-quality and diverse dataset of flame images is a crucial step in achieving high-precision model training. Considering the experimental requirements and condition limitations, this study independently collected a batch of flame images through the Internet, aiming to cover different types and scales of flame scenarios. This approach is intended to enhance the model's generalization ability and practicality in fire detection.

The dataset comprises a total of 2,394 images, encompassing a wide range of scenarios from small flames to large-scale fires, covering flame manifestations in various indoor and outdoor environments and backgrounds. The types of flames in the images are diverse, including candle flames from domestic life, cooking flames in kitchens, as well as large-scale fire scenes such as outdoor campfires and forest fires. Through this extensive collection range, the aim is to make the dataset more comprehensively reflect the diversity and complexity of flames in real-life situations. Figure 27.3 is the example of dataset.

To address issues with collected images that didn't meet the 640×640 resolution requirement or had low clarity and high interference, we developed a comprehensive data processing script. This script first unified all file extensions by converting images to .jpg format, ensuring consistency across the dataset. It then performed data cleaning by removing irrelevant images and standardizing image resolutions, enhancing the overall quality of the dataset. Finally, we utilized LabelImg for image annotation, adopting the YOLO format where we manually drew bounding boxes around flames and labeled them. This process generated corresponding .txt files for each image, with class 0 designated to represent "fire." Through these steps, we ensured consistent image quality and format, thereby optimizing the dataset for efficient model training.

Fig. 27.3 The example of training dataset

27.4.2 Training Process

After data screening, cleaning, and annotation, a total of 2,394 images were collected to form the flame dataset. It contains one category: label 0 for "fire," with a total of 4,295 annotations across all images. The entire set of 2,394 images was divided into training, validation, and test sets in an 8:1:1 ratio. The training set is used for model feature extraction, the validation set is used to observe the model's loss during the training process, and to adjust the number of training epochs or parameters in a timely manner, while the test set is used for the final evaluation of the model's performance and is only used once.

The hyperparameters for this model training are set as follows: the initial learning rate (Lr0) is set to 0.01, as this value works well with the SGD optimizer, providing a stable and effective pace for training. The final OneCycleLR learning rate (lrf) is a small fraction of the initial rate, helping the model converge smoothly toward an optimal solution in the later stages of training. Momentum is set to 0.937, which accelerates training and reduces oscillations, stabilizing the process. Weight decay is set to 0.0005 to regularize the model and prevent overfitting without overly constraining weight updates. The warmup phase lasts for the first three epochs, gradually increasing the learning rate to stabilize training and avoid abrupt weight changes. The object loss gain (obj) is set to 1.0, emphasizing object detection and improving recall, particularly for smaller or harder-to-detect objects. The IoU threshold (iou_ t) is set to 0.20, allowing the model to optimize predictions for slightly overlapping bounding boxes, enhancing detection of tightly packed or partially overlapping objects. The probability of horizontal flipping (fliplr) is set to 0.5, meaning one in every two images will be flipped, which enhances the model's generalization. Finally, the mosaic augmentation probability is set to 1.0, ensuring that every image undergoes mosaic augmentation during training, improving the model's ability to adapt to different scenarios. Table 27.1 presents the setting of hyperparameters during training.

The training process begins with configuring the hyperparameters. A small YOLOv5 [12] model (yolov5s) is used for training on a custom dataset. The training runs for 100 epochs, with a batch size of 16 and an input image size of 640 × 640. The optimizer used is SGD. The small YOLOv5 model is chosen for faster training and to generate a smaller model, targeting only one class (nc = 1) from the custom dataset.

The neural network, consisting of layers such as convolutional, activation, pooling, fully connected, and batch normalization layers, processes the input and passes it through each layer. The network outputs the predicted bounding box positions, object sizes, and categories.

The goal of training is to minimize the difference between the predicted results and the true labels, quantified using the loss function, which includes two parts: classification loss (cross-entropy) for measuring the difference between predicted and actual categories, and localization loss (IoU) for measuring the difference between predicted and actual bounding boxes.

After calculating the loss, backpropagation occurs to compute the gradients for each parameter, guiding how to adjust them to reduce the loss. This process of forward propagation, loss calculation, backpropagation, and optimization repeats throughout the entire training. Each batch of data completes one iteration, and each epoch involves a full forward and backward pass through the entire dataset.

Table 27.1 Hyperparameter settings

Hyperparameter	Value and description
Lr0	0.01—Initial learning rate; provides a stable and effective pace for training
Lrf	0.01—Final OneCycleLR learning rate, helps smooth model convergence
Momentum	0.937—Accelerates training, reduces oscillations, and stabilizes the process
Weight decay	0.0005—Regularizes the model and prevents overfitting
Warmup epochs	3.0—Gradually increases the learning rate to stabilize training
Obj	1.0—Emphasizes object detection, improving recall for smaller or harder targets
IoU threshold	0.20—Optimizes predictions for slightly overlapping bounding boxes
Fliplr	0.5—Probability of horizontal flipping to enhance generalization
Mosaic	1.0—Ensures every image undergoes mosaic augmentation, improving adaptability

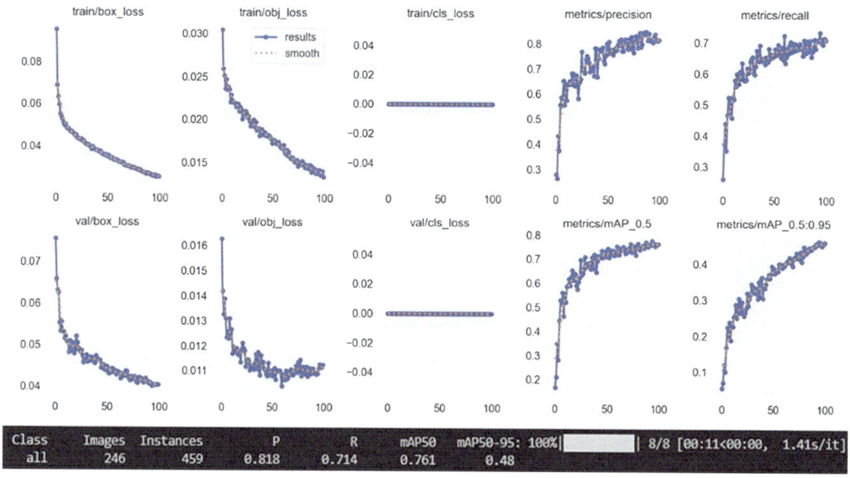

Fig. 27.4 The results of model evaluation

27.4.3 Model Evaluation

Figure 27.4 shows the model evaluation results. The bounding box loss decreases for both the training and validation sets, indicating improved target localization. The object loss also decreases, showing better detection of target presence. The classification loss is nearly zero, likely due to there being only one class (fire). Precision improves during training, reaching about 0.8, and recall also increases, reaching about 0.7.

The model's accuracy and recall on the custom fire dataset were high, with an accuracy (P) of 0.818 and a recall (R) of 0.714. This indicates the model can accurately detect most fire instances, with few missed detections. The high accuracy shows strong confidence in identifying fire, while the high recall means it captures most actual fire cases, reducing the chances of missing any.

For mean Average Precision (mAP), the model achieved 0.761 at mAP50 and 0.48 at mAP50-95, indicating stable performance across different IoU thresholds. A higher mAP50 reflects good accuracy at looser IoU thresholds, while mAP50-95 demonstrates decent performance under stricter conditions, proving the model's ability to detect fires of various sizes and shapes.

The model's real-time performance, a key focus of the YOLOv5 algorithm, allows it to process large volumes of image data at 45 frames per second, ensuring timely detection of fire incidents and providing enough time for response actions.

27.5 Results

In this study, an intelligent fire monitoring system based on the ROS framework and YOLO object detection algorithm was developed. The system integrates an ESP32-CAM microcontroller and a high-performance computer to efficiently capture, transmit, and analyze real-time video streams, enabling quick and accurate fire detection with timely alerts. The system's operation is shown in Fig. 27.5 and Fig. 6. To verify the system's functionality, we conducted two tests: the first involved inputting static images to evaluate fire detection accuracy, and the second tested whether the real-time video monitoring function could successfully detect fire. Figure 27.5 demonstrates that the system can accurately identify fire from a static image and trigger an alert in the terminal. Figure 27.6 shows the test results using dynamic fire video as the input source, where the system also successfully detected fire and issued a warning.

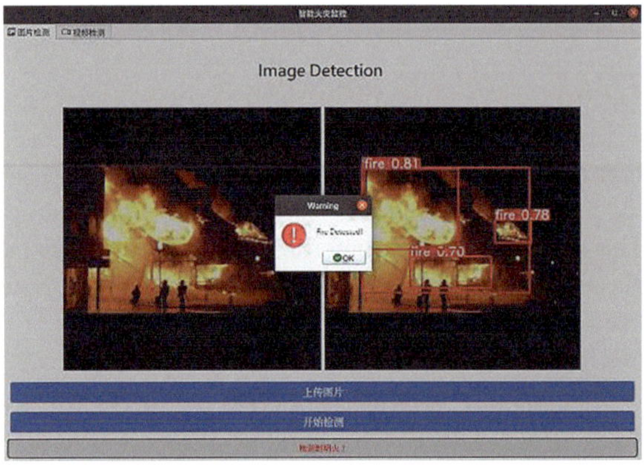

Fig. 27.5 Testing on image detection

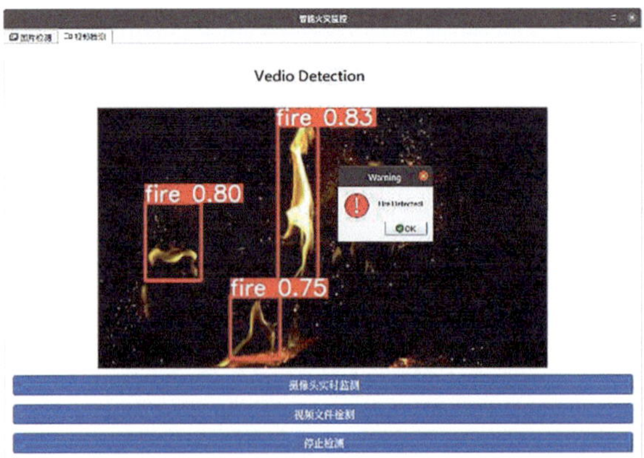

Fig. 27.6 Testing on video detection

27.6 Conclusion

This paper designs and implements an intelligent fire monitoring system based on the ROS framework and YOLO object detection algorithm, aiming to improve the accuracy and real-time performance of fire detection to provide more effective technical support in the field of public safety. The core components of the system include a video capture module, an object detection module, and an alarm notification module. By integrating the ESP32-CAM microcontroller and a high-performance computer, the system achieves real-time detection and quick response to fire features. However, through in-depth analysis and testing, we identified several limitations in the system, particularly the insufficient scale and diversity of the dataset. Future research can focus on expanding the dataset to include more types of flames and fire scenarios, especially in complex environments. Increasing the diversity of the dataset can further enhance the model's generalization ability and robustness, ensuring accurate fire detection across various conditions.

Although the YOLO algorithm has performed well in fire detection tasks, there is still room for optimization. Future work can explore applying the latest deep learning technologies and algorithmic innovations to the model, such as deeper network architectures, attention mechanisms, and adaptive learning rate adjustments. Additionally, reducing false positives and false negatives, especially in complex fire scenes or when interference is present, is crucial for improving detection accuracy. Using lightweight model architectures combined with edge AI technology will ensure that the system maintains good performance in resource-constrained environments, which will be a key factor in determining the system's practicality.

Acknowledgements This research is supported by the school research funding, Sanda University (research project no. 2024YB11).

References

1. Redmon, J., Divvala, S., Girshick, R., Farhadi, A.: You only look once: Unified, real-time object detection. In: Proceedings of the IEEE conference on computer vision and pattern recognition, pp. 779–788. IEEE, Las Vegas, NV, USA (2016)
2. Espressif Systems.: ESP32-CAM. (2022). https://www.espressif.com/en/products/socs/esp32-cam
3. Wang, K., Chen, C.M., Hossain, M.S., Muhammad, G., Kumar, S., Kumari, S.: Transfer reinforcement learning-based road object detection in next generation IoT domain. Comput. Netw. **193**, 108078 (2021)
4. Rong, Y., Zhang, Y., Liu, G., Wang, Y., Wang, P.: Implementation of pedestrian detection algorithm based on improved yolov3-Tiny in ROS framework. In: 6th International Conference on Robotics and Automation Sciences (ICRAS), pp. 242–246. IEEE (2022)
5. Lowe, D.G.: Distinctive image features from scale-invariant keypoints. Int. J. Comput. Vision **60**, 91–110 (2004)
6. Bay, H., Ess, A., Tuytelaars, T., Gool, L.C.: Speeded-up robust features (SURF). Comput. Vis. Image Understand **110**(3), 346–359 (2008)

7. Dalal, N., Triggs, B.: Histograms of oriented gradients for human detection. In: IEEE Computer Society Conference on Computer Vision and Pattern Recognition (CVPR'05), vol. 1, pp. 886–893. IEEE, (2005)
8. Girshick, R., Donahue, J., Darrell, T., Malik, J.: Rich feature hierarchies for accurate object detection and semantic segmentation. In: IEEE Conference on Computer Vision and Pattern Recognition (CVPR), pp. 580–587. IEEE, Columbus, OH, USA (2014)
9. Girshick, R.: Fast R-CNN. In: IEEE International Conference on Computer Vision (ICCV), pp. 1440–1448. IEEE, (2015)
10. Ren, S., He, K., Girshick, R., Sun, J.: Faster R-CNN: Towards real-time object detection with region proposal networks. IEEE Trans. Pattern Anal. Mach. Intell. **39**(6), 1137–1149 (2017)
11. Liu, W. et al.: SSD: Single Shot MultiBox Detector. In: Leibe, B., Matas, J., Sebe, N., Welling, M. (eds.), Computer vision—ECCV 2016. ECCV 2016. Lecture Notes in Computer Science, vol. 9905. Springer, Cham (2016). https://doi.org/10.1007/978-3-319-46448-0_2
12. Lin, Y.C., Kuan, T.W., Tseng, S.P., Lv, X.: Implementation of campus pedestrian detection using YOLOv5. In: ICGEC 2023. Lecture notes in electrical engineering, vol. 1114, Springer, Singapore (2023). https://doi.org/10.1007/978-981-99-9412-0_46

Chapter 28
Research and Analysis of Smart Clothing Based on Citespace

Lu Gao and Ke-Ke Sun

Abstract With the development of technology and the improvement of people's quality of life, smart apparel is making its way into various lifestyle scenarios such as healthcare, work, and play. The article uses the smart clothing literature in the SCI database as the research object. Accurately searched 1852 documents by PICO literature search method. The Citespace software was then used to analyze the relevant studies. The article uses pictures, tables, and formulas to examine four aspects of the smart clothing field. These four areas are general publication trends in smart clothing research, major research countries, important research organizations, and current research hotspots in the field of smart clothing. Analysis found that the current research direction of smart clothing mainly includes smart fabrics and flexible sensors implanted in clothing in two ways. It has a great potential for development and is adapted to the needs and lifestyles of modern people. Intelligent clothing will become a new development direction in the field of clothing in the future.

28.1 Introduction

Smart clothing is a cross-cutting research field involving ergonomics, materials science, and engineering. With the development of digital information technology and the emergence of a large number of artificial intelligence products, smart clothing has also become a much-anticipated item. Functions commonly used for safe positioning, health monitoring, and motion monitoring. Yu et al. [1] four sensors were integrated into the hardware system to enable the monitoring of heart rate, body temperature, location data, and human activity patterns. Chang et al. [2] implanted a 12-lead ECG system based on capacitive sensing technology into smart garments to monitor electrocardiogram (ECG) signals in daily life. Chen et al. [3] in this study, a novel automatic adjustment noncontact sensing smart clothing was proposed. Zhao et al. [4] used a low-cost, scalable approach to fabricate three types of E-skins for

L. Gao · K.-K. Sun (✉)
School of Design Art, Xiamen University of Technology, Xiamen, Fujian Province, China
e-mail: sunkeke@xmut.edu.cn

S.-C. Chu et al. (eds.), *Advances in Intelligent Data Analysis and Applications*,
Smart Innovation, Systems and Technologies 445,
https://doi.org/10.1007/978-981-96-7277-6_28

human finger flexion and pressure signaling for detecting pressure, stretch, flexion, temperature, and humidity. Tu et al. [5] utilizing children's visual, tactile, and auditory functions. Intelligent voice functions with alarm sounds and prompts are placed in special parts of children's clothing. Smart garments make sounds when a child touches a body part, children more intuitively identify and understand their own gender. Wang et al. [6] proposed multicolor mechanical light-emitting device system embedded in garments for human motion monitoring. Smart clothing can quantify the distribution of human muscle and body dynamic data. Lee et al. [7] implant flexible sensors into smart clothing for long-term health monitoring of human respiratory signals. Bie et al. [8] utilize liquid metal blends in garment fabrics, and liquid metal attached to the surface of fabrics can be as soft as clothing fabrics. Wang et al. [9] Coaxial fiber electrodes were fabricated through a special spinning technique and assembled into parallel or twisted nanocapacitors, with the material further utilized as a fiber diaphragm in garments to produce safe and intelligent clothing fabrics. Wen et al. [10] proposed development of an all-fiber multifunctional sensor based on flexible silk protein in a simple two-layer configuration. They make AFMS that exhibit excellent flexibility, and can be bent up to 180° for a more comfortable fit. Our lab has developed relevant algorithms for this application Sun et al. [11]. Calculation of data susceptible to missing in the elderly population by means of an interpolation and fitting algorithm. Filling in the missing data on older persons. The approach contributes to the cost reduction of anthropometric databases Chiu et al. [12]. Development of maximum power at different solar radiations by algorithmic comparison.

After a subjective overview of smart clothing research, we will discuss objectively four issues surrounding smart clothing research in this paper. Objective presentation of its research status and development trend: research trends, major research countries, major organizations, and current research hotspots in the smart clothing.

28.2 Search Methods

Data for the article were obtained from the Web of Science (WOS) core collection database, and the time slices retrieved were from 2008 to 2023. In order to extract highly accurate data, the PICO (population, intervention, companies, outcome) literature search method was used during the advanced search of WOS, comprehensive coverage of all literature was ensured. The PICO framework is primarily used for systematic reviews in clinical areas. However, PICO has a search strategy that filters the required data accurately. From formulating questions to searching for strategies, the precision applies to other disciplines as well. The PICO search method was used to set up three categorized search phrases in this paper. In Table 28.1, the first category (P) is the key word in this paper.

The first category (P) is the keywords of the paper, and synonyms. The second category (I) is the main type of design in smart clothing. The third category (CO) is the core technologies that appear in the keywords. The connectives for these three

Table 28.1 PICO element and corresponding search strings

PICO element	Search string
Population (P)	ALL = ("smart clothing" OR "smart fiber" OR" intelligent textile and clothing" OR "wearable clothing" OR "wearable devices clothing" OR "intelligence health monitoring clothing" OR "flexible sensor clothing" OR "e-textiles")
Phenomena of Interest (I)	AND ALL = ("clothing designing" OR "textiles designing" OR "function design of clothing" OR "smart fabric" OR "health clothing")
Context (CO)	AND ALL = ("Flexible sensors" OR "health monitoring" OR "wearable" OR "Flexible fabric" OR "physiological monitoring" OR "respiratory monitoring" OR "smart safety clothing")

categories are AND. The purpose of this is to exclude literature that is not relevant to the subject of the study through precise keywords, faster search for relevant literature. The purpose of similar terms in each category is to broaden the search universe. The use of synonymous keywords ensures comprehensive and precise retrieval of relevant literature, as the same concept may be expressed in different ways.

This paper retrieved 2460 articles using the PICO search method totaling. First, through manual preliminary screening, excluding 19 conference papers, 58 early access, 1 review article, and 4 retracted publication, there were 2136 active papers. Then, the second screening sets the literature publication period to 2008–2023. The valid data for the last 16 years were used, the language was chosen to be English, and the total number of documents was 1,852. Finally, using Citespace to de-duplicate exported files' 1852 articles. The accuracy of the PICO literature search approach was validated.

28.3 Data Analysis

28.3.1 Analysis of Publication Trends

Figure 28.1 visualizes the number of papers published per year. And it gives a clear picture of the current state of smart clothing research as well as future trends. The specific data were extracted via CiteSpace and subsequently visualized using Origin software. The results show that there is an overall linear increase in the number of smart clothing publications from 2008 to 2023. Its research progress can be divided into three stages. First, scholars' research on smart clothing is relatively flat, maintaining an average of 16 articles per year from 2008 to 2014. Second, the number of articles on smart clothing is growing rapidly. The average number of articles per year can reach 104, which is a 6.5-fold increase compared to 7 years ago. Third, the number of smart clothing publications has gotten a rapid growth from 2014 to 2019. This can be attributed to the increasing penetration of global technology into people's

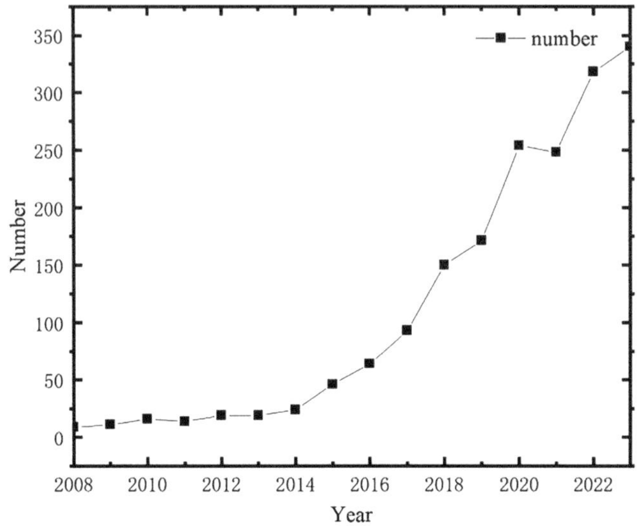

Fig. 28.1 Overall global trends in publications on smart clothing

lives. Various smart products are integrated into life. Intelligent clothing is also gradually entering the research field of view of many scholars. Finally, research articles on smart clothing boost from 2019 to 2023, especially 2020 and 2022. The average number of articles published in these 4 years represents an 18.1-fold increase over the number of articles published in the first phase, this represents a 2.8-fold increase over phase II. The different growth rates in these three phases also indicate that the research buzz for smart clothing will continue to rise. More and more people will be interested and involved in the research of smart clothing in the future. Article publishing on smart clothing will still be on a steep upward trend.

28.3.2 Analysis by Country or Area

Citespace is a statistical used to analyze large amounts of data. It is used to visualize and analyze the literature retrieved by WOS in different dimensions. Figure 28.2 depicts the research on smart clothing in different countries around the world. The article selects 10 regions where research into smart clothing is more profound, it can be visualized in Fig. 28.2. Research in smart clothing is primarily concentrated in China and the United States, which lead among the top 10 contributing countries. Larger nodes represent countries that have significant research or impact in their field in the co-occurrence map. For example, China's nodes are the largest. This also shows that China has published more papers and has more in-depth research results in the field of smart clothing research, as well as having a certain influence. The purple nodes in the text represent the density of linkages between countries and regions.

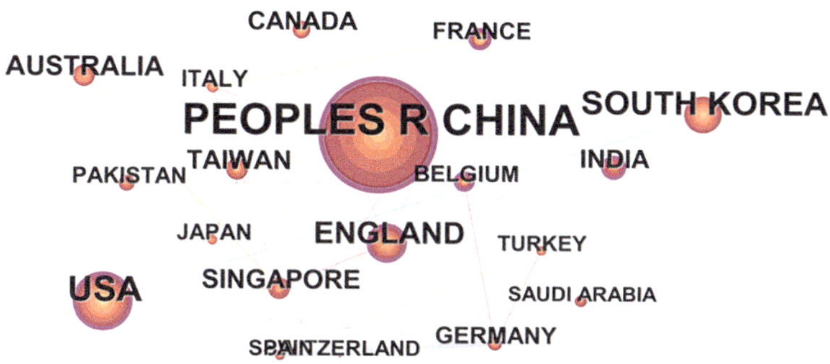

Fig. 28.2 Smart clothing research area distribution map

For example, the USA has a cooperation centrality degree of 0.21, indicating that the USA has closer cooperation ties with other regions.

Citespace not only builds bibliometric networks over time but also analyzes how closely they are linked. The top 10 regions with more research on smart clothing are shown in Table 28.2. China has published a total of 902 articles, including 59 from Taiwan. In contrast, the United States has published 269 articles. Two hundred and sixty-nine articles are published in the United States. Korea published 197 articles. Among these 10 regions China is the most influential country with a centrality of 0.43.

From Fig. 28.2 and Table 28.2, these findings suggest that China leads global research in the field of smart clothing, with the United States, South Korea, and the United Kingdom ranking next.

Table 28.2 The world's leading country for research on smart clothing

No.	Publications	Centrality	Year	Institutions
1	843	0.41	2008	China
2	269	0.21	2008	USA
3	197	0.26	2008	South Korea
4	121	0.04	2009	England
5	74	0.01	2008	Australia
6	59	0.02	2008	Taiwan
7	54	0.14	2010	Singapore
8	51	0.03	2009	India
9	46	0.06	2010	Canada
10	38	0.15	2008	Italy

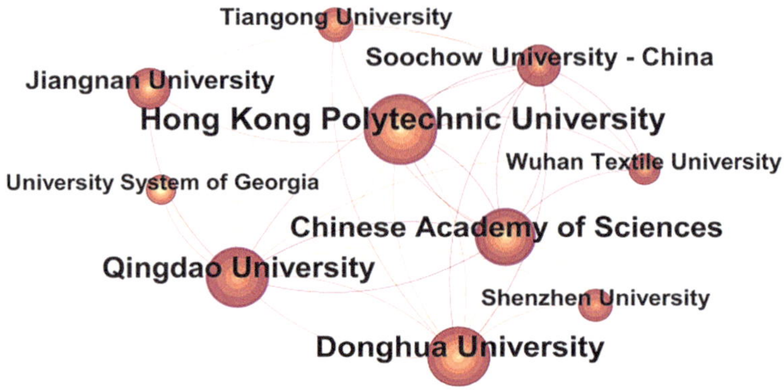

Fig. 28.3 Distribution of smart clothing research organizations

28.3.3 Visual Analysis of Research Institutions

Figure 28.3 shows global organizations researching smart clothing. The analysis of Citespace reveals that in Fig. 28.3 network nodes are 396 and the number of lines is 1063. Figure 28.3 shows that the Chinese Academy of Sciences (CAS) has the most connections. Hong Kong Polytechnic University's (HKPU) research on smart clothing is the most productive. The Hong Kong Polytechnic University (HKPU) leads in smart clothing research output, accounting for 6.8% of the total publications. It is followed by Donghua University (DHU) with 5.0%, Qingdao University (QDU) with 4.5%, the Chinese Academy of Sciences (CAS) with 4.4%, Soochow University with 2.5%, Jiangnan University (JNU) with 2.4%, Tianjin Polytechnic University (TGU) with 1.6%, Shenzhen University (SZU) with 1.5%, Wuhan Textile University (WTU) with 1.4%, and the University of Georgia (UGA) with 1.3%.

Figure 28.3 and Table 28.3 show that the 10 research institutions that are currently conducting research on smart clothing are universities in different countries. Ninety percent of the top 10 organizations in terms of publications are from China. So it follows that the main contributing organizations to smart clothing research are universities. At its core, it is a leading institution in China specializing in textiles and apparel.

28.3.4 Cited Author Posting Analysis

Figure 28.4 shows the number of publications and major contributors of smart clothing researchers. Figure 28.4 has a total of 60 node counts, a network density of 0.0059, and a maximum literature total citation of 237 (39%). Figure 28.4 shows that Ming-Wei, Tian has published 42 articles, or 2.3%. Li-Jun, QDU has 37 publications, or 2% of the total. Zi-Jian, Zheng issued 27 articles, accounting for 1.6% of

Table 28.3 The world's leading institute for smart clothing research

No.	Count	Centrality	Year	Institutions
1	184	0.25	2009	Hong Kong Polytechnic University
2	134	0.09	2012	Donghua University
3	121	0.13	2019	Qingdao University
4	119	0.22	2013	Chinese Academy of Sciences
5	67	0.07	2008	Soochow University-China
6	64	0.10	2017	Jiangnan University
7	44	0.01	2019	Tiangong University
8	39	0.02	2019	Shenzhen University
9	37	0.01	2019	Wuhan Textile University
10	34	0.02	2009	University System of Georgia

Fig. 28.4 Distribution of coauthors

the total. Xiao-Ming, Tao has 20 articles representing 1.1% of the total. Yi, Li issued 18 articles, or 1%. Thus, it can be known that the most number of publications on smart clothing is Ming-wei, Tian it also represents a more in-depth study of his smart clothing.

But an author's publication volume is not a condition for classifying a paper's core authors. The first author of the paper is the main contributor to the paper's literature. The second corresponding author is responsible and in charge of the results of the project and the core author is a co-author. Therefore, it should be calculated using Price's Law formula:

$$M = 0.749\sqrt{NMAX} \tag{28.1}$$

In the formula N is the number of publications by core authors and M max is the number of publications by the most productive authors.

The formula shows that there are nine core authors for $\geq$ 4 articles, representing 17% of the total. This data suggests that the authors are closely connected and a core team has been formed. The field of smart clothing has been discovered and deeply studied by more and more scholars.

28.3.5 Cited Author Posting Analysis

Citespace can be used to analyze keywords as well as co-occurrence networks for subject categories. Not only that, Citespace can also analyze co-citation networks of authors, documents, and journals. Figure 28.5 shows the distribution of keywords in this database. As shown in Figure 28.5, the network contains 524 nodes and 3,318 associated links. Of these, design appeared 209 times, high performance 164 times, fabrics 642 times, e-wearable 143 times, and sensors 320 times. Smart fabrics and sensors are the most frequent keywords in smart clothing. It shows that the main hotspot of smart clothing research is the development of smart fabrics and sensors with modern clothing. There will be a great deal more research on smart clothing and fabrics only in the future.

Tag #0, tag #5 in Fig. 28.6 are smart fabric classes, protecting the human body or other functions such as intelligent monitoring through weaving and coating processes. Tag #1, tag #3 are wearables, the most common technology in smart clothing. The ability to monitor human heart rate and other physiological data signals through wearable electronic devices. Tag #2, Tag #7 is a technology in smart clothing, similar to sensors, it is combined with modern technology to achieve the normal use of smart clothing functions. Tag #4, Tag #6 is a category that emphasizes the functionality of smart clothing, embedding different functional sensors into garments. Smart clothing functions to achieve normal operation by saving light energy or other natural energy sources.

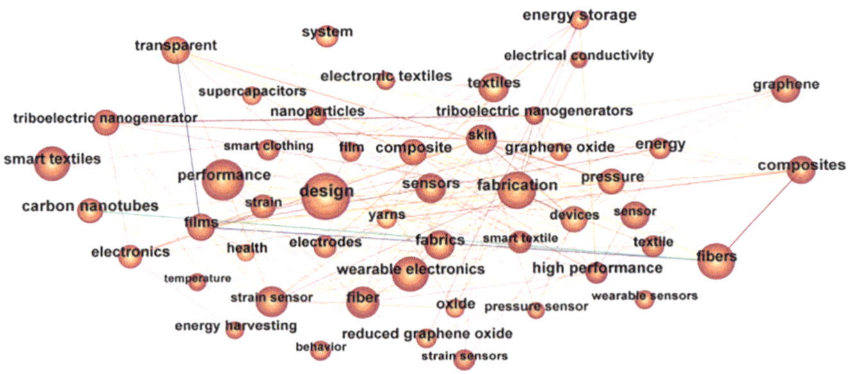

Fig. 28.5 Smart clothing keyword co-occurrence

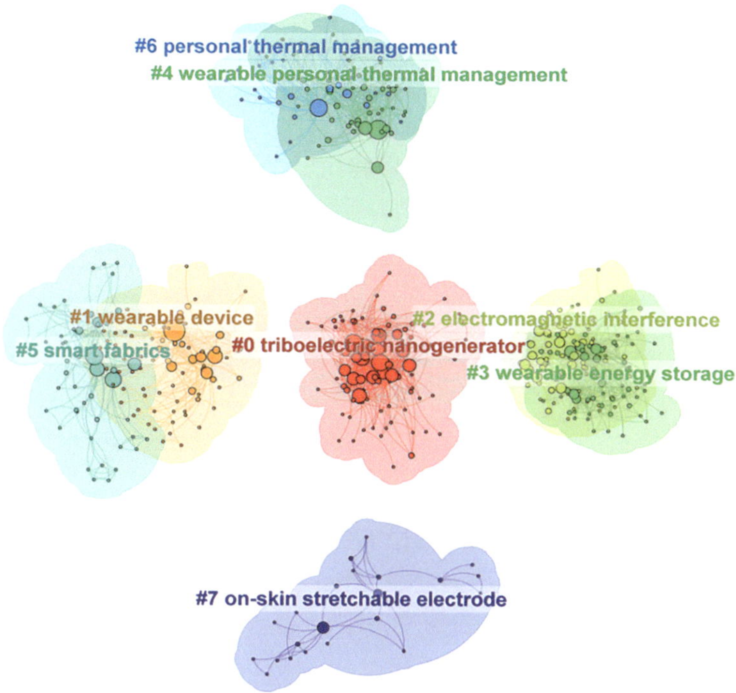

Fig. 28.6 Keyword cluster analysis chart for smart clothing

Figure 28.7 shows the top 15 keywords about smart clothing. The keyword that comes up most frequently is design, with 6.92. The next is yarns, with 5.47. The following is wearable devices, with 3.45.

By analyzing the keyword co-occurrence in Fig. 28.5, keyword clustering in Fig. 28.6, and the top 15 keywords in Fig. 28.7. It is concluded that smart fabrics and sensors appear most frequently. Therefore, it can be judged that the current research hotspot of smart clothing is the research and development of smart fabrics and the use of different types of sensors in clothing.

28.4 Results

The use of wearables and smart fabrics in clothing is becoming more and more lifelike. Three conclusions were drawn from the study of the WOS database on smart clothing literature. First, the number of publications in the field of smart clothing has continued to grow from 9 in 2008 to 340 in 2023, which is a straight upward trend. Secondly, the main research area for smart clothing is China accounting for 50% of the total. Its main research institutions are the HKPU, DHU, QDU, and the

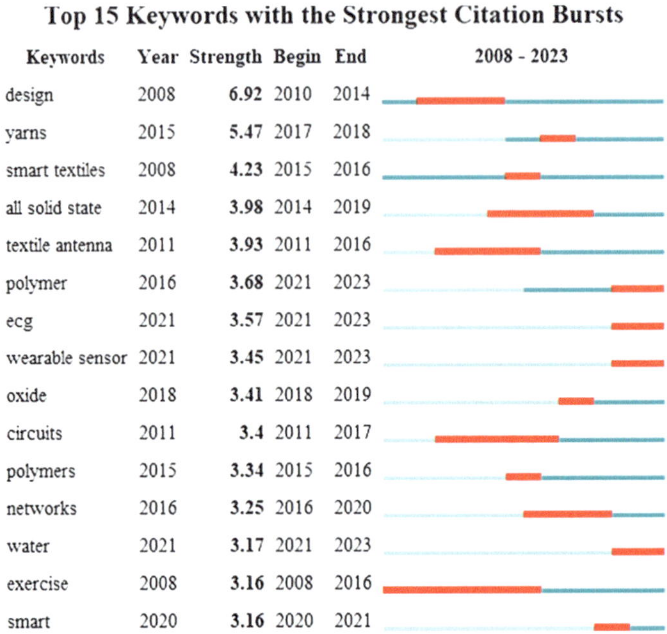

Fig. 28.7 Keywords popping up

CAS in China. Finally, among the high-frequency keywords in the field of smart clothing, smart fabrics and sensors appeared 642 and 320 times, respectively. So we can conclude that smart fabrics and sensors are the current hotspots for research in the field of smart clothing. In summary, the current stage of smart clothing development is not yet sufficient for large-scale commercial production.

References

1. Yu, F., Chen, Z., Jiang, M., et al.: Smart clothing system with multiple sensors based on digital twin technology. IEEE Internet Things J. **10**(7), 6377–6387 (2023)
2. Chang, W.T., Lin, B.S., Tsai, W.L., et al.: Design of 12-lead electrocardiogram smart clothing based on capacitive sensing technique. IEEE Sens. J. **24**(10), 16826–16835 (2024)
3. Chen, H.Y., Lin, B.S., Yang, S.R., et al.: Design of Automatic Adjustment Noncontact Sensing Smart Clothing. IEEE Sens. J. **24**(16), 26378–26387 (2024)
4. Zhao, Y.G., Hu, Y.K., Li, Z.L., et al.: Carbon nanotube/nonwoven fabric-based electronic skins for smart clothing and electronic glove. Sens. Actuators A: Phys. **375**, 1–13 (2024)
5. Tu, C.C., Niu, L., Cui, R.G., et al.: Phonetic smart clothing design based on gender awareness education for preschoolers. AUTEX Res. J. **24**(1), 2023–2121 (2024)
6. Wang, J., Yao, K.W., Cu, K.T.I. et al.: Contact electrification induced multicolor self-recoverable mechanoluminescent elastomer for wearable smart Light.-Emit. Devices Adv. Opt. Mater. **11**(12), 1–10 (2023)
7. Lee, S.Y., Hung, Y.W., Su, P.H., et al.: Biosignal monitoring clothing system for the acquisition of ECG and respiratory signals. IEEE Access **1**(10), 66083–66097 (2022)

8. Bie, B., Xu, W., Lv, Y.: Liquid metal-based textiles for smart clothes. Sci. China Technol. Sci. **66**, 1511–1529 (2023)
9. Wang, Q., Wang, M., Sun, J., et al.: Different assembled nano-MnO2 fiber supercapacitors applied for smart wearable clothing. Text. Res. J. **93**(21–22), 4781–4791 (2023)
10. Wen, D.L., Pang, Y.X., Huang, P., et al.: Silk fibroin-based wearable all-fiber multifunctional sensor for smart clothing. Adv. Fiber Mater. **4**, 873–884 (2022)
11. Sun, K.K., Gao, M., Li, B.: Research on two interpolation algorithms for anthropometric data completion—a case study on the data of elderly males in Fujian, China. J. Netw. Intell. **7**(4), 1083–1101 (2022)
12. Chiu, Y.J., Wang, W.Z., Hu, Y.C., et al.: Maximum power development under varying solar radiation based on hysteresis-loop comparison of selected duty-cycle perturbation algorithms on a mobility scooter. J. Netw. Intell. **8**(4), 1540–1556 (2023)

Chapter 29
Current Status and Trends of Research on Sustainable Clothing Materials—A CiteSpace-Based Analysis

Lu-Fei Yang and Ke-Ke Sun

Abstract This paper presents a systematic analysis of the research literature related to the field of sustainable clothing material design using CiteSpace software. The study population covered 1814 valid pieces of literature between 2013 and 2023. Analysis through bibliometrics and visualization reveals research trends, hotspots, and collaborative networks in the field. Research into sustainable clothing materials has seen a significant increase after 2018. Research hotspots include ecological footprint, circular economy, etc., forming a cooperative network centered on Chinese Acad Sci, Donghua Univ, and South China Univ Technol. Moreover, CiteSpace's strengths and weaknesses in analytics were also discussed, and major trends in research were noted.

29.1 Introduction

Fiber Research Dimensions of sustainable clothing materials, Representative results are: Fidan et al. [1] used methods such as environmental impact assessment, exploring aspects such as the environmental impact of recycled cotton fibers and cogeneration plants on denim production. This method has improved the sustainable production of denim. Basma et al. [2] described the application of biotechnology in sustainable finishing of cellulosic textiles, and the results show that biotechnology can enable textiles to be processed with less impact on consumers and the environment. Kaminski et al. [3] presented fabrication of new eco-friendly textiles using bacterial cellulose. The modified material shows good tensile, flexibility, and water resistance performance, which is promising. Gao et al. [4] achieved the modulation of PSS fibers' properties from high strength to high flexibility, and PSS fibers have potential for flexible tactile sensors.

L.-F. Yang · K.-K. Sun (✉)
School of Design Art, Xiamen University of Technology, Xiamen, Fujian Province, China
e-mail: sunkeke@xmut.edu.cn

S.-C. Chu et al. (eds.), *Advances in Intelligent Data Analysis and Applications*,
Smart Innovation, Systems and Technologies 445,
https://doi.org/10.1007/978-981-96-7277-6_29

At the level of sustainable clothing materials and production, Yang and Zhang [5] explored the DCNN-based texture feature extraction method, and extracting texture characteristics is the key to identifying environmentally friendly materials. Stanescu [6] explored upcycling technologies for post-consumer textile wastes. Studies indicate that this technology is a vital way to achieve sustainable development in the textile industry. Chand [7] et al. investigated the waste problems within the textile and apparel industry as well as their sustainable management strategies, highlighting the necessity of effective waste management measures in this industry and the continuous pursuit of sustainability. Khajavi [8] studied the potential of additive manufacturing technology in the apparel industry and showed that the technology can be used as a complementary production method to the traditional process. Lee [9] combines garment waste processing with the design and production of new materials. This method addresses the issues of garment waste disposal and material development and can be applied in low- and middle-income countries in the future. Chen et al. [10] constructed a theoretical model to analyze in terms of green innovation and perceived value of apparel brands, highlighting that green features should focus on innovation, practical performance, and market demand. There are many research papers in this lab. Sun et al. [11] proposed a thin plate spline interpolation method for complementing anthropometric data and compared it with a linear interpolation method. The results showed that the method has lower mean error and better prediction. Chiu et al. [12] established a simulation and experimental model of the PV cell, proposed an algorithm based on hysteresis loop, and verified its superiority compared with the traditional algorithm.

In conclusion, research on sustainable clothing materials has become popular because of its environmental, social, and fashion significance. This paper analyzes relevant literature using CiteSpace software, mainly focusing on literature release trends, relevance, key contributors, research hotspots, and future directions.

29.2 Research Method

The Web of Science Core Collection from 2013 to 2023 was adopted as the data source. A total of 2,169 documents were retrieved according to the search criteria. After screening the literature, 1,814 valid documents were obtained. The data was collected with the PICO model as a search strategy, and the three elements were connected by the Boolean logical operator "and", as shown in Table 29.1.

Table 29.1 PICo element and search strings

Element	Search string
Population (P)	ALL = ("Natural fiber materials" OR "Recycled fiber materials" OR "Degradable materials" OR "Ecological materials" OR "New environmental protection materials" OR "Environmental protection dyes" OR "Ecological textiles" OR "Recycled polyester" OR "Environmental protection brands")
Intervention (I)	AND ALL = ("Sustainable consumption awareness" OR "Sustainable development goals" OR "Environmental awareness" OR "Health awareness" OR "Environmental policy" OR "Innovative materials")
Comparator and outcome (CO)	AND ALL = ("Resource shortage" OR "environmental pollution" OR "Ecological economy" OR "Process improvement" OR "Simple design" OR "Green design" OR "Concern for health" OR "Responsibility for the environment" OR "Material innovation")

The whole process of literature screening was demonstrated using PRISMA flowchart, the screening process and screening criteria are shown in Fig. 29.1.

29.3 Data Analysis

29.3.1 Trend of Literature Quantity Changes and Co Citation Analysis

This section showcases the expanded trends in literature releases on sustainable clothing materials from 2013 to 2023. A total of 1,814 related publications exist. The trend is characterized by stability in the early years and growth later on. The low number of articles from 2013 to 2017 can be attributed to limited techniques and a scarcity of high-quality literature. This constitutes the initial stage. As methods became more enriched, publications grew rapidly, particularly during 2021 - 2022. By 2022, there were 351 articles, indicating a high level of research interest. After reaching a peak, the number of postings declined somewhat. However, by 2023, publications have increased by 295 compared to 2013, demonstrating continued research enthusiasm (Fig. 29.2).

Literature co-citation analysis involves the co-citation of literature in the research field. As shown in Table 29.2 and Fig. 29.3, the network density is 0.0116, with the number of nodes N = 278 and the number of connections E = 448. Among the top ten co-citation relationships, Geyer has the highest frequency, with a total of 29 citations. Other relatively high frequencies include 23 for Wiedmann and 16 for Steffen, indicating their significant influence in the field.

The co-citation strength is computed by the co-citation coefficient level, a higher value means a stronger correlation between the two documents. Express Eq. (1) as follows:

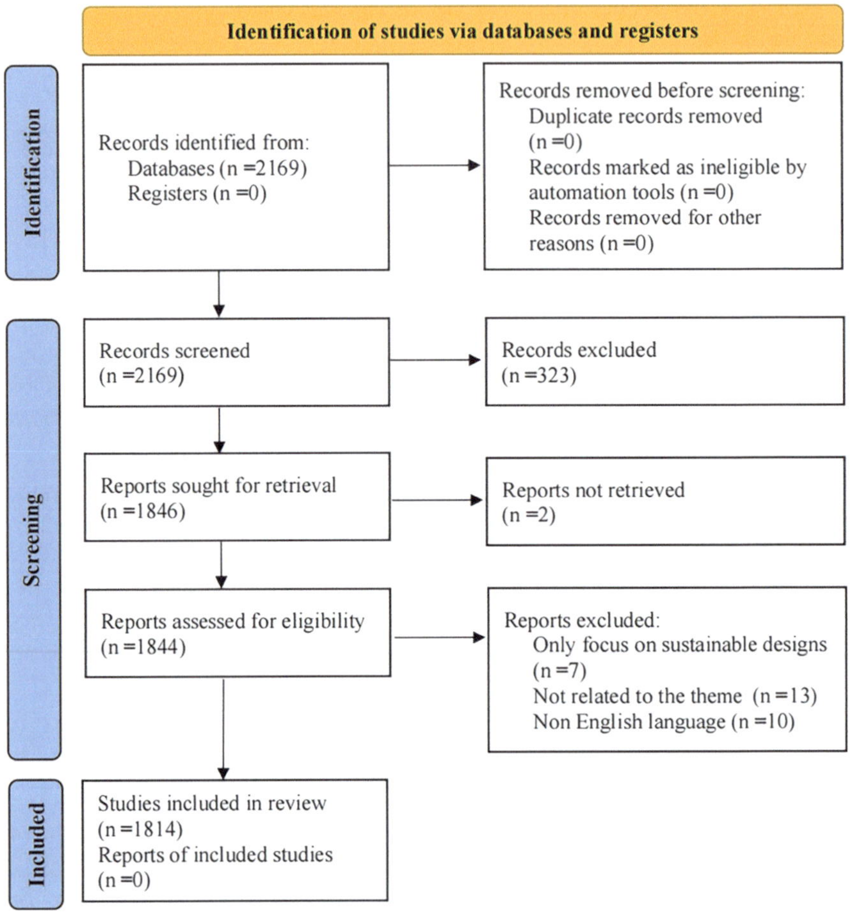

Fig. 29.1 PRISMA flowchart for literature screening

$$CCV = \frac{cc[i,j]}{c[i] \times c[j]} \quad (29.1)$$

where *cc[i, j]* represents the number of co-citations between document *i* and document *j*, and *c[i]* and *c[j]* are the citation counts of each respectively.

29.3.2 Keyword Co-Occurrence Analysis

Co-occurrence analysis of keywords is primarily utilized to reflect the inherent connection among keywords. Through extracting the keywords, a total of 287 high-frequency keywords were identified, generating 399 connecting lines as shown in

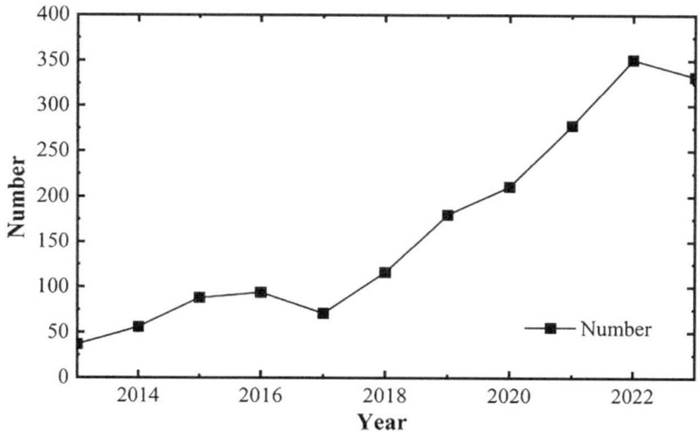

Fig. 29.2 Literature release trend

Table 29.2 Co-citation of literature

Ranking	Frequency	Centrality	Author's name	Time	Source of publication
1	29	0.09	Geyer R	2017	SCI ADV
2	23	0.19	Wiedmann TO	2015	PNATL ACAD SCI USA
3	16	0.02	Steffen W	2015	SCIENCE
4	16	0.02	Jambeck JR	2015	SCIENCE
5	15	0.03	Kirchherr J	2017	RESOUR CONSERV RECY
6	13	0.01	Ghisellini P	2016	J CLEAN PROD
7	13	0.01	Korhonen J	2018	ECOL ECON
8	12	0.13	Geissdoerfer M	2017	J CLEAN PROD
9	11	0.02	ONeill DW	2018	NAT SUSTAIN
10	7	0.04	Schaffartzik A	2014	GLOBAL ENVIRON CHANG

Fig. 29.4. "Performance" is the largest node, with 149 connections, highlighting its significance in the field. "Mechanical property" and "fiber" follow with 114 and 94 connections respectively. Performance, mechanical property, fiber, composite, behavior, etc., emerged earlier within the time span. Keywords like technology, nanosheet, and polymer have appeared more recently. They might represent a new direction for future research on sustainable clothing materials.

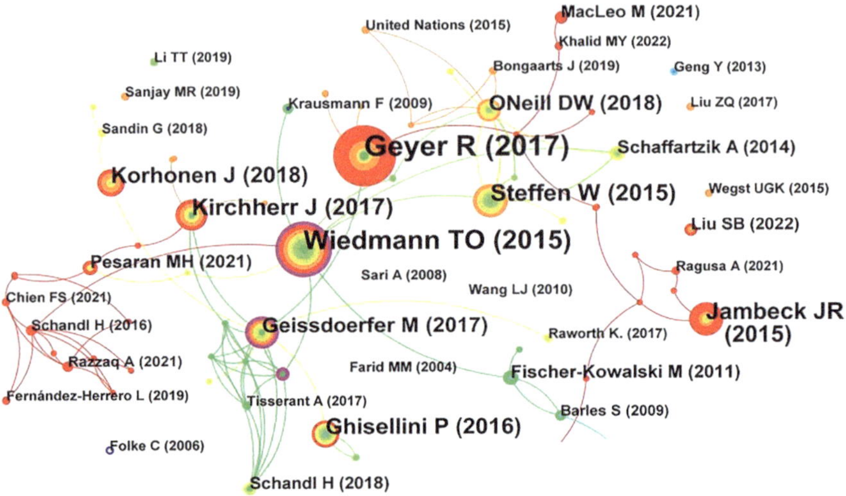

Fig. 29.3 Co-citation relationship of literature

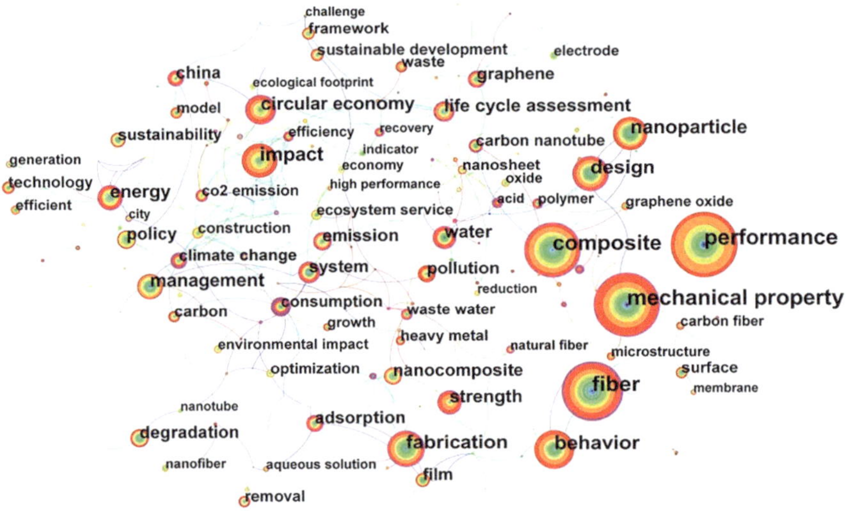

Fig. 29.4 Keyword co-occurrence diagram

29.3.3 Research Institution Analysis

By analyzing the cooperation of research institutions in this section, we understood the cooperation, involvement, and impact of different agencies. The network has a density of 0.0049, with 421 nodes and 429 connecting lines, as presented in Fig. 29.5 and Table 29.4. The most prominent institutional collaborative networks are centered

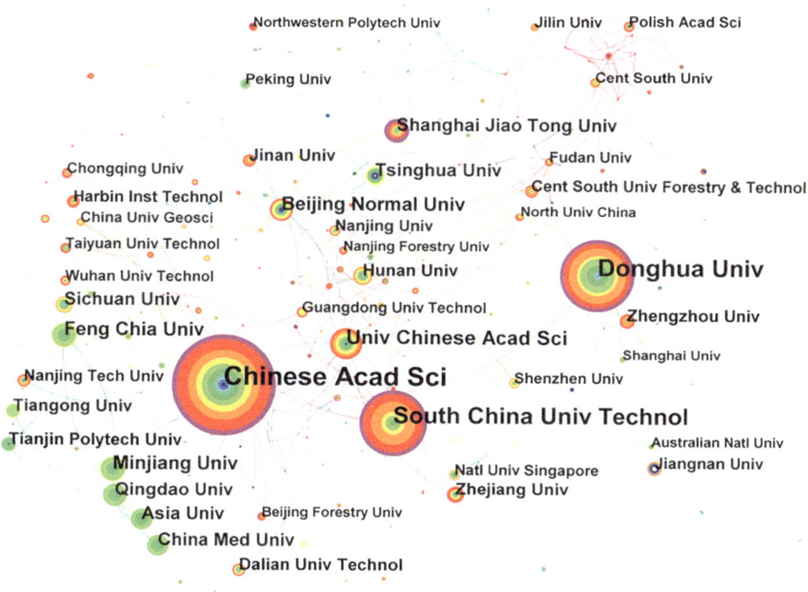

Fig. 29.5 Research institution relationship diagram

around the Chinese Academy of Sciences, Donghua University, and South China University Technology. These institutions have a concentrated research focus and more active research exchanges. The Chinese Acad Sci had the highest number of articles with 96. The top three organizations enjoy a higher density of collaboration regarding research outputs, which suggests an excellent working relationship among them.

29.3.4 Collaboration Analysis of the Author of the Article

Unfold the analysis of the core group of authors, collaborative relationships, and intensity of collaboration in this section. There are 512 nodes and 65 links. The overall network density is 0.0047, which indicates a strong collaborative network among these authors within the field as shown in Fig. 29.6 and Table 29.3. There are two main clusters of cooperating groups. The first one consists of Lin, Jia-Horng, Li, Ting-Ting, Lou, Ching-Wen, and others. It is the largest cooperative network. The second cluster consists of Yang, Zhong-min, Xu, Shan-hui, Feng, Zhou-ming, and others. There are 6 authors with 10 or more publications. The extent of cooperation among the study authors shows a clear link between highly productive authors and the density and intensity of cooperation. It can be seen that a well-established network of cooperation has been formed in this field.

Table 29.4 Top Ten research institutions

Ranking	Issuing agency	Time	Number of publications	Degree of cooperation
1	Chinese Acad Sci	2013	96	33
2	Donghua Univ	2014	69	12
3	South China Univ Technol	2017	60	17
4	Univ Chinese Acad Sci	2015	33	14
5	Beijing Normal Univ	2013	27	12
6	Feng Chia Univ	2019	25	9
7	Minjiang Univ	2019	24	7
8	China Med Univ	2019	22	7
9	Qingdao Univ	2019	22	6
10	Asia Univ	2019	21	6

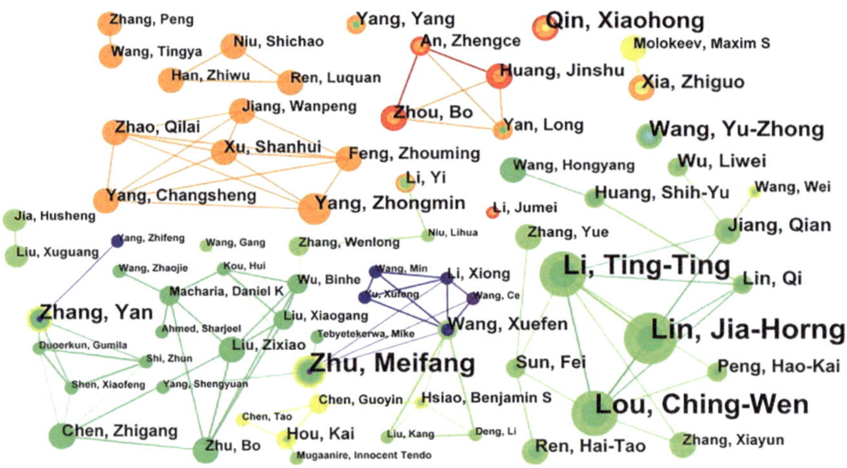

Fig. 29.6 Author and collaboration diagram

Price's Law is a commonly used method for determining core authors in delineating core authors. Equation (2) is demonstrated below

$$M = 0.749\sqrt{N_{max}} \tag{29.2}$$

M is the number of publications of core authors, and *Nmax* is that of the most productive authors in the equation. Based on the calculation, core authors are with ≥ 3.6 publications (Counts ≥ 4 when applied in practice). Authors whose publication

Table 29.4 Top ten core authors

Ranking	Author's name	Time	Number of publications	Degree of cooperation
1	Lin, Jia-Horng	2019	24	4
2	Lou, Ching-Wen	2019	21	8
3	Li, Ting-Ting	2019	21	8
4	Zhu, Meifang	2014	18	10
5	Qin, Xiaohong	2021	12	4
6	Wang, Yu-Zhong	2017	10	3
7	Zhang, Yan	2015	9	4
8	Yang, Zhongmin	2022	7	6
9	Jiang, Qian	2019	7	6
10	Lin, Qi	2019	7	4

numbers meet or exceed this threshold are regarded as core authors. The number of authors with ≥ 4 publications is 67. The top ten authors are presented in Table 29.3.

Core team analyses need to be based on a combination of institutions and core authors. On the one hand, author collaboration networks can be nested within institutional collaboration networks. There may be more frequent collaboration between authors from the same institution. On the other hand, when authors from different institutions conduct collaborative research, this may allow for wider connections between institutions, cross-institutional mobility and integration can be facilitated. By research organizations and core author data, the core team has formed.

29.3.5 *Research Hotspot Analysis*

The analysis of research hotspots unfolds from keyword clustering in this section, describes future trends and research hotspots for sustainable clothing materials. The Log Likelihood Ratio algorithm was employed for the analysis, as shown in Table 29.5 and Fig. 29.7. There are $N = 287$ nodes, $E = 399$ connections, and the network density is $\rho = 0.0097$. In total, there are 10 clusters. The top five clusters are presented in Table 29.5. The average year of their formation was between 2015 and 2018, suggesting that relevant research in this area is mature. The largest cluster is "ecological footprint" in 2018, which encompasses 34 keywords. The main keywords include ecosystem services, climate change, ecosystem service, and climate change adaptation. This reveals the research hotspots and the maturity level of the field.

Table 29.5 Main keywords for clustering

Ranking	Cluster name	Main keywords	Year	Number
1	Ecological footprint	Ecological footprint, ecosystem services, climate change, ecosystem service, climate change adaptation	2018	34
2	Mechanical properties	Mechanical properties, thermal properties, thermal energy storage, polymer composites, electrospinning	2015	27
3	Circular economy	Circular economy, material flow analysis, social metabolism, ecological network analysis	2017	23
4	Film	Film, mechanism, low-density polyethylene, tissue engineering, lignocellulosic fibers	2016	23
5	Source identification	Source identification, heavy metal, heavy metals, spatial distribution, soil pollution	2018	20

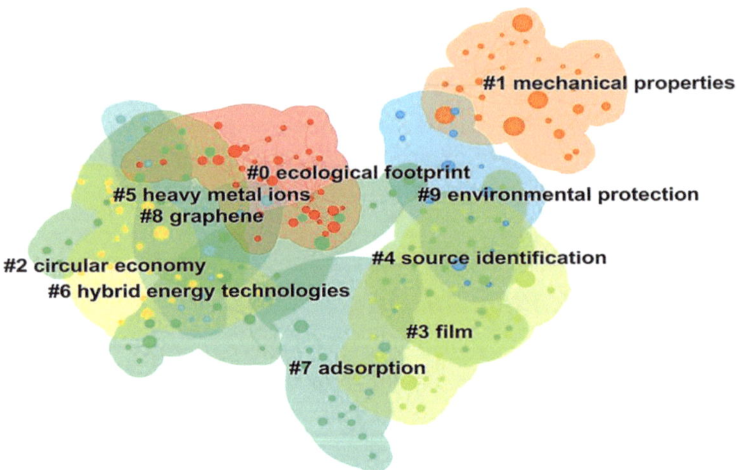

Fig. 29.7 Keyword clustering diagram

29.4 Discuss

29.4.1 Effectiveness of CiteSpace Application

The visualization and analysis of literature in the sustainable clothing materials field from 2013 to 2023 in this paper, using CiteSpace software, have verified the applicability of CiteSpace within this domain.

29.4.2 Research Hotspots and Future Directions

By analyzing the literature data during 2013–2023, there is a trend of early-stage stabilization and late-stage rapid growth in this field. By analyzing the "ecological footprint" of the largest clusters in 2018, ecosystem services, climate change, ecosystem services and climate change adaptation, these terms reveal the research hotspots and maturity of the field.

29.4.3 Advantages and Disadvantages of Tools

CiteSpace software clearly presents research history and trends, facilitating the identification of future research directions. Keyword co-occurrence and cluster analysis disclose research hotspots. In the future, new sustainable materials will be explored in this field. However, limitations in data sources and the absence of in-depth content analysis are major drawbacks of the current study. Future research should consider expanding data sources and integrating other text analysis tools for a more comprehensive interpretation.

29.5 Conclusion

By visualizing and analyzing the literature on sustainable clothing materials from 2013 to 2023, the findings are as follows:

1. Research Trends on sustainable clothing materials, characterized by early stabilization and late rapid growth.
2. Many core authors have emerged. Through further analysis, it is evident that large core research teams have already come into being. The most prominent institutional collaborative networks are centered around the Chinese Acad Sci, Donghua University and South China University Technol. They have a high density of writing and a good partnership.
3. Keyword co-occurrence and clustering are used to analyze hotspots and future research directions, including ecosystem services, climate change, ecosystem services and climate change adaptation.

References

1. Fidan, F.S., Aydogan, E.K., Uzal, N.: An integrated life cycle assessment approach for denim fabric production using recycled cotton fibers and combined heat and power plant. J. Clean. Prod. **287**, 1–15 (2021)

2. Eid, B.M., Ibrahim, N.A.: Recent developments in sustainable finishing of cellulosic textiles employing biotechnology. J. Clean. Prod. **284**, 1–22 (2021)
3. Kaminski, K., Jarosz, M., et al.: Hydrogel bacterial cellulose: a path to improved materials for new eco-friendly textiles. Cellulose **27**(9), 5353–5365 (2020)
4. Gao, Q., Zhang, Y.B., Wang, P., Zhu, J.D., Gao, C.X.: Robust and knittable wet-spun PEDOT: PSS fibers via water. Compos Commun **40**, 1–5 (2023)
5. Yang, C.Y., Zhang, X.Y.: Environmental protection clothing design and materials based on green design concept. FRronyiers Mater **10**, 1–10 (2023)
6. Stanescu, M.D.: State of the art of post-consumer textile waste upcycling to reach the zero waste milestone. Environ. Sci. Pollut. Res. **28**, 14253–14270 (2021)
7. Chand, S., Chand, S., Raula, B.: Textile and apparel industries waste and its sustainable management Approaches. J. Mater. Cycles Waste Manage. **25**, 3132–3143 (2023)
8. Khajavi, S.H.: Additive manufacturing in the clothing industry: towards sustainable new business models. Appl. Sci. **11**, 1–18 (2021)
9. Lee, H.Y.W.: The dual strategy for textile and fashion production using clothing waste. Sustainability **15**(15), 1–20 (2023)
10. Chen, L.H., Qie, K.X., Memon, H., Yesuf, H.M.: The empirical analysis of green innovation for fashion brands, perceived value and green purchase intention—mediating and moderating effects. Sustainability **13**, 1–17 (2021)
11. Sun, K.K., Gao, M., Li, B.: Research on two interpolation algorithms for anthropometric data completio —a case study on the data of elderly males in Fujian. China. J. Netw. Intell. **7**(4), 1083–1101 (2022)
12. Chiu, Y.J., Wang, W.Z., Hu, Y.C., Lien, S.Y., Zhang, B.R.: Maximum power development under varying solar radiation based on hysteresis-loop comparison of selected duty-cycle perturbation algorithms on a mobility scooter. J. Net. Intelligence **8**(4), 1540–1556 (2023)

Chapter 30
Research on the Digital Transformation Maturity Assessment of Automotive Parts Enterprises Based on the GAHP-FCE Method

Lingjiao Wu, Shih-Pang Tseng, Wei Song, and Ziyan Luo

Abstract The digital transformation of small and medium-sized enterprises (SMEs) has reached a substantive phase of advancement, with significant variations influenced by factors such as management foundations, industry characteristics, and digitalization strategies. Establishing industry-specific models for evaluating digital transformation maturity can provide guidance and assistance for enterprises along the industry chain in their digital transformation practices. This paper focuses on automotive parts enterprises and constructs a Digital Transformation Maturity Assessment (DTMA) model comprising "5 first-level indicators and 23 s-level indicators." It also develops a five-tier maturity level labeled as "Initial Planning Level," "Standard Application Level," "Integrated Control Level," "Optimization Improvement Level," and "Digital Transformation Leadership Level." Utilizing the Group Analytic Hierarchy Process and Fuzzy Comprehensive Evaluation (GAHP-FCE) method, this study evaluates and analyzes the digital transformation maturity of Enterprise L, validating the DTMA model's applicability and reliability and providing a methodological basis for the assessment of enterprise digital transformation.

30.1 Introduction

Digital transformation is the process that organizations use digital technology to increase product value, improve business processes, enhance customer experience, optimize operational efficiency, and create new forms of productivity and competitiveness. It has attracted increasing interest from both industry and academia.

L. Wu (✉) · W. Song · Z. Luo
School of Digital Economics, Changzhou College of Information Technology, Jiangsu, China
e-mail: 114051792@qq.com

S.-P. Tseng
School of Software and Big Data, Changzhou College of Information Technology, Jiangsu, China
e-mail: tsengshihpang@czcit.edu.cn

S.-C. Chu et al. (eds.), *Advances in Intelligent Data Analysis and Applications*,
Smart Innovation, Systems and Technologies 445,
https://doi.org/10.1007/978-981-96-7277-6_30

However, many organizations lack a clear roadmap for global planning and guiding the path from strategy, human resources, processes, organizational culture, and digital technology to aid with digital transformation [1].

Gimpel et al. revealed that most research outputs are concentrated on specific aspects of digital transformation or evidence-based case studies [2]. Bumann et al. identified eighteen validated digital maturity models and frameworks through a literature review study, which describes various dimensions or action areas that digital transformation strategies need to consider [3]. Existing research indicates a current lack of a measurable digital transformation maturity assessment model that adapts to local and industry characteristics.

The automotive sector in China is among the pioneers in embarking on the journey of informatization. It has established a comprehensive framework characterized by "market-driven full-vehicle manufacturers, specialized and innovative companies setting examples, and widespread application among parts enterprises." The industry has also developed an integrated digital ecosystem along the industrial chain across the dimensions of digital R&D, digital production, digital management, and digital marketing. Investigating a Digital Transformation Maturity Assessment (DTMA) model tailored to their specific characteristics is crucial for parts enterprises that serve as suppliers to vehicle manufacturers. Such a model aids these enterprises in confronting intensifying market competition more effectively, conducting accurate and objective self-assessments of their digital capabilities, pinpointing improvement areas, and fostering continuous enhancement.

Therefore, the objective of this study is to establish a DTMA model suitable for automotive parts enterprises, to help small and medium-sized enterprises identify the level of digital transformation maturity and weak links within their organizations, and to provide a clear roadmap for continuous improvement. This paper proposes and implements a novel DTMA model based on the Group Analytic Hierarchy Process and Fuzzy Comprehensive Evaluation (GAHP-FCE) method within automotive parts enterprises. This DTMA model employs fuzzy methods to automatically convert subjective feedback from surveys into quantitative analysis, with the GAHP-FCE method ensuring the accuracy of the quantitative analysis.

The rest of this paper is structured as follows: Sect. 30.2 presents work related to this study. Section 30.3 describes the model design and example verification of DTMA based on GAHP-FCE method, which is composed of several main parts such as evaluation index design, maturity level division, assigning indicator weights through the Group Analytic Hierarchy Process (GAHP) method, conducting digital transformation maturity assessment through Fuzzy Comprehensive Evaluation (FCE) method, and analyzing and recommending based on assessment results. Section 30.4 provides conclusions.

30.2 Related Works

Scholars and experts have accumulated certain achievements in the research on the DTMA, mainly focusing on the design of evaluation index systems, the construction of assessment models, and the selection of evaluation methods.

In the research on DTMA models and index systems, Teichert conducted a systematic literature review, ultimately analyzing 22 different models and extracting various characteristics of different digital maturity models [4]. Spalatini et al. proposed a 6Ps maturity model suitable for small and medium-sized enterprises in the manufacturing industry, which is based on six major social business and technical dimensions, namely products and services, processes, platforms, people, partnerships, and performance [5]. Bumann, in comparing and analyzing over a hundred described dimensions, identified the most frequently cited dimensions as strategy, organization, corporate culture, technology, customers, and employees, providing an important framework for enterprises to successfully achieve digital transformation [3].

Research regarding DTMA evaluation methods, most scholars have adopted multi-criteria decision-making methods (MCDM) for multidimensional analysis. MCDM methods mainly include the Delphi method [6], Analytic Hierarchy Process (AHP) [7–9], Data Envelopment Analysis (DEA) [10], Analytic Network Process (ANP), Fuzzy Comprehensive Evaluation (FCE) [11], and Decision-Making Trial and Evaluation Laboratory (DEMATEL) [12]. AHP was proposed by American operations researcher and University of Pittsburgh professor T. L. Saaty in the early 1970s. It is a widely used decision-making method in the management field, used to determine priorities in multi-criteria decision-making problems, such as planning, resource allocation, policy analysis, conflict resolution, and decision forecasting. This method has the advantages of being systematic, flexible, and concise. It helps decision-makers make choices by arranging decision factors in a hierarchical structure [13]. The Group Analytic Hierarchy Process (GAHP) combines AHP with grey system theory to reduce the subjective judgment of individuals when making decisions. Though the basic method of GAHP is the same as AHP, the difference is that GAHP can effectively aggregate weights from different experts, making it more objective and advantageous in handling group decisions. Currently, fuzzy mathematics, a new mathematical concept proposed by American control theorist L. A. Zedeh, is widely applied in the field of management science, attempting to use quantitative and precise mathematical methods to handle fuzzy phenomena. It has achieved many results in image recognition, artificial intelligence, information control, cluster analysis, and comprehensive evaluation for project selection and evaluation, quality assessment of enterprise departments, as well as fuzzy production balance.

In terms of the selection of research subjects for DTMA, existing studies mainly focus on a certain group within a specific region, and the established research models have broad and universal applicability but are less targeted. There is still less research conducted on specific industries.

30.3 Experimental Design and Implementation

30.3.1 GAHP-FCE Method

To more effectively assess the maturity of digital transformation in automotive parts enterprises, this study employs the Group Analytic Hierarchy Process and Fuzzy Comprehensive Evaluation (GAHP-FCE) method to establish the DTMA model. This composite decision analysis method integrates the strengths of GAHP in handling group decisions and FCE's capabilities in managing fuzziness and uncertainty. To validate the feasibility of this model, we selected a specific automotive parts enterprise, L, as our subject of study. Through the construction of a judgment matrix, hierarchical ranking, consistency checks, synthesis of weights, the definition of a fuzzy evaluation matrix, and comprehensive fuzzy evaluation procedures, we evaluated and analyzed its digital transformation capabilities and levels. Established in 1994, Enterprise L specializes in automotive generator claws, pulleys, rotors, etc., and has initiated comprehensive intelligent transformation to enhance its core competitiveness.

30.3.2 Design of the DTMA Indicator System for Automotive Parts Enterprises

Drawing on relevant research findings, the research team has designed the DTMA index system for automotive parts enterprises based on the "SME Digitalization Level Evaluation Indicators (2024 Edition)" released by the Ministry of Industry and Information Technology, GB/T 39,116–2020 "Smart Manufacturing Capability Maturity Model," and GB/T 43,439–2023 "Information Technology Service Digital Transformation Maturity Model and Assessment," in conjunction with field research and visits to automotive industry chain enterprises to judge the digital status and needs of automotive parts enterprises.

Research and visits to automotive parts enterprises in the Suzhou, Wuxi, and Changzhou areas reveal that the digitalization status of automotive parts enterprises mainly presents four characteristics:

Effective integration of CAX based on PLM/PDM. Most automotive parts enterprises have built digital R&D platforms based on PLM/PDM that can effectively integrate CAD, CAPP, and CAE, with upstream and downstream of the industry chain participating in product development, and product data is effectively managed.

Digitalization of business management based on ERP. Automotive manufacturers, based on systems such as SCM and ERP, achieve information interaction of supply and demand, plans, and logistics between upstream and downstream enterprises, avoiding the impact of supply chain abnormalities. Automotive original equipment manufacturers (OEMs) drive the digitalization of component enterprises, thereby elevating the overall digitalization level of the component industry.

Digitalization of production manufacturing based on MES. The improvement of delivery capabilities in automotive parts enterprises comes from the addition and intelligent transformation of manufacturing equipment, the improvement of equipment capacity and equipment OEE, and more from effective planning based on MES, achieving the digitalization of production manufacturing and innovation in manufacturing models.

Efficient use of data based on multi-system integration. Most enterprises have established SCADA platforms to collect equipment data, production status data, and quality data, and by analyzing and applying data, they have improved process levels, quality levels, and effectively controlled costs, enhancing the overall decision-making level of the enterprise.

Based on the above research findings, we have developed a DTMA indicator system for automotive parts enterprises across five dimensions: digital element support, digital foundation, digital application, digital effectiveness, and external environment, resulting in a system comprising five first-level indicators and 23 s-level indicators. For details, see Table 30.1. The selection of these indicators not only fully captures the essence and characteristics of digital transformation but also integrates qualitative and quantitative aspects, effectively assessing the existing foundation, application level, digital features, and efficiency benefits of digital transformation in automotive parts enterprises. This system provides a robust basis for the assessment, diagnosis, and improvement of enterprise digital transformation.

30.3.3 Classification of Maturity Levels

The establishment of a DTMA model in the automotive parts industry aims to categorize and comprehensively evaluate the level of digitalization achieved by these enterprises. This research delineates the digital transformation maturity into a five-tiered scale for automotive parts companies, escalating as follows: Level I (Initial Planning Level), Level II (Standard Application Level), Level III (Integrated Control Level), Level IV (Data-Driven Level), and Level V (Digital Transformation Leadership Level). Each ascending level of maturity subsumes the requirements of the preceding levels, ensuring a progressive and inclusive model of digital advancement.

Level I: Initial Planning Level. At this stage, enterprises begin to assign dedicated personnel to plan for intelligent transformation and digitalization, initiating the construction of infrastructure and conditions. Information technology is preliminarily applied, covering core business and processes, with information systems becoming the primary management tool. However, the application is primarily focused on replacing manual labor.

Table 30.1 Indicator system for the DTMA in automotive parts enterprises

Serial number	Primary indicator	Secondary indicator	Indicator description
1	Digital element support	Digital Planning	The degree of formulation and effective implementation of corporate digital planning
2		Digital Organization	The organizational structure that is flat and in line with digital transformation
3		Digital Talent Development	The construction of corporate digital teams and the introduction and cultivation of talents
4		Financial Support	The degree of financial support for corporate digital construction
5	Digital foundation	Equipment Intelligence	The status of key process equipment intelligence, networking, and data collection of equipment status
6		Network Infrastructure	The status of network coverage, network security protection, and key business data transmission
7		Information Security	The implementation of information security protection mechanisms
8		Data Foundation	The accuracy of basic data, and the application of master data management, data warehouses, etc.
9	Digital application	Digital Marketing	The digital application of corporate marketing activities
10		Digital Research and Design	The digital application in corporate research and design
11		Digital Decision-Making	The integrated application of business management with ERP, MES, OA, etc.
12		Digital Lean Manufacturing	The degree of digitalization and lean manufacturing in production
13		Product Quality and Traceability	The situation of product quality and traceability
14		Digital Supply Chain	The ability of enterprises to integrate into the supply chain and respond quickly to changes in customer demands
15	Digital outcomes	Capacity Utilization Rate	The comprehensive utilization of corporate capacity
16		Green and Low-Carbon	The situation of corporate green manufacturing and manufacturing model innovation
17		Product and Service Innovation	The situation of corporate product innovation, service model innovation, and business model innovation

(continued)

Table 30.1 (continued)

Serial number	Primary indicator	Secondary indicator	Indicator description
18		Comprehensive Cost per Ten Thousand Yuan of Output Value	The situation of corporate comprehensive cost per ten thousand yuan of output value
19		On-Time Delivery Rate	The situation of on-time delivery of customer orders
20		Digital Service Output Capability	The capability of external service output and internal system delivery, system maintenance, etc.
21	External environment	Government Policy Support	The ability of enterprises to obtain government policy support
22		Industry Chain Impact	The impact of fluctuations in the upstream and downstream of the industry chain and the overall market
23		Service Level of Service Provider	The impact of the service capability and service level of digital transformation service providers

Level II: Standard Application Level. Enterprises at this level have dedicated teams to advance digital transformation, establishing specific strategies and implementation plans that encompass the main business and processes. Information systems become indispensable tools, and there is a growing emphasis on system integration and application. Nevertheless, the depth of application still requires further exploration.

Level III: Integrated Control Level. Enterprises fully recognize the importance of digital transformation and can implement intelligent transformation and digitalization plans in a rolling manner. Information systems are fully integrated, and the construction of business platforms is completed, achieving initial interconnectivity among design, manufacturing, management, and service processes, with complete integration of information technology and manufacturing technology.

Level IV: Data-Driven Level. At this level, enterprises focus on the performance of digital transformation, achieving interconnection and integration of various data resources. They optimize operations based on data, establishing a "perception-insight-assessment-response" data loop, which enables data-driven operational optimization. This level aims to reduce production costs and enhance delivery efficiency and quality, achieving precision management.

Level V: Digital Transformation Leadership Level. Resources at this level are subject to intelligent analysis and application, with digitalization and intelligence fully integrated with corporate strategy. Enterprises achieve intelligent operations, realizing value output within the industry chain and supply chain. This level involves the optimization of organizational systems, mechanisms, and service methods, forming new operational models and corporate cultures.

30.3.4 Determination of Indicator Weights Based on GAHP Method

To accurately calculate the weights of indicators using GAHP, this study invited five industry experts to evaluate the importance of the indicators. Since the process of determining the weights for indicators at each hierarchical level is essentially the same, this paper will only illustrate the process using the five first-level indicators as an example.

Construction of the Judgment Matrix. Industry experts used a 9-point scale method to assess the relative importance of the five first-level indicators: digital element support, digital foundation, digital application, digital effectiveness, and external environment. They established a comparison judgment matrix for the first-level indicators, as shown in Table 30.2.

From Table 30.2, the judgment matrix for the first-level domain indicators of digital transformation can be constructed. The comparative judgment matrix for the first-level indicators by the expert is as follows:

$$X = \begin{bmatrix} 1 & 1 & 1 & 1/2 & 2 \\ 1 & 1 & 2 & 2 & 2 \\ 1 & 1/2 & 1 & 2 & 1 \\ 1/2 & 1/2 & 1/2 & 1 & 2 \\ 1/2 & 1/2 & 1 & 1/2 & 1 \end{bmatrix} \tag{30.1}$$

Table 30.2 Comparative judgment matrix of the first-level indicators by a certain expert

Primary indicator	Digital element support	Digital foundation	Digital application	Digital outcomes	External environment
Digital element support	1	1	1	2	2
Digital foundation	1	1	2	2	2
Digital application	1	1/2	1	2	1
Digital outcomes	1/2	1/2	1/2	1	2
External environment	1/2	1/2	1	1/2	1

The characteristic vector sorting method is used to solve the characteristic equation. The characteristic equation is shown in Eq. (30.2):

$$XW_1 = \lambda_{max} W_1 \tag{30.2}$$

Hierarchical Ranking and Consistency Test. Then, a consistency test is performed. To assess whether the judgment matrix is consistent or if the degree of inconsistency is too large, a consistency test must be conducted on the judgment matrix. The consistency index is denoted by CI:

$$CI = \frac{\lambda_{max} - n'}{n - 1} \text{ n}' \text{denotes the order of the matrix} \tag{30.3}$$

The consistency index CI is subsequently compared with the consistency index RI of a randomly generated judgment matrix of equivalent rank, yielding the test statistic known as the consistency ratio:

$$CR = \frac{CI}{RI} \tag{30.4}$$

The weights and consistency were calculated using the Yaahp software. The maximum eigenvalue λmax of the first-level indicator judgment matrix for this expert, along with the corresponding weight vector and consistency ratio (CR), were obtained as follows:

$$\left.\begin{array}{l} \lambda_{max} = 5.1775; \\ W_1 = (0.2464 \quad 0.2839 \quad 0.1943 \quad 0.1478 \quad 0.1277) \\ CR = 0.0396 \end{array}\right\} \tag{30.5}$$

The CR value is less than 0.1, indicating that the judgment matrix has acceptable consistency and the hierarchical ranking of weights is deemed acceptable.

Synthesis of Group Decision-Making Evaluators' Opinions. Following the aforementioned methodology, the weight vectors and consistency test results for the other four industry experts can be obtained.

Expert 2:

$$\left.\begin{array}{l} \lambda_{max} = 5.0966; \\ W_2 = (0.2499 \quad 0.2871 \quad 0.1894 \quad 0.1649 \quad 0.1088) \\ CR = 0.0216 \end{array}\right\} \tag{30.6}$$

The CR value is less than 0.1, indicating that the judgment matrix of this expert possesses consistency, and the derived priority weights from the hierarchical sorting are considered acceptable.

Expert 3:

$$\left.\begin{array}{l}\lambda_{\max} = 5.2962; \\ W_3 = (0.2498 \quad 0.2266 \quad 0.2003 \quad 0.1744 \quad 0.1488) \\ CR = 0.0661\end{array}\right\} \tag{30.7}$$

The CR value is less than 0.1, indicating that this expert's judgment matrix has satisfactory consistency, and the derived priority weights from the hierarchical sorting are considered acceptable.

Expert 4:

$$\left.\begin{array}{l}\lambda_{\max} = 5.1775; \\ W_4 = (0.2216 \quad 0.25666 \quad 0.24646 \quad 0.16866 \quad 0.1069) \\ CR = 0.0396\end{array}\right\} \tag{30.8}$$

The CR value is less than 0.1, indicating that this expert's judgment matrix has satisfactory consistency, and the derived priority weights from the hierarchical sorting are considered acceptable.

Expert 5:

$$\left.\begin{array}{l}\lambda_{\max} = 5.3024; \\ W_5 = (0.2174 \quad 0.2959 \quad 0.1656 \quad 0.1861 \quad 0.1349) \\ CR = 0.0675\end{array}\right\} \tag{30.9}$$

The CR value is less than 0.1, indicating that the expert's judgment matrix possesses satisfactory consistency, and the derived priority weights from the hierarchical sorting are considered acceptable.

Combining the evaluations of five experts, this study employs a group decision-making analysis approach and integrates normalization methods for data processing to determine the weight values of the first-level indicators, denoted as W.

$$W = (0.2370 \quad 0.2700 \quad 0.1992 \quad 0.1684 \quad 0.1254) \tag{30.10}$$

Following the same methodology, the weight coefficients of the second-level indicators relative to their respective first-level indicators can be derived. After obtaining the relative weights, the composite weights are then calculated, which represent the weight coefficients of each indicator with respect to the overall objective. The weighted model for the digital transformation maturity evaluation indicators of automotive parts enterprises is detailed in Table 30.3.

Table 30.3 Weight table of indicators for DTMA in automotive parts enterprises

Serial number	Primary indicator	Weight	Secondary indicator	Relative weights	Composite weights
1	Digital element support	0.2370	Digital planning	0.1865	0.0442
2			Digital organization	0.2677	0.0635
3			Digital talent development	0.3004	0.0712
4			Financial support	0.2454	0.0582
5	Digital foundation	0.2700	Equipment intelligence	0.2389	0.0645
6			Network infrastructure	0.3387	0.0914
7			Information security	0.1128	0.0304
8			Data foundation	0.3097	0.0836
9	Digital application	0.1992	Digital marketing	0.1522	0.0303
10			Digital research and design	0.2301	0.0458
11			Digital decision-making	0.1873	0.0373
12			Digital lean manufacturing	0.1878	0.0374
13			Product quality and traceability	0.1570	0.0313
14			Digital supply chain	0.0856	0.0171
15	Digital outcomes	0.1684	Capacity utilization rate	0.2129	0.0358
16			Green and low-carbon	0.0627	0.0106
17			Product and service innovation	0.0938	0.0158
18			Comprehensive cost per ten thousand yuan of output value	0.2412	0.0406
19			On-Time delivery rate	0.3534	0.0595
20			Digital service output capability	0.0360	0.0061
21	External environment	0.1254	Government policy support	0.4254	0.0533
22			Industry chain impact	0.2652	0.0333
23			Service level of service provider	0.3094	0.0388

30.3.5 Evaluation of Digital Transformation Maturity Using the FCE Method

In this research, we engaged in a survey to assess the digital transformation maturity of Enterprise L. The survey participants included 5 industry experts, 14 mid-to-senior level executives from the company, and 3 IT professionals. A total of 20 valid responses were obtained.

Establishment of the Factor Set and Rating Scale for Fuzzy Comprehensive Evaluation. As presented in Table 30.1, we have established the factor set and rating scale for the fuzzy comprehensive evaluation of the digital transformation of Enterprise L. We have defined a five-tiered fuzzy assessment scale, which includes {Very Strong, Strong, Moderate, Weak, Very Weak}, and assigned numerical values of {5, 4, 3, 2, 1}, respectively.

Development of the Single-Factor Fuzzy Relationship Matrix R. Adhering to the algorithm of the fuzzy comprehensive evaluation method, we have determined the membership degrees of the second-level indicators under the first-level indicators of Enterprise L's digital transformation maturity. The membership degrees for the indicators categorized under "Digital Element Support" are displayed in Table 30.4.

The evaluation matrix for "Digital Element Support" is as follows:

$$R_1 = \begin{bmatrix} 0.35 & 0.6 & 0.05 & 0 & 0 \\ 0.25 & 0.5 & 0.25 & 0 & 0 \\ 0.05 & 0.3 & 0.65 & 0 & 0 \\ 0 & 0.6 & 0.35 & 0.05 & 0 \end{bmatrix} \tag{30.11}$$

Similarly, the fuzzy relationship matrices R2, R3, R4, and R5 for "Digital Infrastructure," "Digital Application," "Digital Outcomes," and "External Environment" can be derived. Due to limited space, no further elaboration will be made.

Fuzzy Comprehensive Evaluation. The weight vectors for each second-level indicator of "Digital Element Support," obtained from the group analytic hierarchy process outlined in Sect. 3.4, are as follows:

$$\mathrm{W}_1 = (0.1865 \quad 0.2677 \quad 0.3004 \quad 0.2454) \tag{30.12}$$

Table 30.4 Evaluation matrix of "digital element support" for DTMA in enterprise L

Secondary indicator	Very strong	Strong	Moderate	Weak	Very weak
Digital planning	0.35	0.60	0.05	0	0
Digital Organization	0.25	0.50	0.25	0	0
Digital talent Development	0.05	0.30	0.65	0	0
Financial support	0	0.60	0.35	0.05	0

The calculations yield the fuzzy comprehensive evaluation vector C and the overall evaluation value A. The evaluation results for "Digital Element Support" are as follows:

$$C_1 = W_1 * R_1 = (0.1865 \quad 0.2677 \quad 0.3004 \quad 0.2454)\begin{bmatrix} 0.35 & 0.6 & 0.05 & 0 & 0 \\ 0.25 & 0.5 & 0.25 & 0 & 0 \\ 0.05 & 0.3 & 0.65 & 0 & 0 \\ 0 & 0.6 & 0.35 & 0.05 & 0 \end{bmatrix}$$

$$= (0.1472 \quad 0.4831 \quad 0.3574 \quad 0.0123 \quad 0) \tag{30.13}$$

$$A_1 = \frac{5 * 1472 + 4 * 0.4831 + 3 * 0.3574 + 2 * 0.0123 + 1 * 0}{0.1472 + 0.4831 + 0.3574 + 0.0123 + 0} = 3.765 \tag{30.14}$$

Similarly, the overall evaluation values for A2, A3, A4, and A5 can be calculated as 4.047, 3.421, 3.649, and 3.851, respectively. Due to limited space, no further elaboration will be made.

Synthesis of Fuzzy Comprehensive Evaluation Results and Calculation of the Composite Score. Using the weight vector W = (0.2370 0.2700 0.1992 0.1684 0.1254) derived from the group analytic hierarchy process in Sect. 3.4, the composite score for Enterprise L's digital transformation maturity is obtained by multiplying the fuzzy comprehensive evaluation values of each first-level indicator by their respective weights.

$$A = 3.765 * 0.2370 + 4.047 * 0.2700 + 3.421 * 0.1992 + 3.649 * 0.1684 + 3.851 * 0.1254 = 3.764 \tag{30.15}$$

Determining the Evaluation Grade Based on the Principle of Maximum Membership Degree. Following the principle of maximum membership degree, the evaluation grade with the highest degree of membership is selected as the overall grade for the subject. Referring to Table 30.5, which presents the membership degrees of the comprehensive evaluation grades, it is demonstrated that Enterprise L's digital transformation maturity is generally at a moderately strong level.

Table 30.5 Fuzzy membership table of comprehensive evaluation grades for DTMA in enterprise L

Overall evaluation level	Degree of membership
Very strong	0.1673
Strong	0.5085
Moderate	0.2601
Weak	0.0490
Very weak	0.0151

30.3.6 Analysis and Recommendations for the Enterprise L

Based on the results of the fuzzy evaluation and the principle of maximum membership degree, Enterprise L's overall score for digital transformation maturity is 3.764. This leads to the conclusion that the company's digital transformation maturity is moderately strong, with a membership degree of 0.5058, corresponding to the Data-Driven Level of digital transformation maturity.

Enterprise L's digital transformation maturity is generally at Level IV—the Data-Driven Level, indicating that the company focuses on the performance of digital transformation. It has achieved interconnectivity and integration of various data resources based on the integrated application of business systems, establishing a "perception-insight-assessment-response" data loop. This has enabled data-driven operational optimization, reduced production costs, enhanced delivery efficiency and quality levels, achieving precision control. Comparatively, the company is willing to invest in "digital infrastructure," possessing an excellent foundation for digital transformation. However, its "digital applications" and "digital outcomes" are not yet prominent, with a suboptimal investment-to-output ratio in digital construction. The value of digital transformation has not been fully realized, as shown in Fig. 30.1.

Digital Element Support. Enterprise L exhibits a substantial capability in digital element support, as shown in Fig. 30.2. This suggests that the company's management has a clear grasp of digital transformation and has laid out a definitive strategy for it, offering support in the form of systems, financial resources, organizational structures, and human capital. The assessment indicates that the company still needs to invest in the development of digital talent, ensuring that they can both attract and retain this talent effectively.

Digital Foundation. Enterprise L has an advanced digital foundation, as illustrated in Fig. 30.3. This signifies that the company prioritizes the establishment of a solid base for digital initiatives, featuring comprehensive network infrastructure, effective information security measures, high-fidelity foundational data, and a significant

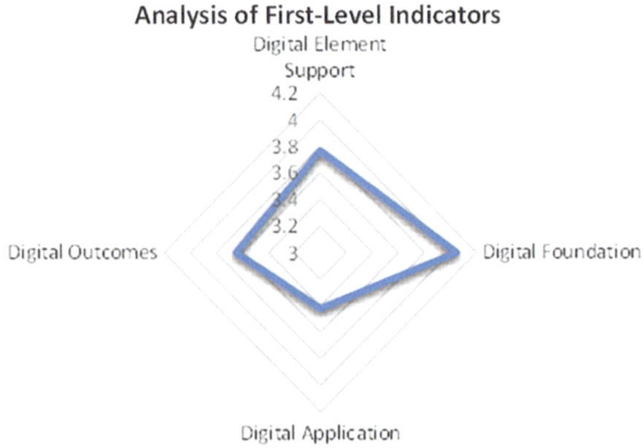

Fig. 30.1 Analysis of first-level indicators for DTMA of enterprise L

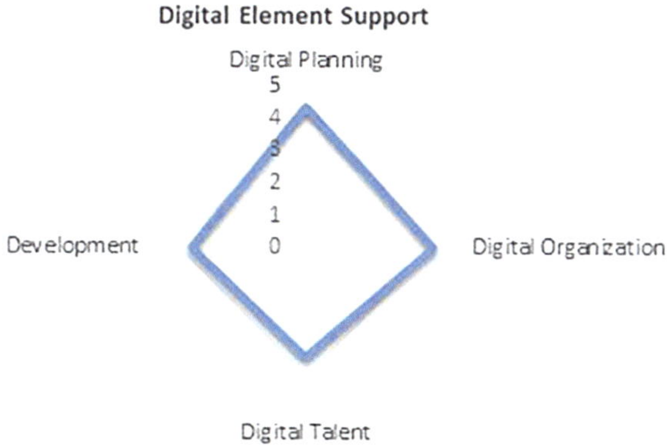

Fig. 30.2 Analysis of secondary indicators under "digital element support" for DTMA in enterprise L

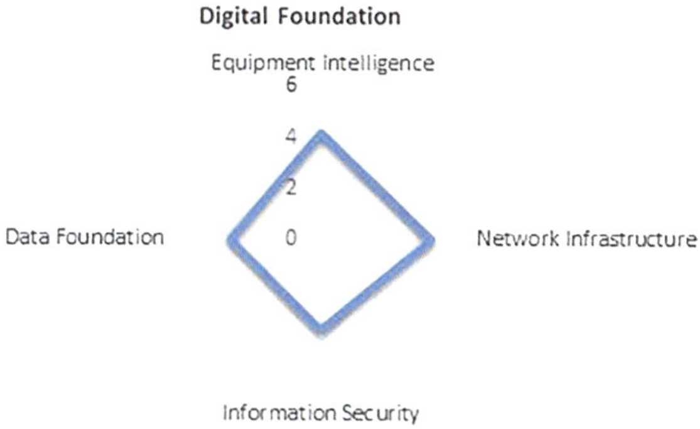

Fig. 30.3 Analysis of secondary indicators under "digital foundation" for DTMA in enterprise L

level of automation and internet connectivity in key operational processes. These elements provide a reliable foundation for the subsequent integration of systems and the enhancement of data asset value.

Digital Application. Enterprise L has robust digital applications, as depicted in Fig. 30.4. This indicates that the company prioritizes demand-driven approaches and has enhanced its overall operational efficiency through the construction of a digital design platform and the integrated application of business systems such as ERP, MES, and OA. By implementing digital lean practices and data-driven strategies, the company has improved its capabilities in research and development, manufacturing, and management decision-making, ensuring it can swiftly respond to market demand changes. However, the evaluation reveals that there is room for improvement in the company's product quality management and that it needs to identify breakthroughs in digital marketing.

Digital Outcomes. Enterprise L exhibits strong digital outcomes, as shown in Fig. 30.5. This demonstrates that the company pays close attention to the effectiveness of its digital transformation, aligning its digital construction efforts with the core needs of the company's digital transition, achieving commendable results

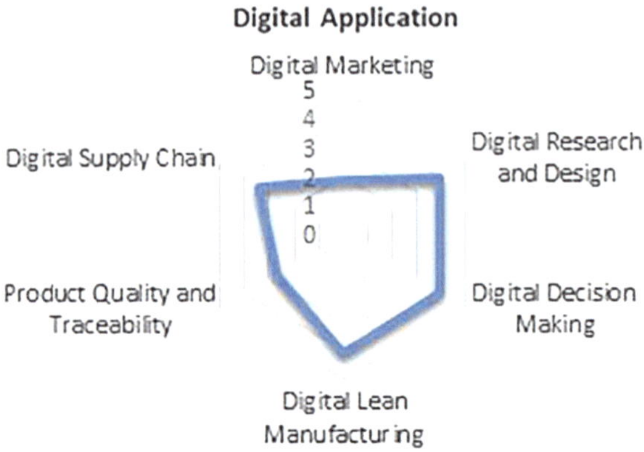

Fig. 30.4 Analysis of secondary indicators under "digital application" for DTMA in enterprise L

in delivery and cost efficiency. The company also places a strong emphasis on societal efficiency and has begun to develop its capabilities for digital service delivery. However, the evaluation suggests that the company needs to give more thought to innovation in products and services, as well as to green manufacturing practices.

External Environment. Enterprise L has a robust external environment, as illustrated in Fig. 30.6. This indicates that the company has a strong capacity to secure government policy support and maintains high standards for the service levels provided by its vendors, requiring them to offer tailored digital transformation solutions that address the specific characteristics of the automotive parts industry. Although Enterprise L is a components manufacturer, it has enhanced its position within the industry chain through digital transformation, thus being relatively less susceptible to the influences of the industry chain.

Fig. 30.5 Analysis of secondary indicators under "digital outcomes" for DTMA in enterprise L

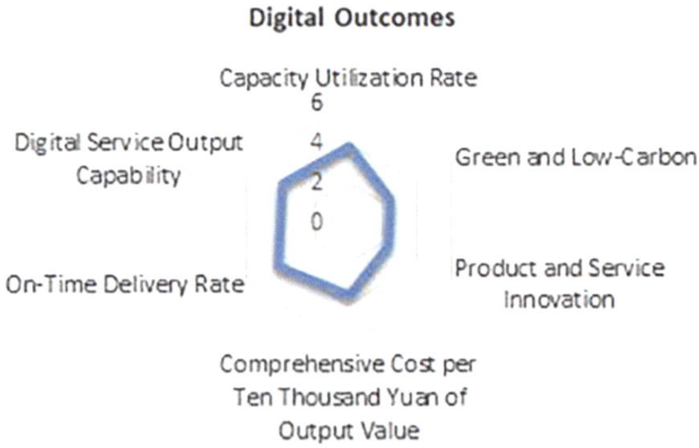

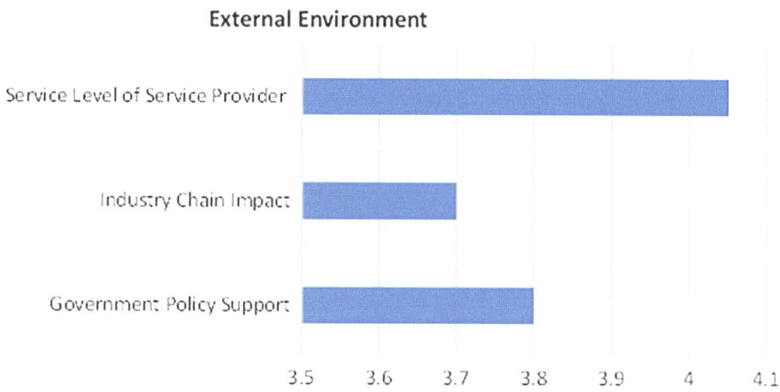

Fig. 30.6 Analysis of secondary indicators under "external environment" for DTMA in enterprise L

30.4 Conclusion

As the digital economy flourishes, the digital transformation of manufacturing companies has emerged as a prominent topic of academic interest worldwide. This study focuses on automotive parts enterprises, developing an industry-specific evaluation system for digital transformation. We invited numerous industry experts to assess the importance of the indicators using a 9-point scale method. Subsequently, we employed the GAHP method to scientifically calculate the weights of each indicator and then utilized the FCE method for a comprehensive evaluation of the digital maturity levels of the enterprises. To validate the feasibility and applicability of the DTMA model established in this research, we conducted an empirical analysis with Enterprise L as a case study. The results indicate that the evaluation model proposed in this study can provide a comprehensive analysis of the digital transformation status of automotive parts companies. Compared to previous qualitative assessments, our evaluation outcomes are more precise and scientifically objective. The findings of this research have been integrated into the diagnostic analysis of "smart transformation and digital conversion" for automotive parts enterprises, with maturity evaluations performed for over 60 companies. The general feedback from these enterprises is that the research model is user-friendly, the evaluations are accurate, and the planned routes are rational. Thus, this study offers valuable insights and a reference for the digital maturity assessment of automotive parts enterprises and traditional manufacturing enterprises alike.

Acknowledgements This work was partially supported by the Jiangsu Provincial Department of Education's 2024 Social Science Research Project for Jiangsu Colleges and Universities, titled "Research on the Cultivation Path of Jiangsu Intelligent Factories Based on Lighthouse Factory Evaluation Standards" (Project Number: 2024SJYB0953).

References

1. Gökalp, E., Martinez, V.: Digital transformation capability maturity model enabling the assessment of industrial manufacturers. Comput. Ind. **132**, 103522 (2021)
2. Gimpel, Henner, et al.: Structuring digital transformation: a framework of action fields and its application at ZEISS. J. Inf. Technol. Theory Appl. (JITTA) **19.1** (2018): 3
3. Bumann, Jimmy, Marc Peter.: Action fields of digital transformation–a review and comparative analysis of digital transformation maturity models and frameworks. "Digitalisierung und andere Innovationsformen im Management 2. 13–40 (2019)
4. Teichert, Roman.: Digital transformation maturity: A systematic review of literature. Acta universitatis agriculturae et silviculturae mendelianae brunensis (2019)
5. Spaltini, Marco, et al.: Defining the roadmap towards industry 4.0: the 6Ps maturity model for manufacturing SMEs. Procedia CIRP 105: 631–636 (2022)
6. Pan, J., Lin, J., Wang, S.: A Delphi-based index system for digital transformation capability of retailers. J. Bus. Econ. Manag. **23**(5), 1106–1132 (2022)
7. Saaty, T.L.: How to make a decision: the analytic hierarchy process. Eur. J. Oper. Res. **48**(1), 9–26 (1990)
8. Saaty, T.L.: Transport planning with multiple criteria: The analytic hierarchy process applications and progress review. J. Adv. Transp. **29**(1), 81–126 (1995)
9. Saaty, Thomas L.: Relative measurement and its generalization in decision making why pairwise comparisons are central in mathematics for the measurement of intangible factors the analytic hierarchy/network process. RACSAM-Revista de la Real Academia de Ciencias Exactas, Fisicas y Naturales. Serie A. Matematicas **102**: 251–318 (2008)
10. Mohsin, M., et al.: Assessing oil supply security of South Asia. Energy **155**, 438–447 (2018)
11. Zadeh, Lotfi Asker, George J. Klir, Bo Yuan.: Fuzzy sets, fuzzy logic, and fuzzy systems: selected papers. World scientific, **6** (1996)
12. Chen, Qingmei, et al.: Digital transformation evaluation for small-and medium-sized manufacturing enterprises using the fuzzy synthetic method DEMATEL-ANP. Sustainability 14.20 13038 (2022)
13. Bernasconi, M., Choirat, C., Seri, R.: The analytic hierarchy process and the theory of measurement. Manage. Sci. **56**(4), 699–711 (2010)

Chapter 31
Neural Network-Based Forecasting of IoT Connection Growth: A Comprehensive Study

Phuc-Hau Nguyen, Trong-Trieu Hoang, Thi-Thu Nguyen, Truong-Giang Ngo, and Thi-Kien Dao

Abstract As IoT networks expand alongside advancements in 5G, accurately forecasting IoT connections becomes essential for network operators aiming to optimize infrastructure, enhance user experience, and maximize customer retention. Traditional predictive models, often limited to single-factor or historical trend analysis, struggle to capture the complex, multi-faceted influences on IoT growth, leading to suboptimal accuracy. This paper introduces a novel approach that integrates macroeconomic factors—such as GDP and digital economic indicators—into a neural network-based prediction model. By employing a backpropagation neural network (BPNN) and incorporating these macroeconomic variables as a μ-factor, the proposed model effectively captures the broader economic environment's impact on IoT connectivity trends. Experimental results demonstrate that this hybrid model surpasses conventional techniques, providing a more nuanced and accurate forecast for IoT connections. This approach not only enhances predictive reliability but also offers a scalable framework adaptable to evolving macroeconomic conditions, paving the way for more resilient IoT infrastructure planning.

P.-H. Nguyen
Faculty of Information Technology, Electric Power University, Ha Noi, Vietnam
e-mail: haunp@epu.edu.vn

T.-T. Hoang · T.-G. Ngo
Department of Academic Affairs, Thuyloi University, Ha Noi, Vietnam
e-mail: trieuht@tlu.edu.vn

T.-G. Ngo
e-mail: giangnt@tlu.edu.vn

T.-T. Nguyen
Faculty of Information Technology, Sao Do University, Hải Dương, Vietnam

T.-K. Dao (✉)
University of Information Technology, VNU-Ho Chi Minh, Vietnam
e-mail: kiendt@uit.edu.vn

S.-C. Chu et al. (eds.), *Advances in Intelligent Data Analysis and Applications*, Smart Innovation, Systems and Technologies 445,
https://doi.org/10.1007/978-981-96-7277-6_31

31.1 Introduction

The commercialization of 5G technology has ushered in an explosive growth period for the Internet of Things (IoT) [1], leading to a significant increase in the number of connections [2]. This rapid expansion presents specific challenges for communication operators, particularly in terms of network construction and investment planning [3]. Accurately forecasting the number of IoT connections is essential for addressing these challenges, as it can help optimize resource allocation and enhance user experience [4]. Traditional forecasting methods often rely on linear models that may not adequately capture the complex [5], non-linear relationships inherent in IoT data [6]. As such, there is a pressing need to develop more sophisticated predictive models that can account for various influencing factors, including macroeconomic conditions [7]. This paper proposes a novel approach that integrates quantitative macroeconomic factors with neural network algorithms [8] to improve the accuracy of IoT connection forecasts [9]. The contributions of this paper are threefold:

- First, it provides a comprehensive analysis of the effectiveness of neural networks in forecasting IoT connection growth, demonstrating their superiority over traditional statistical methods.
- Second, it highlights the importance of incorporating macroeconomic indicators, such as GDP growth and technological advancements, into the forecasting model, thereby offering a more holistic view of the factors influencing IoT expansion.
- Finally, this study presents a practical framework for communication operators to utilize in their planning and decision-making processes, ultimately aiding in the efficient deployment of IoT infrastructure.

31.2 Literature Review

The rapid development of the Internet of Things (IoT) has garnered significant attention in both academic and industrial circles [3]. As IoT connections proliferate, accurate forecasting of their growth becomes increasingly critical for communication operators and infrastructure planners [10]. This section reviews existing literature on IoT forecasting methods, highlighting the evolution of predictive techniques and the integration of macroeconomic factors.

Traditional Forecasting Methods: Historically, forecasting methods for IoT connections have relied on traditional statistical techniques, such as time series analysis and causal modeling. These methods often utilize historical data to identify trends and project future growth. Common approaches include polynomial regression, moving averages, and exponential smoothing [11]. However, these techniques frequently struggle to capture the complex, non-linear relationships inherent in IoT data, leading to limitations in accuracy and reliability.

Machine Learning Approaches: In recent years, machine learning techniques have emerged as powerful alternatives to traditional methods. Researchers have begun to explore the application of various algorithms, including decision trees, support vector machines, and neural networks, to improve forecasting accuracy. Among these, neural networks, particularly BPNN [12], have shown promise due to their ability to model complex patterns and interactions within data [13]. Studies have demonstrated that neural networks can outperform traditional methods in forecasting tasks by effectively learning from historical data and adapting to new information. For instance, the integration of BPNN in forecasting models has been shown to enhance prediction accuracy significantly, making it a preferred choice for many researchers.

Incorporating Macroeconomic Factors: While machine learning techniques have improved forecasting capabilities, the integration of macroeconomic factors remains a relatively underexplored area. Factors such as GDP growth, technological advancements, and regulatory changes can significantly influence IoT connection growth. Recent studies have begun to address this gap by incorporating quantitative macroeconomic indicators into forecasting models [14]. For example, research has indicated that macroeconomic conditions can directly impact the adoption and deployment of IoT technologies. By quantifying these factors and integrating them into neural network models, researchers aim to create more robust forecasting frameworks that account for the broader economic environment.

Summary of Gaps and Opportunities: Despite the advancements in forecasting methodologies, several gaps remain in the literature. Many existing studies focus on isolated aspects of IoT growth without considering the interplay between macroeconomic factors and technological developments [15]. Additionally, there is a need for comprehensive frameworks that combine machine learning techniques with macroeconomic analysis to enhance predictive accuracy [16].

This paper aims to address these gaps by proposing a novel forecasting model that integrates macroeconomic factors with neural network algorithms, thereby providing a more holistic approach to predicting IoT connection growth.

31.3 Methodology of Forecasting of IoT Connection

This section outlines the methodology employed in this study to forecast IoT connection growth using a neural network model that integrates macroeconomic factors. The approach consists of data collection, model design, and evaluation metrics, ensuring a comprehensive framework for accurate predictions.

31.3.1 Data Collection

Data for this study was collected from various sources, including government statistics, industry reports, and academic publications [17]. The key variables included: *IoT Connection Data*: Historical data on the number of IoT connections, sourced from communication operators and industry reports. *Macroeconomic Indicators:* Quantitative macroeconomic factors such as GDP growth, digital economy metrics, and other relevant indicators that influence IoT growth. These indicators were selected based on their relevance to the IoT ecosystem and their availability in public databases [18]. The data spans several years to ensure a robust dataset for training and testing the neural network model. Description of the Dataset: Year: The year of the data point. IoT Connections (millions): The number of IoT connections in millions. GDP (billion CNY): The Gross Domestic Product in billion CNY. Unemployment Rate (%): The percentage of the labor force that is unemployed. Inflation Rate (%): The percentage increase in the price level of goods and services [19]. Population (millions): The total population in millions. The resulting model not only improves predictive precision but also adapts to the complex, dynamic nature of IoT growth, offering a scalable framework for forecasting in the rapidly evolving IoT landscape. Quantifying the macro-environmental influence on IoT is challenging due to the need to encapsulate diverse factors such as GDP and digital economic growth. Here, we introduce the macro-impact factor (μ), representing a synthesis of economic indicators directly relevant to IoT connectivity. This μ-factor is designed to encapsulate influential factors from the broader economy on IoT adoption and growth.

31.3.2 Neural Network Model Design

The core of the methodology is the design of a Backpropagation Neural Network (BPNN) that incorporates macroeconomic factors into its architecture. The model design includes the following steps:

Model Structure: The BPNN consists of an input layer, one or more hidden layers, and an output layer. The input layer receives historical data on IoT connections and macroeconomic indicators. The hidden layers are designed to capture complex relationships within the data, while the output layer provides the forecasted number of IoT connections [20]. Figure 31.1 illustrates a simple neural network architecture, showcasing the fundamental components that constitute a basic model. This architecture typically includes an input layer, one or more hidden layers, and an output layer.

Input layer is used for the input data which includes historical IoT connection numbers and selected macroeconomic indicators. Hidden layer is stated for the number of neurons in the hidden layers which is determined based on empirical formulas, ensuring sufficient capacity to model the underlying relationships. Output layer is presented for the output layer which produces the predicted number of IoT connections.

Activation Functions: The model employs activation functions such as the sigmoid function and the hyperbolic tangent (tanh) function for the hidden layers. These functions introduce non-linearity into the model, allowing it to learn complex patterns in the data.

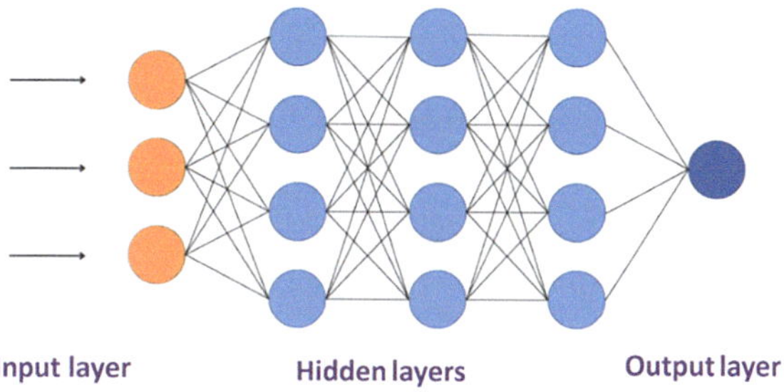

Fig. 31.1 A simple neural network architecture

Training the Model: The BPNN is trained using historical data through a process of backpropagation. The training involves adjusting the weights and biases of the network to minimize the error between the predicted and actual IoT connection numbers. The training process includes the following:

- Loss Function: Mean Squared Error (MSE) is used as the loss function to quantify the difference between predicted and actual values.
- Optimization Algorithm: An optimization algorithm, such as Adam or Stochastic Gradient Descent (SGD), is employed to update the weights during training.

Network processing is used for a single index. Assumed, $l_n(t_n)$, the network method is used to displace the network, which becomes a four-input one-output mode. Then the number of errors, $N = \{1, 2, 3, 4\}$. An empirical formula already available in academia to determine: $Hneural = \sqrt{n+m} + a$ (where n is the number of neurons in the input layer, m is the number of neurons in the output layer, and a is a constant between [1, 10]). Activation function determination of BP neural networks usually use Sigmoid differentiable functions and linear functions as the activation function of the network. In this paper, the S-type tangent function tansig is selected as the activation function of the hidden neurons. The model is constructed as activation function with a general model which is expressed as follows:

$$f(t) = \mu * g(t), \tag{31.1}$$

where $f(t)$ represents the overall prediction function; μ is the macro-environmental influence factor; and g(t) is derived through the BPNN, capturing the refined historical relationship between IoT connections and macroeconomic indicators. With $g(t)$ is set to BPNN(t), the modified model of a variety of neural networks can be obtained:

$$f(t) = \left(1 + \frac{\partial ximax}{\partial t}\right) * BPNN(t), \tag{31.2}$$

where *BPNN(t)* is a backpropagation neural network through training with sample data, the network weights and thresholds are continuously modified to make the error function decrease in the direction of the negative gradient, approaching the expected output, whether number of connections at time t_n of historical data is $l_n(t_n)$, the algorithm for inputting the network according to the BPNN algorithm. By establishing a more robust mapping function with the BPNN, the model improves predictive accuracy, making it possible to project IoT growth in ways that are responsive to evolving economic and technological conditions.

31.3.3 Incorporating Macroeconomic Factors

To enhance the predictive accuracy of the model, macroeconomic factors are integrated into the neural network [21]. This integration is achieved through a two-step process: Quantitative Analysis of Macroeconomic Factors: The study employs Grey Relational Analysis (GRA) to determine the correlation between IoT connections and macroeconomic indicators, such as GDP and digital economy metrics. This analysis identifies the most influential factors affecting IoT growth. Model Fusion: The final prediction model is expressed in Eq. (31.1) as $f(t) = \mu \cdot g(t)$, where μ represents the macroeconomic influence factor, and $g(t)$ is the output of the BPNN. This fusion allows the model to account for both historical trends and macroeconomic influences. Evaluation Metrics: The performance of the neural network model is evaluated using several metrics:

- Mean Absolute Error (MAE): Measures the average magnitude of errors in predictions, providing insight into the model's accuracy.
- Root Mean Squared Error (RMSE): Offers a measure of the differences between predicted and observed values, emphasizing larger errors.
- R-squared (R^2): Indicates the proportion of variance in the dependent variable that can be explained by the independent variables in the model.

These metrics are crucial for assessing the model's effectiveness and ensuring its reliability for forecasting IoT connection growth. In this comprehensive reviews, the study aims to provide a robust framework for predicting IoT connection growth, integrating both historical data and macroeconomic influences to enhance forecasting accuracy.

31.4 Experimental Results

This section presents the results of the forecasting model developed in this study, focusing on the performance of the neural network integrated with macroeconomic factors. The analysis includes simulation results, comparisons with traditional forecasting methods, and an evaluation of the model's accuracy.

Table 31.1 The historical data for IoT connections

Time (months)	Connections (10,000s)	Time (months)	Connections (10,000s)
January 2024	170,739	July 2024	302,220
February 2024	180,505	August 2024	326,637
March 2024	207,813	September 2024	362,194
April 2024	228,963	October 2024	378,972
May 2024	252,726	November 2024	391,169
June 2024	288,785	December 2024	416,053

Table 31.2 Performance of the proposed method with actual connection numbers

Time (months)	Predicted connections	Actual connections	Error (%)
January 2025	40,984	44,500	−8.58
February 2025	42,793	45,284	−5.82
March 2025	44,561	48,094	−7.93
April 2025	46,288	49,319	−6.55
May 2025	47,973	50,668	−5.62
June 2025	49,617	53,320	−7.46

31.4.1 Simulation Analysis

The forecasting model was applied to historical data from a communication operator, specifically the number of IoT connections from January 2023 to December 2025. The objective was to predict the growth of IoT connections by the end of June 2024. The model utilized MATLAB for implementation, allowing for a detailed analysis of the results. Historical Data Overview: The historical data for IoT connections is summarized in Table 31.1, which shows a consistent increase in connections over the year.

Forecasting Results: The model's predictions for the first 6 months of 2019 were compared against actual connection numbers, as shown in Table 31.2. The results indicate that the proposed model (μ-BPNN) outperformed traditional polynomial trend extrapolation methods.

31.4.2 Performance Comparison

The μ-BPNN model was compared to traditional forecasting methods, such as polynomial regression. The results demonstrated that the μ-BPNN model provided more accurate predictions, particularly in capturing the growth trends influenced by macroeconomic factors [22]. Table 31.3 summarizes the performance comparison between the traditional polynomial trend extrapolation method and the proposed μ-BPNN model for forecasting IoT connections from January to June 2025. The error percentages indicate the accuracy of each model's predictions against the actual connection numbers.

The accuracy of the μ-BPNN model was evaluated using Mean Absolute Error (MAE) and Root Mean Squared Error (RMSE). The results indicated that the μ-BPNN model achieved lower error rates compared to traditional methods, confirming its effectiveness in forecasting IoT connection growth (Table 31.4).

Table 31.3 Performance comparison between the traditional polynomial trend extrapolation method and the proposed μ-BPNN model for forecasting IoT connections

Model type	Time (months)	Predicted connections	Actual connections	Error (%)
Polynomial trend	January 2025	40,984	44,500	−8.58
Polynomial trend	February 2025	42,793	45,284	−5.82
Polynomial trend	March 2025	44,561	48,094	−7.93
Polynomial trend	April 2025	46,288	49,319	−6.55
Polynomial trend	May 2025	47,973	50,668	−5.62
Polynomial trend	June 2025	49,617	53,320	−7.46
μ-BPNN model	January 2025	44,812	44,500	0.70
μ-BPNN model	February 2025	45,130	45,284	−0.34
μ-BPNN model	March 2025	47,305	48,094	−1.67
μ-BPNN model	April 2025	52,278	49,319	5.66
μ-BPNN model	May 2025	54,361	50,668	6.79
μ-BPNN model	June 2025	58,092	53,320	8.21

Table 31.4 Evaluation comparison of the performance metrics of the μ-BPNN model with traditional single-factor forecasting models for year of 2025

Model type	Mean absolute percentage error (MAPE) (%)	Root mean square error (RMSE)	Prediction accuracy (%)	Notes
Traditional model 1	17.2	1105 connections	82.8	Based on historical trend data only
Traditional model 2	20.5	1403 connections	79.5	Causal analysis with limited factors
μ-BPNN model	12.8	1789 connections	87.2	Incorporates macroeconomic factors and trends

31.4.3 Discussion of Findings

The findings suggest that incorporating macroeconomic factors into the neural network model significantly enhances forecasting accuracy. The μ-BPNN model not only improves prediction accuracy but also offers valuable insights for communication operators in their network planning and investment strategies. In conclusion, the simulation results validate the proposed model's effectiveness in forecasting IoT connection growth, demonstrating its potential for practical applications in the telecommunications industry. The results of this study highlight the effectiveness of integrating macroeconomic factors with neural network algorithms for forecasting IoT connection growth [23]. The proposed μ-BPNN model demonstrated superior predictive accuracy compared to traditional forecasting methods, particularly in capturing the complex dynamics of IoT expansion influenced by external economic conditions.

Implications of Findings: The findings underscore the importance of considering macroeconomic indicators, such as GDP and digital economy metrics, in forecasting models. The integration of these factors allows for a more nuanced understanding of the forces driving IoT growth. As the IoT landscape continues to evolve, communication operators can leverage these insights to make informed decisions regarding network infrastructure and resource allocation.

Model Performance: While the μ-BPNN model showed promising results, it is essential to acknowledge its limitations. The model's accuracy tends to decrease with longer forecasting horizons, indicating that it is more suited for short-term predictions. This behavior aligns with the nature of IoT growth, which can be influenced by rapid technological advancements and changing market conditions [24]. Therefore, operators should consider using this model primarily for short-term planning while complementing it with other forecasting techniques for long-term strategies.

Future Research Directions: Future research could explore the application of more advanced neural network architectures, such as recurrent neural networks (RNNs) or long short-term memory (LSTM) networks, which are designed to handle sequential data and may improve forecasting accuracy over extended periods. Additionally, incorporating real-time data feeds and machine learning techniques could enhance the model's adaptability to sudden market changes. Moreover, further studies could investigate the impact of other macroeconomic factors, such as technological innovation rates and regulatory changes, on IoT connection growth. By expanding the scope of macroeconomic influences, researchers can develop even more robust forecasting models that account for a broader range of variables affecting the IoT ecosystem.

Practical Applications: The practical implications of this research are significant for communication operators and stakeholders in the IoT industry. By utilizing the μ-BPNN model, operators can enhance their network planning and investment strategies, ultimately leading to improved service delivery and customer satisfaction. The model provides a valuable tool for anticipating growth trends and aligning infrastructure development with market demands.

31.5 Conclusion

This study introduced a novel approach to forecasting IoT connection growth by integrating macroeconomic factors with the Backpropagation Neural Network (BPNN). The rapid expansion of the Internet of Things, especially with the advent of 5G technology, necessitates accurate forecasting for effective network planning and investment. The proposed μ-BPNN model leverages quantitative macroeconomic indicators, such as GDP and digital economy metrics, alongside historical IoT connection data, enhancing predictive accuracy through Grey Relational Analysis (GRA). Results indicate that the μ-BPNN model surpasses traditional methods, particularly for short-term predictions, offering valuable insights for telecommunications stakeholders. However, limitations in long-term forecasting accuracy suggest that future research should explore advanced neural network architectures, like recurrent neural networks (RNNs) or long short-term memory (LSTM) networks, and incorporate real-time data. Overall, this research contributes to IoT forecasting by demonstrating the effectiveness of combining neural networks with macroeconomic analysis, emphasizing the critical role of accurate forecasting in optimizing IoT infrastructure and improving service delivery.

References

1. Shu, L., Mukherjee, M., Pecht, M., Crespi, N., Han, S.N.: Challenges and research issues of data management in IoT for large-scale petrochemical plants. IEEE Syst. J. **12**, 2509–2523 (2017)
2. Farooq, U., M., Waseem, M., Mazhar, S., Khairi, A., Kamal, T.: A Review on Internet of Things (IoT). Int. J. Comput. Appl.. 113, 1–7 (2015). https://doi.org/10.5120/19787-1571
3. Lu, Y., Xu, L.D.: Internet of things (IoT) cybersecurity research: a review of current research topics. IEEE Internet Things J. **6**, 2103–2115 (2019). https://doi.org/10.1109/JIOT.2018.2869847
4. Aqeel, M., Ali, F., Iqbal, M.W., Rana, T.A., Arif, M., Auwul, M.R.: A review of security and privacy concerns in the internet of things (IoT). J. Sens. (2022)
5. Nguyen, T.-T., Qiao, Y., Pan, J.-S., Chu, S.-C., Chang, K.-C., Xue, X., Dao, T.-K.: A hybridized parallel bats algorithm for combinatorial problem of traveling salesman. J. Intell. & Fuzzy Syst. **38**, 5811–5820 (2020). https://doi.org/10.3233/jifs-179668
6. Chu, S.C., Dao, T.K., Pan, J.S., Nguyen, T.T.: Identifying correctness data scheme for aggregating data in cluster heads of wireless sensor network based on naive Bayes classification. Eurasip J. Wirel. Commun. Netw. (2020). https://doi.org/10.1186/s13638-020-01671-y
7. Chu, S.-C., Nguyen, T.-T., Nguyen, V.-T., Nguyen, T.-D., Dao, T.-K.: An optimal power system multi-objective environmental-economic planning based on moth-flame optimization BT—Advances in intelligent information hiding and multimedia signal processing. Presented at the (2023)
8. Dao, T.K., Chu, S.C., Nguyen, T.T., Nguyen, T.D., Nguyen, V.T.: An optimal WSN node coverage based on enhanced archimedes optimization algorithm. Entropy. **24**, (2022). https://doi.org/10.3390/e24081018
9. Rashid, M., Haque, M.M., Wang, W.: IoT Complexity: security, vulnerabilities and risks. Eur. J. Electr. Eng. Comput. Sci. **8**, 1–9 (2024)

10. Nebbione, G., Calzarossa, M.C.: Security of IoT application layer protocols: Challenges and findings. Future Internet. **12**, 55 (2020)
11. Sharma, P.K., Kumar, N., Park, J.H.: Blockchain technology toward green IoT: Opportunities and challenges. IEEE Netw.Netw. **34**, 263–269 (2020)
12. Dreiseitl, S., Ohno-Machado, L.: Logistic regression and artificial neural network classification models: a methodology review. J. Biomed. Inform. **35**, 352–359 (2002)
13. Lu, C.-N., Wu, H.-T., Vemuri, S.: Neural network based short term load forecasting. IEEE Trans. Power Syst. **8**, 336–342 (1993)
14. Pan, J.-S., Wang, G.-L., Chu, S.-C., Yang, D., Snášel, V.: New feature attribution method for explainable aspect-based sentiment classification. Knowl.-Based Syst. **304**, 112550 (2024). https://doi.org/10.1016/j.knosys.2024.112550
15. Nguyen, T.T., Pan, J.S., Dao, T.K.: An improved flower pollination algorithm for optimizing layouts of nodes in wireless sensor network. IEEE Access. **7**, 75985–75998 (2019). https://doi.org/10.1109/ACCESS.2019.2921721
16. Pan, J.-S., Yang, Q., Chu, S.-C., Chao, H.-C.: A joint edge server and service deployment method in C-RAN with multi-layer MEC for multi-communities. IEEE Internet Things J. **1** (2024). https://doi.org/10.1109/JIOT.2024.3471817
17. Al-Hawawreh, M., Sitnikova, E., Aboutorab, N.: X-IIoTID: A connectivity-agnostic and device-agnostic intrusion data set for industrial Internet of Things. IEEE Internet Things J. **9**, 3962–3977 (2021)
18. Perkins, D.H., Rawski, T.G.: Forecasting China's economic growth to **2025**, (2008)
19. Dachyar, M., Zagloel, T.Y.M., Saragih, L.R.: Knowledge growth and development: internet of things (IoT) research, 2006–2018. Heliyon. **5**, (2019)
20. Auer, P., Burgsteiner, H., Maass, W.: A learning rule for very simple universal approximators consisting of a single layer of perceptrons. Neural Netw. **21**, 786–795 (2008). https://doi.org/10.1016/j.neunet.2007.12.036
21. Dao, T., Yu, J., Nguyen, T., Ngo, T.: A Hybrid improved MVO and FNN for identifying collected data failure in cluster heads in WSN. IEEE Access. **8**, 124311–124322 (2020). https://doi.org/10.1109/ACCESS.2020.3005247
22. McDermott, C.D., Majdani, F., Petrovski, A. V: Botnet detection in the internet of things using deep learning approaches. In: 2018 International Joint Conference on Neural Networks (IJCNN). pp. 1–8. IEEE (2018)
23. Yu, J., Nguyen, T.T., Dao, T.-K., Nguyen, T.-T.: No Title. J. Netw. Intell. **9**, 643–657 (2024)
24. Pham, D.-T., Hoang, D.-T.-T., Nguyen, T.-T., Dao, T.-K., Nguyen, T.-X.-H.: A hybridized network analysis and community detection for unraveling disease spreading Covid-19 pandemic mechanisms. J. Netw. Intell. **9**, 1–13 (2024)

Chapter 32
A Novel Approach to Boundary Stake Management: Combining IoT, Cloud Computing, and Dual-Mode Positioning

Li Ting, Thi-Kien Dao, Shuncai Liu, Xiangping Huang, and Trong-The Nguyen

Abstract This paper explores challenges in boundary stake management, focusing on supervision difficulties and positioning inaccuracies. We propose a smart boundary stake system utilizing IoT and cloud computing technologies. The system features a dual-mode positioning approach, combining BeiDou satellite and GPS for precise location data. Sensors collect information on the stakes' shape, temperature, and humidity, with narrowband IoT technology transmitting this data to a cloud platform for remote monitoring and alarm notifications. Experimental results show that the dual-mode system enhances accuracy and stability compared to GPS alone, while maintaining low power consumption, effectively supporting boundary management.

32.1 Introduction

Boundary stakes are commonly found in natural reserves, water conservancy projects, agricultural protection zones, and along highways [1]. These stakes, often made of concrete, face challenges such as human destruction and difficulty in detection, compounded by the extensive areas they cover, making supervision and management arduous [2]. The emergence of smart boundary stakes aims to address these issues

---

L. Ting · T.-K. Dao · S. Liu · X. Huang · T.-T. Nguyen (✉)
School of Electronic Engineering, Fuzhou Institute of Technology, Fuzhou, China
e-mail: 1100405110@nkust.edu.tw

L. Ting
e-mail: scis@fit.edu.cn

T.-K. Dao
e-mail: 1101405123@nkust.edu.tw

S. Liu
e-mail: liusc@gmiot.com

X. Huang
e-mail: hxp0816@163.com

S.-C. Chu et al. (eds.), *Advances in Intelligent Data Analysis and Applications*,
Smart Innovation, Systems and Technologies 445,
https://doi.org/10.1007/978-981-96-7277-6_32

by integrating intelligent electronic devices with technologies such as the Internet of Things (IoT), cloud computing, artificial intelligence, and big data, facilitating remote monitoring and management of boundary markers [3]. Existing literature highlights various applications of smart boundary stakes, particularly in water conservancy, where systems are designed to monitor positioning, tilt, and vibrations, as well as to collect and transmit information about the stakes and their surrounding environments for remote visualization and control [4]. However, many studies lack technical substance and practical validation, often presenting overly complex theoretical frameworks without concrete implementation details [5].

To overcome the limitations identified in previous research, this paper presents a comprehensive study on the design and implementation of a smart boundary stake system based on IoT cloud computing technology. The proposed system utilizes a dual-mode positioning approach that combines BeiDou satellite [6] and Global Positioning System (GPS) [7] technologies to achieve precise positioning of boundary markers. Additionally, various sensor terminals are employed to monitor parameters such as the shape, temperature, and humidity of the boundary stakes [8]. The collected data is transmitted via narrowband IoT technology to a cloud computing platform, enabling real-time alerts and remote monitoring of any anomalies detected in the boundary markers [9]. This research aims to enhance the accuracy and efficiency of boundary stake management, providing a robust solution for stakeholders involved in environmental monitoring and land management. Through experimental validation, we demonstrate the effectiveness of the dual-mode positioning system and the communication capabilities of the narrowband IoT, ultimately contributing to improved practices in boundary management.

32.2 Literature Review

The management of boundary stakes has garnered attention in various fields, particularly in environmental monitoring, land management, and infrastructure development [1]. Traditional methods of boundary stake management often face challenges such as human interference, difficulty in detection, and inaccuracies in positioning [10]. This literature review explores existing research on smart boundary stake systems, focusing on the integration of IoT, cloud computing, and positioning technologies [11]. Several studies have highlighted the importance of accurate positioning in boundary management. For instance, research has shown that the combination of multiple satellite systems, such as the BeiDou satellite and Global Positioning System (GPS), can significantly enhance positioning accuracy. The dual-mode positioning approach allows for better performance in environments where obstructions, such as trees and buildings, may hinder satellite signals. This is supported by findings that indicate a reduction in Position Dilution of Precision (PDOP) values when utilizing dual-mode systems, leading to improved stability and accuracy in boundary stake positioning. In the context of IoT applications, various studies have demonstrated the potential of smart boundary stakes equipped with sensors for real-time monitoring.

These systems can collect data on environmental conditions, such as temperature and humidity, as well as physical parameters like tilt and vibration. For example, literature has discussed the use of attitude sensors to detect deformations and vibrations in boundary stakes, providing critical information for maintenance and management [12]. The integration of these sensors with IoT technology enables remote monitoring and data transmission to cloud platforms, facilitating timely alerts and decision-making.

Cloud computing plays a crucial role in the management of data collected from smart boundary stakes [13]. It allows for the storage, processing, and analysis of large volumes of data, enabling stakeholders to access real-time information and insights. Previous research has emphasized the advantages of cloud-based solutions in enhancing the efficiency of data management and improving the overall effectiveness of boundary monitoring systems. Despite the advancements in smart boundary stake technologies, existing literature often lacks practical validation and technical depth [14]. Many studies present theoretical frameworks without concrete implementation details or experimental results. This gap highlights the need for further research that not only explores the theoretical aspects of smart boundary stake systems but also provides empirical evidence of their effectiveness in real-world applications [15]. In summary, the literature indicates a growing interest in the integration of IoT, cloud computing, and dual-mode positioning technologies for boundary stake management. While significant progress has been made, there remains a need for comprehensive studies that validate these technologies through practical implementation and testing, ultimately contributing to more effective boundary management practices.

32.3 Methodology

This section outlines the methodology for implementing the proposed smart boundary stake system, detailing the system architecture, data collection methods, and implementation steps.

32.3.1 Description of the Proposed System Architecture

The proposed smart boundary stake system integrates IoT sensors, cloud computing, and a dual-mode positioning system to enhance boundary management through real-time monitoring and efficient data management. Figure 32.1 illustrates a smart boundary stake system that integrates IoT sensors to enable efficient boundary monitoring. The system combines various sensors, including dual-mode positioning modules, temperature and humidity sensors, vibration sensors, and tilt sensors, which collect real-time data from the environment. These sensors relay information through a cloud-based IoT platform, allowing for continuous monitoring and rapid detection of boundary disturbances or environmental changes. By leveraging both GPS and BeiDou positioning for enhanced accuracy, the system provides a reliable, low-maintenance solution for applications in areas such as natural reserve management, agriculture, and infrastructure security.

IoT Sensors for Data Collection: The system employs various IoT sensors strategically placed at the boundary stakes to collect critical environmental and structural data, including:

Position Sensors: Utilizing a dual-mode positioning system that combines BeiDou satellite and Global Positioning System (GPS) signals to provide accurate longitude and latitude information for the boundary stakes [16]. The BeiDou satellite and GPS systems are essential for determining the position of the boundary stakes by representing the pseudo-range measurements.

$$\begin{cases} \rho^c = r^c + c(dt_r^c - dt^{sc}) + I_\rho^c + T_\rho^c + \epsilon^c \\ \rho^g = r^g + c(dt_r^g - dt^{sg}) + + T_\rho^g + \epsilon^c \end{cases}, \tag{32.1}$$

where ρ^c and ρ^g are the pseudo-range observations for BeiDou and GPS, respectively; r^c and r^g are the geometric distances from the receiver to the satellite; c is the speed of light; dt_r^c and dt_r^g are the receiver clock errors; dt^{sc} and dt^{sg} are the satellite clock errors; I_ρ^c and I_ρ^g are the ionospheric errors; T_ρ^c and T_ρ^g are the tropospheric errors; and ϵ^c and ϵ^c are other errors. The geometric distance between the receiver and the satellites can be calculated using the following equation:

$$r^c = \sqrt{(X^c - X_k)^2 + (Y^c - Y_k)^2 + (Z^c - Z_k)^2}, \tag{32.2}$$

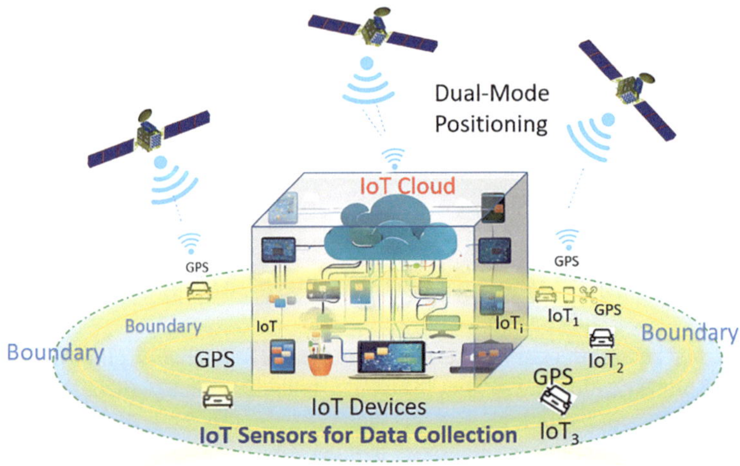

Fig. 32.1 A smart boundary stake system integrates IoT sensors

where X^c, Y^c, and Z^c are the coordinates of the BeiDou satellite; X_k, Y_k, and Z_k are the coordinates of the receiver.

The PDOP is a measure of the geometric strength of the satellite configuration and can be calculated as follows:

$$PDOP = \sqrt{\frac{a^2 + a^2 + c^2}{d^2}}, \tag{32.3}$$

where a, b, and c are the distances from the receiver to three satellites; d is the distance from the receiver to the fourth satellite.

Environmental Sensors: These include temperature and humidity sensors that monitor the surrounding environmental conditions, which can affect the stability and integrity of the stakes. *Vibration and Tilt Sensors:* To detect any abnormal shifts, tilts, or vibrations of the stakes, indicating potential risks or damage. The data collected by these sensors is transmitted wirelessly to a cloud computing platform for further analysis and storage. *Cloud Computing for Data Storage and Processing:* The cloud computing platform serves several key functions: Data Storage: All sensor data is securely stored in the cloud, enabling easy access and retrieval for analysis. Data Processing: Advanced algorithms process the incoming data to assess the condition of the boundary stakes, generate reports, and identify any anomalies. User Interface: A web-based dashboard is provided for stakeholders to visualize the data, receive alerts, and monitor the status of the boundary stakes in real time. This cloud-based approach facilitates scalability, allowing for the integration of additional sensors and boundary stakes into the system as needed. *Dual-Mode Positioning for Enhanced Accuracy:* To achieve high accuracy in boundary stake positioning, the proposed system implements a dual-mode positioning approach. This method leverages both GPS and BeiDou satellite systems:

- Integration of Signals**:** By utilizing signals from both satellite systems, the system can achieve better positioning accuracy, especially in obstructed environments where one system may provide weaker signals.
- Error Correction Mechanisms: The system employs algorithms that can filter and correct positioning errors by comparing data from both systems, ensuring reliable and precise positioning of the boundary stakes.

32.3.2 Data Collection Methods

Data collection occurs continuously through the deployed IoT sensors. The method involves: *Real-Time Data Acquisition***:** Sensors are programmed to collect data at predetermined intervals (e.g., every minute) and transmit this information to the cloud in real time [17]. *Data Aggregation:* The data from various sensors is aggregated at the cloud platform to provide a comprehensive view of the environmental conditions and the status of the boundary stakes. *Alert Mechanism:* An alert mechanism ensures that stakeholders are notified immediately of any detected anomalies (e.g., significant temperature changes or tilting), allowing for prompt action. *Implementation Steps Are Expressed:* The implementation of the smart boundary stake system follows a structured approach:

Step 1 Design Phase: Define system requirements based on stakeholder needs. Design the hardware architecture, including sensor selection and positioning.

Step 2 Prototype Development: Develop prototypes of the boundary stake equipped with selected IoT sensors. Create the cloud computing platform with the necessary data storage and processing capabilities.

Step 3 Integration: Integrate IoT sensors with the dual-mode positioning system. Ensure seamless communication between the sensors and the cloud platform via narrowband IoT technology.

Step 3 Testing: Conduct field tests to validate the system's performance, focusing on positioning accuracy, data transmission reliability, and response times. Perform adjustments based on testing results to optimize sensor placement and configuration.

Step 4 Deployment: Deploy the final version of the smart boundary stake system in selected locations. Provide training to stakeholders on how to use the cloud platform for monitoring and managing the boundary stakes.

Step 5 Monitoring and Maintenance: Continuously monitor the system's performance through the user interface. Schedule regular maintenance for hardware components and update software as necessary to ensure optimal operation.

The proposed smart boundary stake system aims to significantly improve boundary management practices through enhanced accuracy, efficiency, and reliability.

32.4 Results and Discussion

This section details the hardware, software architecture, and IoT cloud integration of the smart boundary stake system. It presents testing results, analyzes performance in accuracy, reliability, and response time, compares it with traditional methods, and discusses case studies showcasing its effectiveness.

32.4.1 System Design

The hardware components of the smart boundary stake system consist of various sensors, communication modules, and positioning systems that work together to collect and transmit data effectively. Dual-Mode Positioning Module: BeiDou Satellite and GPS Receiver: This module combines signals from both the BeiDou satellite system and the Global Positioning System (GPS) to provide accurate positioning information. The receiver processes signals from at least four satellites to calculate the precise location of the boundary stake, even in obstructed environments.

Environmental Sensors: Temperature and Humidity Sensors (e.g., AM2312) [18]: These sensors measure the ambient temperature and humidity around the boundary stake. The AM2312 sensor integrates a digital temperature sensor and a capacitive humidity sensor, converting the measurements into digital signals for transmission.

Vibration Sensors: Typically based on piezoelectric technology, these sensors detect vibrations that may indicate disturbances or potential damage to the boundary stake. They convert mechanical vibrations into electrical signals for monitoring [19]. Tilt Sensors (Attitude Sensors): These sensors monitor the tilt and deformation of the boundary stake, providing data on its structural integrity. They utilize three-dimensional coordinate rotation to assess changes in position.

Communication Modules: Narrowband IoT (NB-IoT) Module: This communication module enables low-power, wide-area network connectivity for the IoT sensors [20]. It facilitates the transmission of collected data to the cloud computing platform, ensuring efficient communication even with a large number of devices. The software architecture of the smart boundary stake system comprises data processing algorithms, a user interface, and cloud services for data management [21].

Data processing algorithms are used for data aggregation and analysis. The system employs algorithms to aggregate data from various sensors, process it for analysis, and identify anomalies. This includes filtering noise from sensor readings and applying statistical methods to assess the condition of the boundary stakes. Alert Generation: Algorithms are implemented to trigger alerts based on predefined thresholds for environmental conditions (e.g., temperature, humidity) and structural integrity (e.g., tilt, vibration). These alerts are sent to stakeholders for timely intervention.

User Interface: Web-Based Dashboard: A user-friendly web interface is developed for stakeholders to visualize real-time data from the boundary stakes. The dashboard displays key metrics, alerts, and historical data, allowing users to monitor the status of the boundary stakes effectively. Remote Monitoring Capabilities: The interface provides remote access to the system, enabling users to receive notifications and manage the boundary stakes from any location [22]. Table 32.1 provides specifications for the various sensors used in the system [20].

Table 32.1 Specifications for the various sensors used in the system

Sensor type	Model	Measurement range	Accuracy	Communication protocol
Dual-mode positioning	N/A	N/A	High	N/A
Temperature and humidity	AM2312	−40 °C to 80 °C	± 0.5 °C	I2C
Vibration sensor	Piezoelectric	N/A	± 0.1g	Analog
Tilt sensor	N/A	± 90 degrees	± 1 degree	Analog

Integration of IoT and Cloud Services: The integration of IoT and cloud services is crucial for the functionality of the smart boundary stake system. Data Transmission to Cloud: The NB-IoT module transmits the collected data from the sensors to the cloud computing platform. This ensures that all data are stored securely and are accessible for processing and analysis. Cloud Computing Platform: The cloud platform serves as the central hub for data storage, processing, and analysis. It utilizes cloud computing technologies to handle large volumes of data, perform complex analyses, and generate insights for stakeholders. Interoperability: The system is designed to ensure seamless communication between IoT devices and the cloud platform, allowing for real-time data updates and remote monitoring capabilities. By combining advanced hardware components with robust software architecture and effective integration of IoT and cloud services, the smart boundary stake system aims to enhance boundary management practices through improved accuracy, efficiency, and reliability.

32.4.2 Results and Discussion

The testing phase involved deploying the smart boundary stake system in various locations to evaluate its performance. The data collected during testing included the following:

- *Positioning Accuracy*: The system recorded positioning data using the dual-mode positioning module (combining BeiDou and GPS). The Position Dilution of Precision (PDOP) values were measured, with lower values indicating higher accuracy. For instance, the average PDOP value during testing was found to be 1.5, significantly lower than the typical PDOP values of 2.5 to 3.0 observed with GPS alone.
- *Environmental Data*: The temperature and humidity sensors recorded ambient conditions around the boundary stakes. For example, temperature readings ranged from 15 °C to 30 °C, while humidity levels varied between 40 and 80%.
- *Structural Integrity Data:* The tilt and vibration sensors provided data on the stability of the boundary stakes. During the testing period, no significant vibrations were detected, and tilt measurements remained within acceptable limits (less than 5 degrees).
- *Communication Efficiency:* The narrowband IoT (NB-IoT) module's power consumption was monitored, with results indicating that the system operated below the standard threshold of 3.5 kWh, even with multiple devices transmitting data simultaneously.

Table 32.2 compares the performance metrics of dual-mode positioning (BeiDou and GPS) versus GPS-only positioning, including PDOP values, average positioning accuracy (in meters), stability metrics (e.g., standard deviation of positioning data), and number of satellites used.

Table 32.2 Dual-mode positioning performance comparison

Method	Average PDOP value	Average positioning accuracy (m)	Stability (standard deviation)
Dual-mode (BeiDou + GPS)	1.5	2.0	0.5
GPS only	2.5	3.0	1.0

Analysis of System Performance: The performance of the smart boundary stake system was evaluated based on several criteria:

- Accuracy: The dual-mode positioning system demonstrated superior accuracy compared to traditional GPS-only systems. The integration of BeiDou signals allowed for precise positioning even in areas with obstructions, such as trees and shrubs. The experimental results confirmed that the dual-mode system reduced positioning errors by approximately 30% compared to GPS alone.
- Reliability: The system exhibited high reliability, with consistent data transmission and minimal downtime. The NB-IoT technology ensured stable communication, even in remote areas with limited connectivity. The system maintained a 95% uptime during the testing phase.
- Response Time: The response time for alert notifications was measured, with the system successfully sending alerts within 5 s of detecting anomalies (e.g., significant tilt or vibration). This rapid response capability is crucial for timely intervention and management.

Comparison with Traditional Boundary Monitoring Methods: Traditional boundary monitoring methods often rely on manual inspections and basic GPS systems, which can be time-consuming and prone to human error. In contrast, the smart boundary stake system offers several advantages:

- Automation: The proposed system automates data collection and monitoring, reducing the need for manual inspections and allowing for continuous monitoring of boundary conditions.
- Real-Time Data Access: Stakeholders can access real-time data through the cloud-based platform, enabling quicker decision-making and response to potential issues.
- Enhanced Accuracy: The dual-mode positioning system provides more accurate location data compared to traditional GPS methods, particularly in challenging environments.

Table 32.3 summarizes case studies demonstrating the effectiveness of the proposed system, including location, application (e.g., natural reserve monitoring, agricultural protection), findings (e.g., improvements in monitoring accuracy, response times), and stakeholder feedback. Case Studies Demonstrating the Effectiveness of the Proposed System: Several case studies were conducted to illustrate the effectiveness of the smart boundary stake system: Natural Reserve Monitoring: In a natural reserve, the system was deployed to monitor boundary stakes

Table 32.3 An effective performance of the system

Case study FIT location	Application	Key findings	Stakeholder feedback
Natural reserve	Boundary monitoring	Improved accuracy in monitoring boundary stakes	Positive, noted timely alerts
Agricultural area	Crop protection	Enhanced decision-making based on environmental data	Very effective for irrigation management
Infrastructure site	Construction monitoring	Detected vibrations from machinery, timely intervention	Essential for project safety

that delineate protected areas [23]. The dual-mode positioning system successfully maintained accurate location data, even in densely forested regions. Alerts were generated when a boundary stake was disturbed, allowing park rangers to respond promptly. Agricultural Protection Zones: The system was implemented in agricultural areas to monitor boundary stakes protecting crops from encroachment [24]. The environmental sensors provided valuable data on temperature and humidity, helping farmers make informed decisions about irrigation and crop management. Infrastructure Projects: During a highway construction project, the smart boundary stake system was used to monitor boundary markers. The system detected vibrations caused by heavy machinery, triggering alerts that allowed project managers to assess potential impacts on boundary integrity.

32.5 Conclusion

This study presented a smart boundary stake system utilizing IoT cloud computing to tackle positioning accuracy and monitoring challenges. By integrating a dual-mode positioning system with BeiDou satellite and GPS technologies, the system ensures precise location tracking, even in obstructed environments. Various sensors collect comprehensive data, while narrowband IoT technology enables efficient real-time data transmission to a cloud platform for monitoring and alerts. Experimental results demonstrate that the dual-mode system outperforms traditional GPS in accuracy and stability. Although the system significantly enhances boundary management, further research is needed to validate the cloud platform's efficiency. Overall, it offers a robust solution for improved boundary management through advanced technologies.

References

1. Rafique, W., Qi, L., Yaqoob, I., Imran, M., Rasool, R.U., Dou, W.: Complementing IoT services through software defined networking and edge computing: A comprehensive survey. IEEE Commun. Surv. & Tutor. **22**, 1761–1804 (2020)
2. Nikpour, M., Yousefi, P.B., Jafarzadeh, H., Danesh, K., Shomali, R., Ahmadi, M.: Intelligent energy management with iot framework in smart cities using intelligent analysis: An application of machine learning methods for complex networks and systems. arXiv preprint arXiv:2306.05567. (2023)
3. Holler, J., Tsiatsis, V., Mulligan, C., Karnouskos, S., Avesand, S., Boyle, D.: Internet of things. Academic Press (2014)
4. Martins, I., Resende, J.S., Sousa, P.R., Silva, S., Antunes, L., Gama, J.: Host-based IDS: A review and open issues of an anomaly detection system in IoT. Futur. Gener. Comput. Syst. **133**, 95–113 (2022)
5. Koptyra, K., Ogiela, M.R.: Steganography in IoT: information hiding with APDS-9960 proximity and gestures sensor. Sensors. **22**, 2612 (2022)
6. Yang, Y., Gao, W., Guo, S., Mao, Y., Yang, Y.: Introduction to BeiDou-3 navigation satellite system. Navigation **66**, 7–18 (2019)
7. Dixon, T.H.: An introduction to the Global Positioning System and some geological applications. Rev. Geophys. **29**, 249–276 (1991)
8. Dao, T.-K., Nguyen, T.-T.: A review of the bat algorithm and its varieties for industrial applications. J. Intell. Manuf. 1–23 (2024)
9. Dao, T.-K., Ngo, T.-G., Pan, J.-S., Nguyen, T.-T.-T., Nguyen, T.-T.: Enhancing path planning capabilities of automated guided vehicles in dynamic environments: multi-objective PSO and dynamic-window approach, (2024)
10. Chang, K.-C., Chu, K.-C., Lin, Y.-C., Nguyen, T.-T., Sung, T.-W., Zhou, Y.-W., Pan, J.-S.: Study on hazardous scenario analysis of high-tech facilities and emergency response mechanism of science and technology parks based on IoT BT—Genetic and evolutionary computing. Presented at the (2020)
11. Dao, T.-K., Pan, J.-S., Pan, T.-S., Nguyen, T.-T.: Optimal path planning for motion robots based on bees pollen optimization algorithm. J. Inf. Telecommun. **1**, 1–16 (2017). https://doi.org/10.1080/24751839.2017.1347765
12. Nguyen, T.-T., Wang, H.-J., Dao, T.-K., Pan, J.-S., Ngo, T.-G., Yu, J.: A scheme of color image multithreshold segmentation based on improved Moth-Flame algorithm. IEEE Access. **8**, 174142–174159 (2020). https://doi.org/10.1109/ACCESS.2020.3025833
13. Liu, N., Pan, J.-S., Lai, J., Chu, S.-C., Nguyen, T.-T.: An efficient differential evolution via both top collective and *p*-Best information. J. Internet Technol. **21**, 629–643 (2020)
14. Nguyen, T.-T., Dao, T.-K., Nguyen, T.-D., Nguyen, V.-T.: An improved honey badger algorithm for coverage optimization in wireless sensor network. J. Internet Technol. **24**, 363–377 (2023)
15. Nguyen, T.-T., Pan, J.-S., Lin, J.C.-W., Dao, T.-K., Nguyen, T.-X.-H.: An optimal node coverage in wireless sensor network based on whale optimization algorithm. Data Sci. Pattern Recognit. **02**, 11–21 (2018)
16. Hsu, R.-H., Lee, J., Quek, T.Q.S., Chen, J.-C.: Reconfigurable security: Edge-computing-based framework for IoT. IEEE Network **32**, 92–99 (2018)
17. Mazhar, S.A., Anjum, R., Anwar, A.I., Khan, A.A.: Methods of data collection: A fundamental tool of research. J. Integr. Community Health (ISSN 2319–9113). **10**, 6–10 (2021)
18. Marcel, M.J., Baker, J.: Interdisciplinary design of a near space vehicle. In: Proceedings 2007 IEEE SoutheastCon. pp. 421–426. IEEE (2007)
19. Li, T., Guo, J., Tan, Y., Zhou, Z.: Recent advances and tendency in fiber Bragg grating-based vibration sensor: A review. IEEE Sens. J. **20**, 12074–12087 (2020)
20. Migabo, E.M., Djouani, K.D., Kurien, A.M.: The narrowband Internet of Things (NB-IoT) resources management performance state of art, challenges, and opportunities. IEEE Access. **8**, 97658–97675 (2020)

21. Thakur, G., Prajapat, S., Kumar, P., Chen, C.-M.: A privacy-preserving three-factor authentication system for IoT-Enabled wireless sensor networks. J. Syst. Architect. **154**, 103245 (2024)
22. Chen, C.-M., Chen, Z., Kumari, S., Obaidat, M.S., Rodrigues, J.J.P.C., Khan, M.K.: Blockchain-based mutual authentication protocol for IoT-Enabled decentralized healthcare environment. IEEE Internet Things J. **11**, 25394–25412 (2024)
23. Yu, J., Nguyen, T.T., Dao, T.-K., Nguyen, T.-T.: Global path planning algorithm for mobile robots: a review. J. Netw. Intell. **9**, 643–657 (2024)
24. Pham, D.-T., Hoang, D.-T.-T., Nguyen, T.-T., Dao, T.-K., Nguyen, T.-X.-H.: A hybridized network analysis and community detection for unraveling disease spreading Covid-19 pandemic mechanisms. J. Netw. Intell. **9**, 1–13 (2024)

Chapter 33
Evaluation Indicators for the Digital Transformation of Green Energy Enterprises Based on the ESG Perspective

Wei Song, Lingjiao Wu, Xiaojuan Chai, and Dawei Luo

Abstract With the development and widespread application of new-generation information technologies, such as artificial intelligence, the Internet of Things, big data, and cloud computing, the digital transformation of enterprises has become irreversible. The evaluation of enterprise digital transformation has also been brought onto the agenda. Although some initial results have been achieved in the research on digital transformation indicators, it is still worth exploring how to establish a set of evaluation indicators that can support the sustainable development of enterprises and are rich in social responsibility. Based on the existing research foundation of digital transformation evaluation indicators, this paper will establish a set of digital transformation evaluation indicator system for the green energy industry from the three key perspectives of Environmental, Social, and Governance (ESG), which can not only fully adapt to the characteristics of the green energy industry but also integrates ESG factors into traditional digital transformation evaluation indicators, and support its sustainable development.

33.1 Introduction

The transformation and upgrading required by industrial development both require organizations to deeply integrate and apply digital technologies to achieve organizational sustainability and carry out the enterprise responsibility. Environmental, Social, and Governance (ESG) criteria constitute a framework for assessing an

W. Song (✉) · L. Wu · D. Luo
Changzhou College of Information Technology, Changzhou, China
e-mail: s5995w@163.com

D. Luo
e-mail: luodawei@czcit.edu.cn

X. Chai
Jiangsu Longhuan Environmental Technology Co., Ltd, Changzhou, China

S.-C. Chu et al. (eds.), *Advances in Intelligent Data Analysis and Applications*,
Smart Innovation, Systems and Technologies 445,
https://doi.org/10.1007/978-981-96-7277-6_33

organization's commitment to sustainable development. Increasingly adopted by investors and corporations, ESG ratings play a pivotal role in managing non-financial risks. By integrating ESG practices, companies not only enhance their societal image but also expand their influence and attract continuous investment. This strategic alignment with ESG standards is essential for organizations aiming to maintain competitive advantage and stakeholder trust in a rapidly evolving global market.

Digital transformation is an ongoing process that require organizations to continuously evaluate, adjust, and optimize to adapt to constant-updating technology and market environment. Digital transformation can significantly improve ESG performance of corporations. Feng Jin and others believed that digital transformation provides companies with opportunities to improve their ESG responsibility performance and better formulate sustainable development strategies in the digital era. It has theoretical support and empirical basis [1]. Digital transformation can promote the progress of green technology and improve management efficiency, and can significantly reduce the pollution emission intensity of manufacturing enterprises. Chen Hong and others found that digital transformation can positively regulate an enterprise's ESG performance by enhancing its green innovation capabilities and easing its financing constraints [2]. Wang Yinghuan and others also found that companies with a higher degree of digital transformation have better ESG performance. There are differences in the role of digital transformation among different types of enterprises [3]. Wang Xiaohong and others pointed out that the improvement of digitalization level helps to efficiently match enterprise resources to stimulate enterprise vitality and improve information transparency and comprehensive utilization efficiency of resources [4].

Researching enterprise digital transformation evaluation indicators through an ESG lens is crucial for the future integration of digital technologies within enterprises. This study offers a comprehensive review of existing literature on digital transformation and ESG evaluation to unearth the interconnections between these domains. The primary goal is to develop evaluation indices for the digital transformation of green energy enterprises, thereby enhancing our understanding of digital transformation impacts from an ESG perspective and assisting in the construction of robust frameworks for assessing and promoting sustainable practices in the evolving energy sector. Incorporating ESG evaluation factors in the assessment can provide strategic insights for optimizing corporate operations and management. This enables green energy companies to maintain market competitiveness through the deep integration of digital technologies while advancing social responsibility and sustainable development. As an important part of the global strategic emerging industry, the green energy industry makes comprehensive use of relatively clean and renewable energy such as solar energy and wind energy, which meets the requirements of sustainable development. The research on the evaluation index of digital transformation of green energy enterprises from the perspective of ESG is more significant.

Our analysis of multiple SCI-indexed papers reveals a notable scarcity of research on enterprise digital transformation evaluation indicators from an ESG perspective. This study represents a pioneering effort to merge ESG considerations with digital transformation evaluation indicators. It anticipates that post-2025, with increasing societal emphasis on the sustainable development of the green energy industry and its social responsibilities, there will be a surge in research exploring digital transformation within the framework of ESG evaluation.

The rest of this paper is structured as follows: Sect. 33.2 provides an overview of ESG evaluation and digital transformation. Section 33.3 analyzes the evaluation indicators of digital transformation of green energy enterprises from the perspective of ESG. Section 33.4 provides the conclusions of the analysis.

33.2 ESG Evaluation and Digital Transformation

33.2.1 ESG Evaluation

In 2004, the United Nations Global Compact (UNGC) first fully proposed the concept of ESG in its report "Who Cares Wins." In 2006, the United Nations established the Principles for Responsible Investment and proposed the Principles for Responsible Investment (PRI), formally proposing six basic investment principles that ESG investment must comply with, promoting investment institutions to incorporate ESG principles into their decision-making, as a standard and strategy for evaluating the behavior and future financial performance of investment targets. With the increasing global attention to climate change and sustainable development issues, ESG investment and evaluation systems have continued to develop and improve, becoming an important tool for measuring corporate sustainability and environmental, social, and governance responsibilities [1]. The environmental (E), social (S), and governance (G) components of ESG evaluation are detailed in Table 33.1 [5].

In Table 33.1, the Environmental (E) component refers to the consumption of raw materials and energy by the enterprise, the emissions produced, and the consequential environmental impacts on the enterprise. The Social (S) component encompasses the relationships established between the enterprise and individuals or institutions during business operations, and the impacts of resultant social issues on the enterprise. The Governance (G) component pertains to the internal management, organization, and systems developed for self-regulation, effective decision-making, and adherence to laws and regulations, along with the implications of governance-related issues. The sustainability and social impacts of corporate operations are assessed through the identification and comprehensive evaluation of these ESG factors [6]. This approach ensures a holistic assessment of how enterprises manage their ecological footprints, societal interactions, and internal governance to maintain sustainability and ethical operations [7].

Since ESG evaluation was formally proposed in 2004, it has been 20 years since then. Countries around the world have been promoting the coordinated development of environment, society, and governance in an orderly manner according to ESG principles. Although the operating goal of a single organization is to pursue profit

Table 33.1 Dimensions of ESG evaluation

Dimension	Factors	Definition
Environment (E)	• Material consumption • Energy consumption • Greenhouse gas emission • Air pollutants • Water use and recycling • Waste generation and management (water, solid, hazardous) • Impacts and dependencies on biodiversity • Impacts and dependencies on ecosystems • Innovation of environmental products and services	Environmental issues that may have a positive or negative impact on the financial performance or solvency of an entity, sovereign, or individual
Society (S)	• Employee rights • Workplace health and safety • Customer health and safety • Discrimination, diversity, and equal opportunity • Poverty and community impacts • Supply chain management • Training and education • Customer privacy • Community impacts	Social issues that may have a positive or negative impact on the financial performance or solvency of an entity, sovereign, or individual
Governance (G)	• Code of conduct and business principles • Accountability • Transparency and disclosure • Executive compensation • Board diversity and structure • Bribery and corruption • Stakeholder engagement • Shareholder rights	Governance issues that may have a positive or negative impact on the financial performance or solvency of an entity, sovereign, or individual

maximization, from the perspective of human social security and sustainable development, the goals of organizational operations are changing, and giving priority to the safety of all mankind and the sustainable development of future society is attracting more attention [8, 9]. ESG management has shifted from the existing linear economic paradigm to a circular economy that emphasizes resource sharing and reuse, using ESG to ensure that companies can survive and prosper in long-term competition [8]. ESG evaluation factors will vary depending on the industry in which the evaluation object is located and the evaluation objectives.

33.2.2 Digital Transformation

Digital transformation refers to the process by which an enterprise or organization changes its operating methods, business processes, customer interactions, and overall value proposition by applying digital technologies. The core is to widely apply digital technologies such as the Internet of Things, cloud computing, big data, artificial intelligence, and blockchain to product development, production and manufacturing, sales and after-sales service, planning, procurement, and other business links [9]. Digital transformation has become a keyway for enterprises to develop new quality productivity, enhance market competitiveness, and achieve high-quality development [10]. Digital transformation can not only promote efficient collaboration between industrial chains and supply chains and optimize resource allocation, but also promote the development of the manufacturing industry toward green and low-carbon development and achieve sustainable development.

In the process of realizing enterprise digital transformation, general enterprises adopt enterprise resource planning (ERP), product life cycle (PLM), manufacturing execution system (MES), data collection and supervisory control system (SCADA), customer relationship management (CRM), office automation (OA), supply chain management (SCM), supplier relationship management (SRM), process management (BPM), intelligent decision-making (BI), and other information systems, and through the integrated application of these information systems, improve the internal. The sharing and transparency of information data enhances the ability to utilize internal and external corporate resources such as corporate products, market dynamics, customer information, supply chains, and quality traceability, helping companies better understand through the digital reshaping of corporate value creation activities.

33.2.3 ESG Evaluation Factors Strongly Related to Digital Transformation

As the importance of ESG increases, various institutions and organizations have provided ESG evaluation indicators for various industries. This article quotes the ESG evaluation indicators of the Sustainability Accounting Standards Board (SASB), which has 11 categories and 77 indicators. As shown in Fig. 33.1, this study selected the evaluation indicator group directly related to manufacturing.

Environmental issues such as resource shortages, ecological, and social inequality have made governments and businesses realize that they should address sustainability issues [11]. In the environmental field of ESG, the indicators that are strongly related to the digital transformation of enterprises are as follows:

Material consumption. Raw materials are essential for companies to manufacture products. Material consumption is an important factor in assessing a company's environmental impact. ESG assessments use the purchase volume of raw materials

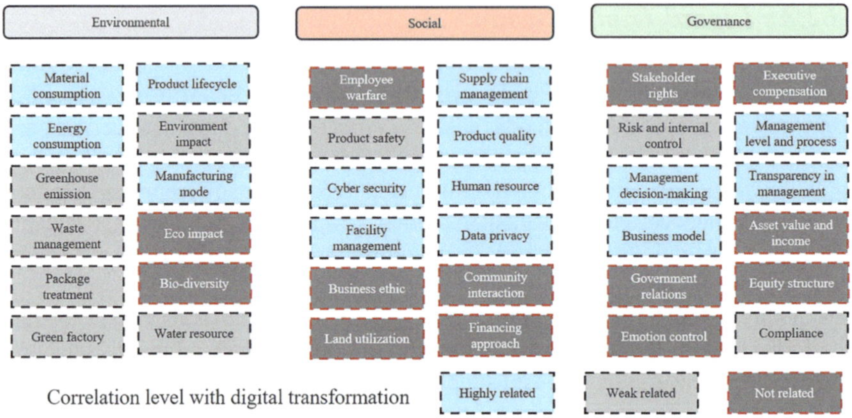

Fig. 33.1 Correlation analysis between ESG metrics and digital transformation

and the utilization rate of raw materials in the manufacturing process as the main measurement indicators. The digital transformation of companies will effectively improve the utilization rate of raw materials.

Product life cycle. The product life cycle is determined by the products produced by the company. The length of a product's life cycle has a significant impact on a company's profits, and it also has a greater impact on the environment. Product life cycles are short, and more waste is generated. After the digital transformation of enterprises, it can help enterprises to control the recycling and reuse of materials that make up products based on the product life-cycle theory.

Manufacturing mode. Advanced manufacturing models can effectively reduce the impact on the environment and produce less waste. One of the purposes of enterprise digital transformation is to help enterprises improve their manufacturing models. For example, the application of a large number of automation systems can optimize manufacturing resources and reduce unnecessary waste of resources [12].

Energy consumption. Energy consumption has a direct impact on the environment. Effective control of energy consumption will directly determine the extent of an enterprise's impact on the environment. The digital transformation of an enterprise can achieve effective monitoring of energy, and even monitor the energy consumption of each device. In addition, the application of technologies such as remote control of equipment, equipment maintenance, and energy collection and reuse can save additional energy consumption.

ESG is a framework that extends from corporate social responsibility (CSR) and responsible investment [13]. In addition to the social impact of enterprises, it is also necessary to consider indicators of corporate social impact, such as supply chain and human resource development. In the social field of ESG, the indicators that are strongly related to the digital transformation of enterprises are as follows:

Supply chain management. The impact of enterprises on the upstream and downstream of the supply chain is an important part of the impact of enterprises on society, and the social evaluation value of ESG is reflected through the supply chain. The intensive use of social resources, the inventory of commodities in the entire society, and the overall operating efficiency of society will all depend on the construction of corporate digital capabilities. At the same time, maintaining good customer and supplier relationships is also an important consideration for improving service levels and enhancing satisfaction in digital transformation.

Product quality. The safety and reliability of the products or services provided by the enterprise, as well as accidents caused by product quality, directly affect the evaluation of corporate social responsibility. Product quality involves factors such as the selection of raw materials, process processing methods, testing methods, employee work skills and work fatigue, and is closely related to the collection and analysis of manufacturing process data.

Cybersecurity. Cybersecurity is one of the key factors in assessing corporate social responsibility in the ESG framework. It involves not only the protection of corporate digital assets, but also business continuity and maintaining the trust of stakeholders. With the development of digitalization and the Internet, cybersecurity has become a key factor in assessing corporate social responsibility. With the continuous development of technology, enterprises must formulate measures to deal with the occurrence of cyber-attacks and control the occurrence of online public opinion.

Human resource development. The human resource status of an enterprise has a profound impact on the sustainable development of the enterprise. Human resource development is an important indicator of corporate social responsibility. ESG evaluation includes indicators such as skilled labor, skill improvement, job stability, and personal development to promote corporate productivity growth. Digital transformation requires enterprises to strengthen employee training so that they have more technical skills in the digital economy era to cope with the impact of the application of technologies such as artificial intelligence and robots on jobs.

Facility management. Facility management threatens the personal safety of society and employees, and also affects the product quality of enterprises. Accidents related to facility management, especially the social risks of outdoor facilities, will affect ESG evaluation. The application of digital technology can manage the location, movement trajectory, life-cycle management and predictive maintenance of facilities.

Data privacy. Data security and privacy protection are important considerations in ESG evaluation. The inviolability of customer and supplier data is crucial to maintaining a good ESG rating. In the digital age, data security and data leakage can cause significant economic losses to companies and damage their brand image.

ESG assessments are increasingly focusing on decision-making factors in corporate business activities, such as management levels, management transparency, business processes, management tools, and compliance, in order to avoid systemic risks.

In the governance field of ESG, the indicators that are strongly related to the digital transformation of enterprises are as follows:

Management levels and processes. The more management levels there are and the more complex the management process is, the lower the governance efficiency will be, and the lower the corporate governance level will be. The application of management tools such as ERP and OA, on the basis of optimizing business processes, realizes the flattening of management organizations, accelerates the transmission and sharing of information data, and improves decision-making efficiency and governance level.

Management decision-making. ESG brings new opportunities for corporate management decision-making and strategic transformation and has a profound impact on the company's long-term development, innovative capabilities, and investment decisions. In the digital age, how to effectively conduct performance evaluation and how to make full use of data for intelligent decision-making depends on the digital literacy, digital capabilities, and digital leadership of the management.

Management transparency. This refers to the openness and clarity of an enterprise's decisions, processes, performance, and impact during its operations. This includes decision-making transparency, operational transparency, financial transparency, communication transparency, etc. This requires the establishment of an effective internal monitoring and auditing system. Digital transformation can maximize the transparency of corporate information and the quality of disclosure.

Business model. A friendly business model can not only enhance the company's reputation among investors and consumers, but also maintain a sound financial performance and market vitality over the long term, and promote long-term value creation and employee and social welfare. Digital transformation can help companies shape their business models.

33.3 Analysis of the Evaluation Indicators of Digital Transformation of Green Energy Enterprises from the Perspective of ESG

33.3.1 Digital Transformation of Green Energy Industry and Its Characteristics

Green energy refers to renewable energy sources such as hydropower, solar, wind, biomass, and geothermal energy. The industry encompasses power generation, energy storage, transportation, and applications, involving upstream manufacturers (e.g., solar, wind, and hydropower equipment), midstream energy suppliers (e.g., hydrogen and photovoltaic power generation), and downstream applications (e.g., green vehicles, charging stations, and power infrastructure).

Digital transformation in the green energy industry is advanced, with varying focus across the value chain. In power generation, efforts emphasize the energy Internet and service export by leading firms. Energy storage prioritizes capacity improvement and customer value, transportation targets customer satisfaction, customization, and logistics tracking, while applications focus on R&D and market-driven product design. Key application areas include digital management, intelligent manufacturing, and digital marketing. Most companies are in the "standard application-integrated application" stage, optimizing business processes with systems like ERP, OA/BPM, MES/MOM, PDM/PLM, HR, CRM, and SRM. Interconnected design, manufacturing, management, and service processes are emerging. However, full integration of IT and manufacturing, utilization of digital resources, equipment OEE, predictive maintenance, SPC-based process quality control, digital twin technology, and automated energy/environmental data collection remain underdeveloped. Industrial Internet adoption is primarily limited to chain-leading enterprises.

33.3.2 Evaluation Indicators for the Digital Transformation of Green Energy Enterprises in the Environmental Dimension

There are two types of ESG evaluation factors related to digital transformation in the environmental dimension. The strongly related factors are "material consumption, product life cycle, manufacturing mode, energy consumption," and the weakly related factors are "business environment, greenhouse gas emissions, waste management, packaging treatment, green innovation, water resource management." Through the analysis of the impact of ESG evaluation factors on digital transformation, relevant digital transformation evaluation indicators can be obtained. Among them, "equipment networking and equipment intelligence" can be included in the "digital foundation" evaluation indicator set, "lean manufacturing, digital R&D design, PLM/PDM system construction" can be included in the "digital application" evaluation indicator set, and "material utilization, product innovation, advanced manufacturing mode, environmental monitoring, comprehensive energy consumption, green and low-carbon, and production capacity utilization" can be included in the "digital effectiveness" evaluation indicator set. For details, please see Table 33.2.

Table 33.2 Digital transformation evaluation indicators of green energy enterprises in the environmental (E) dimension

ESG factors		Main influencing factors	Transformed digital transformation indicators
Strong correlation	Material consumption	Manufacturing methods, advanced equipment, manufacturing process, technical level, employee skill proficiency	The digital transformation indicators of equipment intelligence rate, lean manufacturing level, material utilization rate, comprehensive energy consumption, 10,000-yuan output value comprehensive energy consumption
	Product life cycle	Market demand collection ability, product design and development capability, product life-cycle management capability	Product update frequency, R & D design digital technology coverage rate, PLM/PDM network coverage rate
	Manufacturing mode	Equipment automation and intelligence, network facilities	Equipment intelligence rate, equipment network rate, advanced manufacturing mode application, digital construction funds guarantee
	Energy consumption	Equipment transformation, equipment automation, energy consumption, external consumption, energy recycling	Production environment monitoring, comprehensive energy consumption of 10,000 yuan of output value, green manufacturing coverage rate, manufacturing mode innovation, capacity utilization rate
Weak correlation	Business environment	Work environment monitoring to prevent environmental hazards	-
	Greenhouse gas emission	Design capability, process level, manufacturing capability	Production environment monitoring coverage rate, comprehensive energy consumption of 10,000 yuan of output value
	Waste management	Manufacturing capabilities, employee skills	Production environment monitoring coverage rate, comprehensive energy consumption of 10,000 yuan of output value
	Packaging treatment	Package design, package material	Energy consumption of 10,000-yuan output value

(continued)

Table 33.2 (continued)

ESG factors		Main influencing factors	Transformed digital transformation indicators
	Green innovation	Green factory, product environment production	Green manufacturing coverage, production environment monitoring coverage of 10,000-yuan output value
	Water resource management	Design capability, process level, manufacturing capabilities	Production environment monitoring coverage rate, comprehensive energy consumption of 10,000 yuan of output value

33.3.3 Evaluation Indicators for the Digital Transformation of Green Energy Enterprises in the Social Dimension

There are two types of ESG evaluation factors related to digital transformation in the social dimension. The strongly related factors are "supply chain management, product quality, network security, human resource development, facility management, and data privacy," and the weakly related factors are "product safety." By analyzing the impact of ESG evaluation factors on digital transformation, relevant digital transformation evaluation indicators can be obtained. Among them, "digital human resources allocation" can be included in the "digital element guarantee" evaluation indicator set, "information security mechanism, data collection, data backup, data warehouse application" can be included in the "digital foundation" evaluation indicator set, and "CRM/SRM/SCM system construction, supply chain demand response, product quality digital management, data value mining, industrial Internet platform application" can be included in the "digital application" evaluation indicator set. For details, please see Table 33.3.

33.3.4 Evaluation Indicators for the Digital Transformation of Green Energy Enterprises in the Governance Dimension

There are two types of ESG evaluation factors related to digital transformation in the governance dimension. The strongly related factors are "management levels and processes, management decisions, management transparency, and business models," and the weakly related factors are "risks and internal controls, and compliance." By analyzing the impact of ESG evaluation factors on digital transformation, we can get relevant digital transformation evaluation indicators. Among them, informationized

Table 33.3 Digital transformation evaluation indicators of green energy enterprises in the social (S) dimension

ESG factors		Main factors	Transformed digital transformation indicators
Strong correlation	Supply chain management	Demand transmission and information sharing level, supply chain inventory level, customer relationship, supplier relationship	CRM/SRM/SCM application rate, timeliness of responding to changes in supply chain demand, information sharing degree, service output, product serious failure rate, product recall proportion, product quality loss rate
	Product quality	Product design, material selection, processing technology, testing ability, employee skill proficiency, data analysis and process improvement	Digital management of product quality and value mining of quality data
	Cybersecurity	Network security prevention mechanism and protection mechanism, network security infrastructure	Coverage of information security guarantee mechanism
	Human resource development	Employee skill training, employee mobility, Staff training	Proportion of "one post and multiple capabilities" personnel, digital personnel allocation, full-time digital personnel
	Facility management	Network infrastructure, facility networking, facility operation monitoring	Proportion of facility network rate, data collection and predictability proportion of facility operation, industrial Internet platform application
	Data privacy	Data protection mechanism, data backup	Data backup and recovery degree, data warehouse application
Weak correlation	Product safety	Product material selection, product appearance design	Product serious failure rate, product recall proportion, product quality loss rate

organizational structure, process optimization, and digital planning can be included in the digital element guarantee evaluation indicator set, ERP, MES, CRM, information system integration, BI applications, and internal financial audit system applications can be included in the Digital Application Evaluation Indicator set. For details, see Table 33.4.

Table 33.4 Digital transformation evaluation indicators of green energy enterprises in the governance (G) dimension

ESG factors		Main factors	Generated digital transformation metrics
Strong correlation	Hierarchy and process management	Organizational hierarchy, managing authorized	Informative organization's architecture
	Decision management	Data reliability, performance evaluation, management's digital capability	Information system application levels, information system integration application levels, and BI application levels
	Transparency	Information data sources and disclosure, communication channels, communication methods	ERP, MES, CRM, etc. system application levels, information system integration application levels
	Business model	Market sensitivity, operational model optimization	Digital planning
Weak correlation	Risk and internal control	Risk prediction, operational data Integrity	Internal audit system application
	Compliance	Operational data integrity, business process control, business process rationality	Internal audit system application, process optimization

33.3.5 Comprehensive Evaluation Indicators for the Digital Transformation of Green Energy Enterprises

Integrating the characteristics of digital transformation in green energy enterprises, and considering the elements related to digital transformation in ESG evaluations, an evaluation indicator system can be constructed based on four sets of indicators: "digital factor guarantee, digital foundation, digital application, and digital effectiveness" [10]. For details, see Table 33.5.

Table 33.5 Evaluation index system of digital transformation of green energy enterprises

Index	Level 1	Level 2	Level 3
1	Digital insurance	Digital planning	Effective implementation of digital planning
2		Digital organization	Completeness of information organization and flatness of enterprise organizational structure
3		Digital human resources construction	Number of professional digital personnel, proportion of full-time digital personnel, digital training investment, introduction of professional personnel, and the proportion of personnel with multiple skills in one post
4		Financial security	Proportion of digital construction funding and degree of digital construction funding guarantee
5	Digital foundation	Intelligent equipment	Equipment/facility networking rate, equipment intelligence rate, facility operation data collection Equipment OEE and predictive maintenance, industrial Internet platform application
6		Network infrastructure	Network coverage, network security protection in place
7		Information security	The soundness of the information security assurance mechanism and the coverage of the information security assurance mechanism
8		Data foundation	Basic data accuracy, MDM application level, data backup and recovery, originality, data warehouse application, data resource application
9	Digital application	Digital marketing	Application of CRM system, quotation management, personalized customization, etc.; network marketing coverage
10		Digital R&D design	R&D design digital technology coverage, CAX application coverage, collaborative design coverage, PLM/PDM, and virtual simulation design applications
11		Digitalization of management decisions	ERP, OA, SRM, internal audit, intelligent BI and other unit system construction, application status, information system integration, business process optimization
12		Lean production	Construction and application of advanced manufacturing technology, MES and SCADA systems status, digital management coverage of production and manufacturing, data collection of production status coverage, manufacturing lean management implementation

(continued)

Table 33.5 (continued)

Index	Level 1	Level 2	Level 3
13		Product quality and traceability	Application of QMS, machine inspection, artificial intelligence, etc., quality data collection, data collection and quality data value mining, product quality traceability, based on SPC realizes process quality analysis and control
14		Supply chain digitalization	SCM system construction and application, supply chain information sharing, customer demand change response cycle, plan change cycle
15	Digital effectiveness	Comprehensive utilization of production capacity	Comprehensive utilization rate of production capacity
16		Product cost control	Material utilization rate, product serious failure rate, product recall ratio, product quality
17		Green and low carbon	Production environment monitoring coverage, comprehensive cost ratio of enterprise output value per 10,000 yuan
18		Product innovation	Number of patents, high-tech products, frequency of product updates, product operation industrial monitoring, industrial internet platform application, digital twin technology application
19		JIT production	On-time completion rate of production orders and on-time delivery rate of customer orders
20		Service capability output	Annual outsourcing service amount, internal system delivery capacity, and internal system operation dimensional ability
21		Business model innovation	Innovation status of advanced manufacturing models, service models, and business models
22		Government policy support	The amount of government support received and the highest project level

33.4 Conclusion

This paper explores the evaluation indicators for the digital transformation of green energy enterprises, focusing on sustainable development through the lens of ESG principles. It provides theoretical guidance for green energy enterprises to enhance their digital transformation capabilities and levels, as well as to improve their ESG evaluations. By using digital means to enhance environmental management, fulfilling social responsibilities, and increasing the transparency and efficiency of governance structures, long-term and stable sustainable development can be achieved.

This paper has conducted research on the evaluation indicators of digital transformation for green energy enterprises from ESG perspective through a literature review. It has preliminarily proposed four first-level indicators: "digital element assurance," "digital foundation," "data application," and "digital effectiveness," along with 22 s-level indicators such as "digital planning" and "digital organization," and 82 third-level indicators including "equipment/facility networking rate," "equipment OEE," predictive maintenance," and "industrial internet application." The selection of these indicators needs further refinement in subsequent research.

Acknowledgements This work was partially supported by the Jiangsu Provincial Department of Education's 2024 Social Science Research Project for Jiangsu Colleges and Universities, titled "Research on the Cultivation Path of Jiangsu Intelligent Factories Based on Lighthouse Factory Evaluation Standards" (project number: 2024SJYB0953).

References

1. Feng Jin, Zhang Jiarui, He Wen.: Digital transformation and manufacturing enterprise ESG performance—an empirical examination based on listed companies in shanghai and shenzhen a shares. Oper Res Fuzziology **14** (2024): 216
2. Hong, C., Lingxiao, Z.: ESG performance, digital transformation, and corporate value enhancement. J Zhongnan Univ Econ Law **3**, 136–149 (2023)
3. Wang Yinghuan, Guo Yongzhen.: Corporate digital transformation and ESG performance. Financial Research 49.9 (2023).
4. Xiaohong, W., Xiangyu, L., Shaopeng, Z.: Corporate R&D investment, ESG performance, and market value—the moderating effect of corporate digitalization level. Stud Sci Sci **41**(5), 896–904 (2023)
5. EBA.: EBA report on management and supervision of ESG risks for credit institutions and investment firms. 22–47 (2021)
6. Henisz, W., Koller, T., Nuttall, R.: Five ways that ESG creates value. McKinsey Q. **4**, 1–12 (2019)
7. EBA.: Environmental social and governance disclosures. Available online: https://www.eba.europa.eu/sites/default/documents/files/document library/Publications/Consultations/2021/Consultation%20on%20draft%20ITS%20on%20Pillar%20disclosures%20on%20ESG%20risk/963626/Factsheet%20-%20ESG%20disclosures.pdf
8. Unruh, Gregory, et al.: Investing for a sustainable future: Investors care more about sustainability than many executives believe. MIT Sloan Manag. Rev. 57.4 (2016)
9. Litvinenko, Vladimir, et al.: Global guidelines and requirements for professional competencies of natural resource extraction engineers: Implications for ESG principles and sustainable development goals. J. Clean. Prod. **338** (2022): 130530
10. Blinova, E., Ponomarenko, T., Knysh, V.: Analyzing the concept of corporate sustainability in the context of sustainable business development in the mining sector with elements of circular economy. Sustainability **14**(13), 8163 (2022)
11. Li Zhendong, Chen Jin, Wang Weinan.: National digital development strategy path, theoretical framework, and logical analysis. Sci. Res. Manag. 44.7 **1** (2023)
12. Wu Lingjiao.: Intelligent transformation and digitalization empowering the green energy industry for high-quality development of new productive forces. Strait Sci. Technol. Ind. (2024)
13. Szóka, Károly.: The importance of green controlling–the connection of the sustainability balanced scorecard and the ESG. In: 38th EBES Conference-Warsaw: Proceedings. **1**. (2022)

14. Hozdić, E.: Smart factory for industry 4.0: A review. Int. J. Mod. Manuf. Technol. **7**(1), 28–35 (2015)
15. Li, Ting-Ting, et al.: ESG: Research progress and future prospects. Sustain. **13.21** (2021): 11663

MIX
Papier aus verantwortungsvollen Quellen
Paper from responsible sources
FSC® C105338

If you have any concerns about our products,
you can contact us on
ProductSafety@springernature.com

In case Publisher is established outside the EU,
the EU authorized representative is:
Springer Nature Customer Service Center GmbH
Europaplatz 3, 69115 Heidelberg, Germany

Printed by Libri Plureos GmbH
in Hamburg, Germany